S0-BUA-050

INTERFACE CONVERSION
for Polymer Coatings

Published Symposia Held at the General Motors Research Laboratories, Warren, Michigan

	Editors
FRICTION AND WEAR, 1959	ROBERT DAVIES
INTERNAL STRESSES AND FATIGUE IN METALS, 1959	GERALD M. RASSWEILER and WILLIAM L. GRUBE
THEORY OF TRAFFIC FLOW, 1961	ROBERT HERMAN
ROLLING CONTACT PHENOMENA, 1962	JOSEPH B. BIDWELL
ADHESION AND COHESION, 1962	PHILIP WEISS
CAVITATION IN REAL LIQUIDS, 1964	ROBERT DAVIES
LIQUIDS: STRUCTURE, PROPERTIES, SOLID INTERACTIONS, 1965	THOMAS J. HUGHEL
APPROXIMATION OF FUNCTIONS, 1965	HENRY L. GARABEDIAN
FLUID MECHANICS OF INTERNAL FLOW, 1967	GINO SOVRAN
FERROELECTRICITY, 1967	EDWARD F. WELLER
INTERFACE CONVERSION FOR POLYMER COATINGS, 1968	PHILIP WEISS and G. DALE CHEEVER

INTERFACE CONVERSION
for Polymer Coatings

PROCEEDINGS OF THE SYMPOSIUM ON
INTERFACE CONVERSION FOR POLYMER COATINGS,
GENERAL MOTORS RESEARCH LABORATORIES,
WARREN, MICHIGAN, 1967

Edited by

PHILIP WEISS

and

G. DALE CHEEVER

Polymers Department
General Motors Research Laboratories
Warren, Michigan

AMERICAN ELSEVIER PUBLISHING COMPANY, Inc.

NEW YORK

1968

AMERICAN ELSEVIER PUBLISHING COMPANY, INC.
52 VANDERBILT AVENUE
NEW YORK, N.Y. 10017

ELSEVIER PUBLISHING CO. LTD.
BARKING, ESSEX, ENGLAND

ELSEVIER PUBLISHING COMPANY
335 JAN VAN GALENSTRAAT
P.O. BOX 211, AMSTERDAM, THE NETHERLANDS

Standard Book Number 444-00047-X

Library of Congress Card Number 68-29883

COPYRIGHT © 1968 BY AMERICAN ELSEVIER PUBLISHING COMPANY, INC.

ALL RIGHTS RESERVED. THIS BOOK OR ANY PART THEREOF
MUST NOT BE REPRODUCED IN ANY FORM WITHOUT THE WRITTEN
PERMISSION OF THE PUBLISHER, AMERICAN ELSEVIER PUBLISHING
COMPANY, INC., 52 VANDERBILT AVENUE, NEW YORK, N.Y. 10017

PRINTED IN THE UNITED STATES OF AMERICA

PREFACE

This book includes the papers and subsequent discussions presented at the Symposium on Interface Conversion for Polymer Coatings which was held at the General Motors Research Laboratories, Warren, Michigan, on October 2 and 3, 1967. This symposium considered the substrate (metallic and nonmetallic) and how it may be treated or converted to make it an effective acceptor for a polymer coating. To cover interface conversion in greater depth than is possible in a two-day conference, papers in specific disciplines, such as zinc phosphate crystal formation, surfactant adsorption, and oxide nucleation, were submitted by invitation, and these are included as part of the symposium proceedings. This eleventh annual symposium once again reflected the general theme of creating an atmosphere for an active exchange of ideas, an inventory of the state of the art, and future considerations required to understand fully the science of interface conversion.

The interdisciplinary nature of interface conversion brought forth the need to consider the total concept of substrate, polymer coating, and role of the interface between the substrate and the polymer coating. Of importance in this consideration is molecular structure, surface heterogeneities, diffusion and wetting in a capillary matrix, formation mechanisms of phosphate coatings, absorption, and polymer-solvent criteria at interfaces.

To characterize the symposium, it was decided to create a symbol which would depict the subject under discussion, namely interface conversion. It was an easy matter to represent an interface pictorially, but the question arose, How do we indicate that an interface is being changed or converted? In looking at electron microscope replicas of microcrystals of zinc phosphate, a replica of a 4-micron-long crystal growing across a scratch in the steel substrate seemed to show the theme

of the symposium. The original micrograph shown here was chosen as the Interface Conversion Symposium symbol.

Thanks are due Mr. R. L. Scott for the handling of local arrangements. Special recognition must go to Mr. S. R. Tiderington for his efficient and alert operation of the projection and recording facilities. We also appreciate the efforts of Miss S. Y. Hoffner and Mrs. V. H. Hartshorn for handling the secretarial functions associated with the symposium. Special thanks go to the session chairmen L. Onsager, F. M. Fowkes, P. H. Mark, and H. Burrell for creating a scientific atmosphere leading to active discussions.

PHILIP WEISS
G. DALE CHEEVER

CONTENTS

Preface v

Attendance List ix

Session I—Chairman: L. Onsager
Molecular Structure and Electrostatic Interactions at Polymer–Solid Interfaces
    J. C. Bolger and A. S. Michaels 3
Diffusion and Reaction of Polymers in a Capillary Matrix
    H. F. Mark 61
Interfacial Turbulence: Spontaneous Emulsification and Evaporative Convection
    A Contributed Discussion by D. T. Wasan 83

Session II—Chairman: F. M. Fowkes
Epitaxy and Corrosion Resistance of Inorganic Protective Layers on Metals
    A Contributed Discussion by A. Neuhaus *and* M. Gebhardt . 91
Surface Finishes of Metals from the Electrochemical Point of View .
    A Contributed Discussion by V. Cupr *and* B. Cibulka . . 120
The Kinetics of the Formation of Phosphate Coatings
    W. Machu 128
Wetting of Phosphate Interfaces by Polymer Liquids
    G. D. Cheever 150
The Role of Oxide Films in the Zinc Phosphating of Steel Surfaces
    J. V. Laukonis 182

Session III—Chairman: P. H. Mark
A Collective Viewpoint of Surfaces
    A Contributed Discussion by P. H. Mark 205
The Physical Chemistry of Surfaces and Surface Heterogeneities
    A. C. Zettlemoyer 208
Localized Oxidation Processes on Iron
    A Contributed Discussion by E. A. Gulbransen . . . 238
Interface Conversion of Polymers by Excited Gases
    R. H. Hansen 257

Dynamics of Ionic Adsorption on Polymers
   E. G. BOBALEK 289

Session IV—Chairman: H. Burrell
Polymer Adsorption on Substrates
   R. R. Stromberg 321
Adsorption of Surfactants at Polymer Interfaces
   A Contributed Discussion by R. G. GRISKEY 338
Factors in Interface Conversion for Polymer Coatings
   F. R. EIRICH. 350

AUTHOR INDEX 379

SUBJECT INDEX 385

ATTENDANCE LIST

(*Denotes Session Chairman; **denotes speaker)

W. A. ALBERS, JR.
Research Laboratories, GMC†

R. G. ALDRICH
Syracuse University
Syracuse, New York

R. D. ANDREWS
Stevens Institute of Technology
Hoboken, New Jersey

W. K. ASBECK
Union Carbide Corporation
South Charleston, West Virginia

W. J. BAXTER
Research Laboratories, GMC

S. E. BEACOM
Research Laboratories, GMC

P. BECHER
Atlas Chemical Industries, Inc.
Wilmington, Delaware

G. R. BELL
GM Institute, GMC
Flint, Michigan

H. S. BENDER
Research Laboratories, GMC

R. J. BLACKINTON
Interchemical Corporation
Detroit, Michigan

**E. G. BOBALEK (pp. 289–317)
University of Maine
Orono, Maine

**J. C. BOLGER (pp. 3–60)
Amicon Corporation
Lexington, Massachusetts

D. B. BRUCE
Pinchin, Johnson & Associates, Ltd.
Coventry, Warks, England

R. BULT
Paint Research Institute
Delft, The Netherlands

*H. BURRELL
Interchemical Corporation
Clifton, New Jersey

L. R. BUZAN
Research Laboratories, GMC

C. J. CAMPBELL
Hercules Research Center
Wilmington, Delaware

J. M. CAMPBELL
Research Laboratories, GMC

W. R. CAMPBELL
Oldsmobile Division, GMC

J. D. CAPLAN
Research Laboratories, GMC

H. CHARLES
Chemfil Corporation
Troy, Michigan

† GMC = General Motors Corporation.

**G. D. CHEEVER (pp. 150–181)
Research Laboratories, GMC

P. F. CHENEA
Research Laboratories, GMC

J. E. COWLING
U.S. Naval Research Laboratory
Washington, D.C.

D. W. CRIDDLE
Chevron Research Corporation
Richmond, California

R. E. CUTHRELL
Sandia Corporation
Albuquerque, New Mexico

R. DAVIES
Research Laboratories, GMC

A. DOUTY
Amchem Products, Inc.
Ambler, Pennsylvania

**F. R. EIRICH (pp. 350–377)
Polytechnic Institute of Brooklyn
Brooklyn, New York

E. EUSEBI
Research Laboratories, GMC

W. F. FLANAGAN
Research Laboratories, GMC

L. L. FLECK
Research Laboratories, GMC

*F. M. FOWKES
Lehigh University
Bethlehem, Pennsylvania

W. F. FOWLER, JR.
Eastman Kodak Company
Rochester, New York

Z. G. GARDLUND
Research Laboratories, GMC

R. W. GIBSON
Research Laboratories, GMC

M. S. GLASER
Midland Industrial Finishes
Company
Waukegan, Illinois

R. G. GRISKEY (pp. 338–349)
Newark College of Engineering
Newark, New Jersey

J. E. GUILLET
University of Toronto
Toronto, Canada

E. GUTH
Oak Ridge National Laboratory
Oak Ridge, Tennessee

G. L. GUTSCHER
Buick Motor Division, GMC

L. R. HAFSTAD
Research Laboratories, GMC

**R. H. HANSEN (pp. 257–288)
J. P. Stevens & Co. Inc.
Garfield, New Jersey

D. R. HAYS
Research Laboratories, GMC

W. HELLER
Wayne State University
Detroit, Michigan

J. J. HERMANS
Chemstrand Research Center
Durham, North Carolina

G. A. ILKKA
Research Laboratories, GMC

ATTENDANCE LIST

E. L. Jacks
Research Laboratories, GMC

H. H. G. JELLINEK
Clarkson College of Technology
Potsdam, New York

S. KATZ
Research Laboratories, GMC

D. V. KELLER, JR.
Syracuse University
Syracuse, New York

H. J. KIEFER
The Glidden Company
Cleveland, Ohio

J. N. Koral
American Cyanamid Company
Stamford, Connecticut

A. KUHLKAMP
Farbwerke Hoechst AG
Frankfurt, Federal Republic of Germany

C. A. KUMINS
A-M Corporation
Mt. Prospect, Illinois

**J. V. LAUKONIS   (pp. 182–202)
Research Laboratories, GMC

L. H. LEE
Dow Chemical Company
Midland, Michigan

G. L. LEITHAUSER
Research Laboratories, GMC

L. L. LEWIS
Research Laboratories, GMC

H. K. LIVINGSTON
Wayne State University
Detroit, Michigan

**W. MACHU   (pp. 128–149)
Technische Hochschule
Vienna, Austria

**H. F. MARK   (pp. 61–82)
Polytechnic Institute of Brooklyn
Brooklyn, New York

J. E. MARK
University of Michigan
Ann Arbor, Michigan

*P. H. MARK   (pp. 205–207)
Princeton University
Princeton, New Jersey

E. MATIJEVIĆ
Clarkson College of Technology
Potsdam, New York

J. I. MAURER
Hooker Chemical Company
Detroit, Michigan

D. G. McCULLOUGH
Pontiac Motor Division, GMC

**A. S. MICHAELS   (pp. 3–60)
Amicon Corporation
Lexington, Massachusetts

A. L. MICHELI
Research Laboratories, GMC

W. E. MITCHELL
Fisher Body Division, GMC

T. O. MORGAN
Research Laboratories, GMC

N. L. MUENCH
Research Laboratories, GMC

R. R. MYERS
Kent State University
Kent, Ohio

*L. ONSAGER
Yale University
New Haven, Connecticut

S. PARTINGTON
Metasurf Corporation
Detroit, Michigan

T. C. PATTON
The Baker Castor Oil Company
Bayonne, New Jersey

A. PETERLIN
Research Triangle Institute
Research Triangle Park,
North Carolina

E. A. PRASCHAN
Research Laboratories, GMC

G. M. RASSWEILER
Research Laboratories, GMC

A. E. RHEINECK
North Dakota State University
Fargo, North Dakota

R. A. RICHARDSON
Research Laboratories, GMC

A. A. ROBERTSON
Pulp & Paper Research Institute of
Canada, Pointe Claire, Quebec,
Canada

M. ROSOFF
Columbia University
New York, New York

F. G. ROUNDS
Research Laboratories, GMC

G. G. SCHURR
The Sherwin-Williams Co.
Chicago, Illinois

A. M. SCHWARTZ
Gillette Research Institute
Washington, D.C.

R. L. SCOTT
Research Laboratories, GMC

L. H. SHARPE
Bell Telephone Laboratories
Murray Hill, New Jersey

R. W. SMITH
GM Institute, GMC
Flint, Michigan

M. STAND
Sealectro Corporation
Mamaroneck, New York

**R. R. STROMBERG (pp. 321–337)
National Bureau of Standards
Washington, D.C.

J. D. THOMAS
Research Laboratories, GMC

R. F. THOMSON
Research Laboratories, GMC

S. R. TIDERINGTON
Research Laboratories, GMC

J. W. TOMECKO
University of Waterloo
Waterloo, Ontario, Canada

A. F. UNDERWOOD
Research Laboratories, GMC

L. VALENTINE
Paint Research Station
Teddington, Middlesex, England

H. VAN OLPHEN
National Research Council
Washington, D.C.

ATTENDANCE LIST

W. von Fischer
Switzer Brothers, Inc.
Cleveland, Ohio

**D. T. Wasan (pp. 83–88)
Illinois Institute of Technology
Chicago, Illinois

P. Weiss
Research Laboratories, GMC

H. L. Williams
University of Toronto
Toronto, Canada

M. Wismer
Pittsburgh Plate Glass Co.
Springdale, Pennsylvania

J. J. Wojtkowiak
Research Laboratories, GMC

P. E. Wright
Research Laboratories, GMC

**A. C. Zettlemoyer (pp. 208–237)
Lehigh University
Bethlehem, Pennsylvania

J. W. Zimmerman
Chevrolet Motor Division, GMC

W. S. Zimmt
Marshall Laboratory DuPont
Philadelphia, Pennsylvania

SESSION I

Chairman
L. ONSAGER

Molecular Structure and Electrostatic Interactions at Polymer–Solid Interfaces

J. C. BOLGER and A. S. MICHAELS

Amicon Corporation
Lexington, Massachusetts

Introduction

This paper will attempt to focus on the chemical, physical, and geometric characteristics of certain types of high-energy solid surfaces and to propose some generalizations regarding specific mechanisms whereby these surfaces can interact with polymeric materials commonly used as coatings and adhesives. In particular, this paper is concerned with the surface structure and reactivity of metals, metal oxides, and silicates which are of broad commercial and practical importance.

In addition, this paper will consider which types of specific interfacial interactions should, in theory, best serve to prevent or retard bond disruption after exposure to water, humid air, or other important environmental conditions.

During the past decade, a major body of literature has been generated which raises the question of whether either specific chemical interactions or the formation of covalent or ionic bonds is necessary to provide good adhesion between an adherend and an adhesive of dissimilar composition. The authors who raise this question (*1–4*) have maintained, and have accompanied their theoretical arguments with extensive experimental data, that the formation of specific interfacial bonds across an adhesive–adherend interface is neither necessary nor essential to form a good bond. According to this school of thought, every material adheres to every other material, provided that good interfacial contact is achieved. This close contact alone should always be sufficient to yield an adhesive bond strong enough to cause failure of a joint, under mechanical loading, in either the adhesive or the adherend phase, rather than at the interface.

Bikerman, for example, has long advocated that, if complete wetting between adhesive and adherend is achieved, interfacial separation is impossible. Bikerman argues (*2*) partially from a probability standpoint—that is, that once failure initiates, the plane of failure may propagate along or away from the interface in either direction, so that after a finite area of separation has been generated, the probability becomes small that the loci of separation will be entirely along the interface. Primarily, however, Bikerman has postulated that interfacial attractive forces between phases 1 and 2 never deviate appreciably from the approximation

$$A_{12} \approx \sqrt{A_{11}A_{22}} \tag{1}$$

where A_{11} and A_{22} are the intermolecular attractive forces in phases 1 and 2, and A_{12} is the intermolecular attractive force across the interface.

If this approximation were universally true, then indeed, whenever A_{11} is different from A_{22}, A_{12} would always be larger than one or the other of the cohesive force terms.

The criteria for achieving the degree of wetting and spreading needed to provide the desired degree of high interfacial contact at the interface have been developed primarily by Zisman and his co-workers (5–7). Zisman *et al.* have developed the useful concept of critical surface tension for wetting (γ_c), which can be regarded as a fundamental property of an adherend surface. Liquids of surface tension less than γ_c will spontaneously wet, and spread upon, the solid surface, giving a contact angle of zero, and should therefore provide the maximum degree of interfacial contact. When liquids having γ_1 greater than γ_c are applied to the surface, contact angles greater than zero are produced.

Sharpe and Schonhorn (4, 8) have proposed that the best criterion for selecting an adhesive is that its γ should be less than γ_c for the adherend surface. They maintain that this criterion must be satisfied to have the maximum possible wetting and performance and that, when wetting is complete, interfacial separation cannot occur. The maximum reversible work of adhesion in this case, if we assume that failure occurs in the adhesive layer, equals $2\gamma_1$. Theoretically, of course, the force necessary to cause failure under such conditions is very large, even for low-energy adhesives. For example, if we assume that failure occurs after a separation in the adhesive layer of x, approximately 4 Å, the maximum bond strength calculated for a low-energy adhesive such as polyethylene is extremely high:

$$F_{\max} \approx \frac{2\gamma_1}{x} = \frac{2 \times 31}{4 \times 10^{-8}} = 1.55 \times 10^9 \frac{\text{dynes}}{\text{cm}^2} = 22{,}800 \text{ psi} \tag{2}$$

Thus, the adhesive forces attainable via simple adsorption—that is, without requiring specific covalent or other interfacial forces across the interface—are indeed theoretically higher than the strength of existing polymeric materials.

Recently, Huntsberger (9, 10) analyzed the adsorption theories of adhesion mentioned above and pointed out that limiting the selection of coating or adhesive compositions to those that satisfy the sole criterion of $\gamma_1 < \gamma_c$ is tantamount to limiting the selection to low-energy adhesives which will fail cohesively. Huntsberger also pointed out that Zisman's equations derived for maximum work of adhesion lead to the prediction that the work of adhesion can be greater than $2\gamma_c$ in certain cases where an adhesive of γ_1 larger than γ_c is used, despite the fact that interfacial contact may be incomplete in such cases. Moreover, Huntsberger showed that the spreading rate for most liquids having γ_1 greater than γ_c is generally still positive and that, given sufficient time, good interfacial contact should be produced unless the solidification rate for the adhesive is very rapid or $\gamma_1 - \gamma_c$ is unusually large.

The present authors agree that the energetics involved in simple physical adsorption of polymeric materials on high-energy surfaces can, theoretically, be more than sufficient to form joints which are stronger than the cohesive strength of existing coating and adhesive materials, and that, given good interfacial contact, adhesive joints essentially never fail at the interface under mechanical loading. After testing of several hundred adhesive formulations in this laboratory, by a variety of tensile and peel strength measurements on carefully made joints, close examination (under high microscope magnifications where necessary) has never failed to show small quantities of either the adhesive or the adherend phase remaining on one or the other of the fractured surfaces. But, as Zisman points out (7), specific chemical bonding can offer major advantages other than joint strength—for example, in retention of adhesion after exposure to heat, water, or other chemical environments.

Because of the complexity of the analysis, most previous investigators interested in specific interfacial forces have been forced to study relatively simple adhesive–adherend compositions, such as polyethylene on steel or other combinations where either the adherend or the adhesive is nonpolar, and where interfacial forces are due primarily to London force interactions. The fact remains, however, that essentially all polymeric materials used as coatings or adhesives in combination with metallic or oxide surfaces not only are relatively polar in nature, but consist of a complex mixture of poorly characterized ingredients. For these polar complex mixtures, it is extremely difficult to interpret the results of adhesion tests, and particularly of fracture experiments, in meaningful quantitative terms. Notwithstanding these difficulties, the present authors feel that, since these polar polymeric compositions are of such overwhelming practical importance, an attempt (albeit qualitative) should be made to provide some useful generalizations on how such compositions can interact with metal or oxide surfaces by forces other than simple London (dispersion) interactions.

Methods for Estimation of Interfacial Forces

Consider the system shown in Fig. 1, wherein a polymeric material, initially in intimate interfacial contact with an inorganic solid substrate, is exposed to an aqueous environment (either bulk water or humid air) such that the

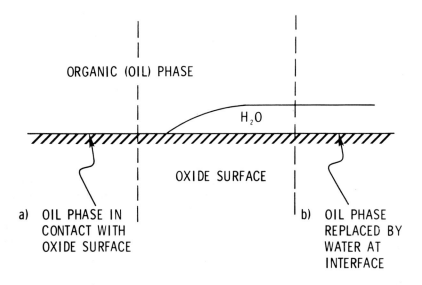

FIG. 1. Three-phase oil–water–solid oxide system.

water is able to penetrate to the interface either by diffusion through the polymeric film (the usual case for coatings), or by surface diffusion in from the exposed edges (as in most adhesive joints), or by any combination of the two. For our purposes, we shall consider displacement by water to occur if an interfacial water film more than a few molecules thick (Fig. 1b) accumulates at the interface. An interfacial separation of this type is known, from experience, to destroy the utility of a protective coating and the mechanical properties of an adhesive joint. Obviously, the free energy change in going from Fig. 1a to Fig. 1b—that is, the free energy change to create an organic–water and a substrate–water interface and to eliminate the original organic–substrate interface is

$$\Delta F = \gamma_{OW} + \gamma_{SW} - \gamma_{OS} \tag{3}$$

If ΔF is negative, the organic phase has a thermodynamic tendency to be displaced by water. This alone, however, is a necessary but not a sufficient requirement for predicting water displacement. This is a classic example of a process which is highly rate-dependent. It is probably safe to state that a major portion of the coatings and adhesives in successful commercial use today would not be able to resist loss of adhesion in contact with water if equilibrium conditions were to prevail at the interface. Rather, such materials retain their useful adhesive lifetimes because of rate-limiting factors which include:

1. Rate of diffusion of water molecules through the organic phase to the interface.
2. Solubility of water in the organic phase.
3. The elastic modulus of the organic phase—particularly in the immediate vicinity of the interface.
4. Surface topography—scale of surface irregularities and roughness.
5. Initial degree of approach to complete organic–solid interfacial contact.

Several of these factors are, of course, interrelated. Diffusivity and solubility both generally decrease, and elastic modulus increases, as the cross-link density of the resin phase increases. Unfortunately, an adequate treatment of these kinetic factors not only is beyond the scope of the present discussion of interfacial bonding mechanisms but also requires considerably more experimental evidence than is presently available. It is hoped that subsequent investigators will be able to couple the present treatment of equilibrium factors with predictive correlations which take into account the chemical, mechanical, and geometric parameters which govern the kinetic factors.

For the present, we address ourselves to a discussion of the methods for estimating the relative magnitudes of the interfacial tensions across the organic–water, substrate–water, and organic–substrate interfaces to form some generalizations regarding the thermodynamic tendency of a film to resist or to yield to displacement by water, and we shall illustrate these generalizations by some experimental measurements where rate factors can be eliminated. To do this, we must first have some method of computing the interfacial tension terms given in equation 3. Over the past decade, two (somewhat related) computational methods have evolved—the method of Good, Girifalco, *et al.*, and that of Fowkes. Both methods are based on attempts to adapt the earlier successful solubility theories of Hildebrand, Scott, and Hirschfelder (*11*, *12*) to the problem of estimating force interactions across an interface, and both are based on the following two physical concepts:

1. Additivity of Forces. The cohesive energy density (C.E.D.) of a given liquid or solid material is defined as the sum of all energies of intermolecular interaction per unit volume, and is equal to the work needed to separate all interacting sites from their original equilibrium distance to infinite separation. The additivity principle states that if these interactions are due to a number of different types of forces—that is, to force types i, j, etc.—then

$$\text{C.E.D.} = \sum n_i a_i + n_j a_j + \cdots + \tag{4}$$

where n_i is the number of attractive forces of type i, and a_i is the magnitude of the individual attractive forces of type i.

TABLE I

BOND TYPES AND TYPICAL BOND ENERGIES (13, 14)

Type	Bond Energy (kcal/mole)
1. Ionic	140–250
2. Covalent	15–170
3. Metallic	27–83
4. Permanent dipole–dipole interactions	
Hydrogen bonds involving fluorine	Up to 10
Hydrogen bonds excluding fluorine	5–8
Other dipole–dipole (excluding hydrogen bonds)	1–5
5. Dipole–induced dipole	Less than 0.5
6. London (dispersion) forces	0.02–10

2. Inverse-Square Relationship. First postulated by Berthelot for van der Waals' forces, this relationship states that for a given type of intermolecular force

$$A_{12} = \sqrt{A_{11} A_{22}} \tag{5}$$

where A_1 and A_2 are the interaction energies in components 1 and 2, respectively, and A_{12} is the resulting magnitude of the interaction energy between component 1 and component 2.

Regarding the types of attractive forces to be used in these relationships, Pauling (*13*) distinguishes three general types of chemical bond: metallic, covalent, and electrostatic. Since the various subtypes of electrostatic attraction are of major significance to the present study, these are listed in their generally recognized subvariations and are shown with estimates of the range of magnitude of bond energies in Table I.

In adapting these solubility concepts to estimation of interfacial forces, both Good and Fowkes assume that

$$\text{Total interfacial forces} = \sum \text{London forces} + \text{dipole–dipole forces} + \cdots + \quad (6)$$

Girifalco and Good (*15*) and Good and later co-workers (*16–19*) assume that not all forces in material 1 can interact with all forces in material 2, for several reasons. Some force components in material 1 have no counterpart in material 2 (for example, when a metal and a nonmetal are in contact). In addition, because of poor lattice fit at the interface—for example, because of large differences in molar volumes in the two phases—the number and intensity of interfacial force interactions may be further decreased. Good expresses these inefficiencies by means of an "interaction efficiency" parameter, Φ, so that the effective interaction term is

$$A_{12} = \Phi\sqrt{A_{11}A_{12}} \quad (7)$$

and the corresponding expression for the interfacial tension is

$$\gamma_{12} = \gamma_1 + \gamma_2 - \Phi\sqrt{\gamma_1\gamma_2} \quad (8)$$

where

$$\Phi = \Phi_A \Phi_V \quad (9)$$

$$\Phi_V = \frac{4(V_1 V_2)^{1/3}}{(V_1^{1/3} + V_2^{1/3})^2} \quad (10)$$

V_1, V_2 = molar volumes of phases 1 and 2

Φ_A = ratio of force components which can interact across the interface to the total internal force components

All Φ terms must, of course, be equal to or less than unity: Φ_V is the interfacial mismatch term; Φ_A is a considerably more complex term which can be written in closed form and solved numerically only for a few special cases—for example, for pure dispersion–force interaction, or for several simple combinations of dipole–dipole, dipole–induced dipole, and London interactions. Note, however, that the Good approach has many useful implications. For example, for any given series of homologous liquids on a given surface, Zisman's critical surface tension for wetting (γ_c) is given by

$$\gamma_c = \Phi^2 \gamma_{so} \quad (11)$$

That is, γ_c is always less than γ_{so} (the surface energy of the solid in vacuum) by the square of Good's efficiency factor, reflecting the fact that only a portion of the forces in the solid are able to interact with molecules across the interface.

Fowkes (20, 21) first extended the original Good and Girifalco equations to certain cases where the internal forces in two different materials were of unlike types and showed that the dispersion energy components of these internal forces could be deduced from relatively simple contact angle measurements. When dispersion (London) forces *only* are responsible for interfacial attraction, the Fowkes equation is

$$A_{12} = \sqrt{A_1^d A_2^d} \tag{12}$$

where A_1^d is the dispersion component of the internal forces in material 1.

The interfacial tension in this case, therefore, is

$$\gamma_{12} = \gamma_1 + \gamma_2 - \sqrt{\gamma_1^d \gamma_2^d} \tag{13}$$

The advantage of the Fowkes approach is that the dispersion force component for a large number of liquids and solids can be measured and tabulated, and these values can then be used to compute the interfacial tension for any pair wherein only London forces interact. For example, from a series of interfacial tension measurements of water versus hydrocarbons, Fowkes calculates (22) that the surface tension of water is equal to the sum of a term attributable to internal London interactions, plus a term attributable to hydrogen bonding:

$$\gamma_{H_2O} = \gamma_{H_2O}^d + \gamma_{H_2O}^h \tag{14}$$
$$71.8 = 21.8 + 50.0 \text{ dynes/cm}$$

and for mercury (internal London plus metallic forces):

$$\gamma_{Hg} = \gamma_{Hg}^d + \gamma_{Hg}^m \tag{15}$$
$$484 = 200 + 284 \text{ dynes/cm}$$

From these values Fowkes can estimate the interfacial tension between water and mercury, since only the dispersion force components can interact between these two substances. For this case,

$$\gamma_{H_2O/Hg} = \gamma_{H_2O} + \gamma_{Hg} - \sqrt{\gamma_{H_2O}^d \gamma_{Hg}^d} = 424.8 \text{ dynes/cm} \tag{16}$$

which agrees well with the best experimental values of 426 to 427 dynes/cm.

Where direct comparisons can be made, the Fowkes approach and the Good equations are in relatively good agreement. For water versus nonpolar organic surfaces such as polyethylene, for example, Good's equations for Φ generally give a value of about 0.5 to 0.6; the coupling efficiency factor computed via the Fowkes approximation gives about the same result:

$$\frac{\sqrt{\gamma^d_{H_2O}\gamma^d_{PE}}}{\sqrt{\gamma_{H_2O}\gamma_{PE}}} = \frac{\sqrt{21.8 \times 31}}{\sqrt{71.8 \times 31}} = 0.55 \quad (17)$$

Unfortunately, these theories are not able to yield quantitative predictions in the more complex cases where a relatively polar coating or adhesive bonds to a hydrophilic, polar surface through a combination of forces (possibly including primary valence bonds) in addition to dispersion force interactions. These equations are also presently unable to predict whether a polar solid will be able to resist displacement by a third (liquid) component, such as water.

Zisman and co-workers (23–25) have studied extensively the factors leading to displacement on hydrophilic surfaces of one liquid, either water or oil, by a second "displacing" liquid. These studies have been aimed primarily at developing special classes of displacing liquids which can cause removal (for example, for salvaging purposes) of undesired aqueous or oil films on steel and other surfaces. It was found that the general requirements for the displacement of liquid A, by the displacing liquid B, are:

1. A low value of γ_B. Fluorinated alcohols, other long-chain fluorocarbons, and low-molecular-weight dimethyl silicones, having γ_B as low as 14 to 20, were found to be the most effective displacing liquids.

2. A high-equilibrium spreading pressure (that is, a maximum value of $\gamma_A - \gamma_B$). As liquid B is applied on top of, and spreads rapidly on, liquid A, part of liquid B's displacing action is due to removal of the underlying liquid by viscous drag effects.

3. High solubility of liquid B in liquid A.

4. An ability of liquid B to depress the surface tension of liquid A.

5. High volatility of liquid B when *rapid rates* of displacement are required.

Most of the conditions pertinent to the work of Zisman and co-workers do not, however, pertain to the present problem of displacement of a solid, polar, organic material from an oxide surface via a high-energy liquid such as water. In this case, there are no viscous drag forces, $\gamma_A - \gamma_B$ is generally negative, volatility is not an important factor, and very little cosolubility of the aqueous and organic phases results. Under Zisman's conditions, specific solid–liquid interactions are not important except where special ionic surfactants are added to liquid B to deposit on the cleaned surface to prevent rewetting by the displaced liquid. For long-term retention of adhesion of a solid, high-molecular-weight polymeric material to hydrophilic surfaces in the presence of water, it is essential to consider the types and relative strengths of possible ionic, covalent, and other specific interactions at the polymer–substrate interface. To do this, one must first attempt to develop a general model for the nature and structure of the adhered surfaces of interest and

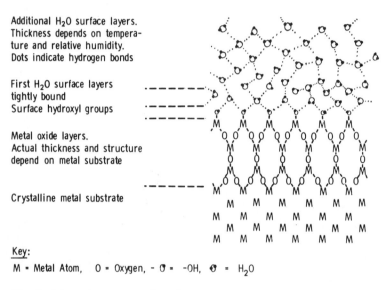

FIG. 2. Schematic representation of water and oxide layers on metal surface.

then explore modes by which either water or a polymeric solid may form attachments to this surface.

Physicochemical Nature of High-Energy Metal, Metal Oxide, or Silicate Surfaces

This discussion is restricted to the subject of adhesion to nonporous metal, metal oxide, or silicate surfaces and excludes consideration of porous, cellulosic, polymeric, or other organic adherends where either mechanical interpenetration or interdiffusion, of the type studied by Voyutski (26) or Stromberg (27), is a major factor in providing adhesion.

Figure 2 shows a schematic representation of the basic model assumed for the hydrated oxide surface of the metals, oxides, and silicates of interest in this study. The essential features of this model are, of course, not novel, but rather are based on the results of a large number of prior studies of adherend surfaces (*28–33*). The three basic assumptions for this model are as follows:

1. A metal surface is, in fact, a metal oxide surface. Surface oxide films may be thin for mercury or for noble metals (gold, platinum) but thick enough (40 to 80 Å or more) for metals and alloys of primary commercial importance (nickel, iron, aluminum, copper, zinc, magnesium, etc.) to eliminate, for purposes of estimating attractive forces, interaction effects due to metal atoms underlying the oxide layer.

2. While the oxide surface can be dehydrated and dehydrogenated at elevated temperatures, at normal ambient bonding conditions the outermost surface oxygens hydrate to form a high density of hydroxly groups. Brooks (*34, 35*), Zettlemoyer (*36, 37*), and Ter-Minassian-Saraga (*38*), for example, have estimated that about one silanol group is present per 60 Å^2 on the surface of a variety of glasses, silicas, montmorillonites, and aerosils.

3. The hydroxyl-rich surface adsorbs, and strongly retains, several molecular layers of bound water.

To cite a few of the many experiments supporting this model: Debye and Van Beek (*39*) found that silica powder, washed with water and then dried at 100°C, retained an adsorbed surface layer several hundred angstroms in thickness, and that this adsorbed water layer could not be removed completely at temperatures below 500°C. Bowden and Throssell (*40*) found that, for smooth aluminum, iron, and SiO_2 surfaces at ambient temperatures and humidities, up to twenty molecular layers of water were present on "dry" surfaces, although with special precautions to remove the hygroscopic surface contaminants normally present, the surface H_2O film could be reduced to about two molecular layers in thickness at low humidities. Bowden and Tabor (*41, 42*) report similar water film thicknesses on platinum surfaces.

Shafran and Zisman (*43*) have also reported on contact angles measured for nonpolar liquids on carefully cleaned silica as a function of water activity in the vapor phase. At normal temperatures and humidities, the contact angles measured on silica approach those predicted for a bulk water surface. Shafran and Zisman find that, as water activity in the vapor phase decreases, the contact angles decrease (that is, the effective surface energy of the substrate increases), although either very high temperatures or very low partial pressures (about 10^{-6} mm) are required to approach the high surface energies expected for dry silica.

Similarly, many investigators have verified the influence of oxide films on wetting of one metal by another metal. Keller (*44*), for example, studied the wetting of liquid gallium on smooth nickel surfaces and found that wetting is prevented unless the nickel oxide surface film and all adsorbed gases are removed. When this is done—for example, by ion bombardment in high vacuum—metallic bonds can be established between the gallium and nickel, and the gallium spreads.

If, in light of the foregoing evidence, one accepts that the model shown in Fig. 2 is correct, and that virtually all metal oxides and silicates have a similar hydrated oxide surface structure, does this imply that all such materials should behave identically in surface interactions with water and with organic coatings and adhesives? In certain respects this is true to a first approximation. Fowkes (22) analyzed data of Boyd and Livingston (45, 46) on heats of immersion for a series of different inorganic powders in water and showed that the excess interaction energy (that is, the free energy of adsorption in excess of that predicted by London forces) is nearly the same in all cases studied. Similarly, Wade and Hackerman (47) measured enthalpies of adsorption for water, methanol, and hexane on three powdered oxides (SiO_2, Al_2O_3, and TiO_2). While the values vary widely for any single liquid on any single oxide, owing primarily to particle size effects, the range of values for each of the three liquids is approximately the same on all three surfaces. Water gave a range of about 250 to 600 ergs/cm², while hexane gave a range of about 70 to 140 ergs/cm² for all three oxides.

Generally speaking, however, our practical experience is that different metals, oxides, or silicates vary widely in their surface interactions with organic adhesives. To understand why this is so, one must examine the *differences* between the hydrated oxide surfaces on such materials. These differences include:

1. Geometric Factors. The spacing and regularity of the surface hydroxyls generally correlate with the spacing of the metal atoms in the oxide. Oxides on metals above grain boundaries or above other amorphous regions exhibit a different surface energy from that shown by oxides above crystalline regions, as illustrated by Wade and Hackerman's (47) heat of wetting measurements on amorphous versus crystalline silica, and by Cheever's (48) nucleation experiments described later in this Symposium.

2. Rate of Hydration and Hydroxyl Formation. Many experiments have shown that the rate of adsorption of water is very rapid; for example, Irwin's (49) fracture experiments with glass and Orowan's (50) and Bowden and Tabor's (41) cleavage experiments with mica show that the crack propagation velocities in glass and the energy to cause cleavage in mica depend strongly on the activity of water in the surrounding vapor phase. Orowan (50), in cleaving mica in high vacuum, estimates energies of the freshly exposed surfaces to be about 4500 ergs/cm², whereas cleavage in air with water vapor present yields the much lower values of 375 to 500 ergs/cm², thereby illustrating both the extreme rapidity with which surfaces hydrate and the profound ability of these adsorbed surface layers to reduce the surface energy of a high-energy solid.

For most metals and oxides studied (*33, 51–53*), however, the rate of the hydroxyl formation reaction via

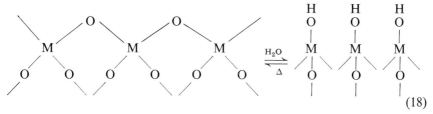
(18)

is slow relative to the rates for simple surface adsorption of H_2O.

Estimates of the surface population, and rate of formation, of hydroxyl groups on silica and silicates have usually been based on infrared measurement of the absorption bands at about 2.6 to 2.9 microns, which are characteristic of ionic hydroxyls, and at the slightly larger wavelengths characteristic of hydrogen-bonded hydroxyl groups (*28*). Such measurements indicate that the number of hydroxyl groups per unit surface area depends on oxide composition, level and type of impurities, prior thermal history, time and temperature of exposure to water vapor, oxide crystallinity and particle size, and other factors. Nevertheless, even for those surfaces that hydroxylate slowly, such as freshly calcined TiO_2, Al_2O_3, and Fe_2O_3, reaction 18 is rapid enough to validate the assumption that a significant fraction of the metal atoms at the surface exist as —MOH groups under virtually all practical bonding conditions. At room temperature, for example, freshly calcined alumina hydrates in 1 to 2 days to yield a surface having the properties of $Al(OH)_3$, while Fe_2O_3 hydrates somewhat more rapidly to give a surface resembling —FeOOH (*47*).

3. *Activity of the Surface Hydroxyl Groups.* The attractive forces attributable to the surface hydroxyl groups shown in Fig. 2 depend strongly on the electron density of the oxygen atoms in the group —MOH. If the electron density on the oxygen atom is low, the strength of the hydrogen bonds which can be formed with polarized hydrogen atoms in an adhesive phase is reduced, and the probability of ionization via the following dissociative surface reaction is increased:

$$\text{—MOH} + H_2O \rightleftarrows \text{—M}\overline{O} + H_3O^+ \tag{19a}$$

Not all oxides acquire a surface charge by dissociation of the surface —MOH group. As described below, some oxides are believed to ionize via a (nondissociative) adsorption process for which the potential determining ions are H^+, OH^-, $[M^{z+}(OH)_{z-1}]^+$, or $[M^{z+}(OH)_{z+1}]^-$; that is,

$$-MOH + OH^- \rightleftarrows -MOH \cdots OH^- \tag{19b}$$

Similarly, a high electron density on the surface oxygen atoms increases its tendency to bind protons, either to form hydrogen bonds with compounds containing polarized hydrogen atoms, or to ionize via the dissociation reaction:

$$-MOH + H_2O \rightleftarrows -M\overset{+}{O}H_2 + OH^- \tag{20a}$$

while for the corresponding nondissociative adsorption reaction:

$$-MOH + H_3O^+ \rightleftarrows -M\overset{\overset{H}{|}}{O} \cdots H_3O^+ \tag{20b}$$

The equilibria for the above reactions depend both on the solid and on the electrolyte in which it is immersed. Values for the isoelectric point (IEPS) of solid oxides and silicates have traditionally been measured in water by a variety of techniques used to estimate the pH at which the net surface charge is zero. Analysis of the factors that influence the IEPS of various oxides in water should, therefore, also serve as a qualitative guide to the factors that govern the ability of these surfaces to lose, donate, attract, or bind protons in contact with nonaqueous liquids and solids.

Parks (54) has compiled and has critically reviewed the literature on prior measurements of the isoelectric points for a large number of solid oxides and hydroxides. The values which Parks considers most reliable for silicon, iron, and aluminum oxides are shown in Table II; they illustrate several points. For a given metal oxide such as iron or aluminum, the IEPS can vary over a relatively broad range, reflecting:

1. The composition and crystalline (or amorphous) form of the oxide.
2. Lattice substitution and impurities.
3. The degree of hydration and hydroxylation of the surface.
4. The state of oxidation or reduction.

Oxides representing the maximum degree of oxidation (highest cation valence) give the lowest IEPS—for example, the ferrous versus the ferric oxides in Table II. Similarly, Parks shows that for the various oxides of molybdenum the IEPS decreases as the molybdenum valence increases, from a value in excess of 11.5 for Mo_2O to a value below 0.5 for Mo_2O_5 and MoO_3. Fully hydrated surfaces give a higher IEPS than freshly calcined surfaces (Table III), reflecting the time dependence of the hydroxyl formation reactions (equation 18).

TABLE II
Isoelectric Points for Solid Oxides in Water (54)

Oxide	IEPS Range
α-Al_2O_3	6.6–9.2
γ-Al_2O_3	7.4–8.6
AlOOH (boehmite)	6.5–8.8
AlOOH (diaspore)	5.0–7.5
$Al(OH)_3$ (bayerite)	5.4–9.3
Amorphous Al_2O_3	7.5–8.0
$Fe(OH)_2$	11.5–12.5
Natural Fe_2O_3 (hematite)	5.4–6.9
Synthetic Fe_2O_3	6.5–8.6
Fe_2O_3	6.5–6.9
FeOOH (goethite)	6.5–6.9
FeOOH (lepidocrocite)	7.4
Hydrous Fe_2O_3 and amorphous hydroxides	8.5
Fe_3O_4	6.3–6.7
SiO_2 (quartz)	2.2
SiO_2 sols and gels	1.8

Parks has attempted to correlate the IEPS values for a large number of oxides by a simple electrostatic model involving cationic size and charge. For the equilibrium reaction between positive and negative surface sites written as

$$\text{—MO}^- \text{ (surface)} + 2H^+ \rightleftarrows \text{—MOH}_2^+ \text{ (surface)} \quad (21)$$

Parks first estimates the electrostatic work gained by the approach of two protons from an initial infinite separation to a final distance (R) equal to the radius of the oxalated cation and then equates this work term to the free energy change for equation 21, for which

$$-\Delta F = RT \ln \frac{[MOH_2^+]}{[MO^-][H^+]^2} \quad (22)$$

TABLE III
Effect of Hydration on IEPS (54)

Solid	IEPS Freshly Calcined	IEPS After Hydration	Difference
Fe_3O_4	6.7	8.6	1.9
Al_2O_3	6.7	9.2	2.5
TiO_2	4.7	6.2	2.5

At the IEPS, $[MOH_2^+] = [MO^-]$, and

$$-\Delta F = RT \ln \frac{1}{[H^+]^2} \tag{23}$$

After rearrangement, Parks' simplified electrostatic model leads to the equation

$$\text{IEPS} = C_1 - C_2\left(\frac{Z}{R}\right) \tag{24}$$

where C_1 and C_2 are constants characteristic of the electrolyte.

Parks' plot of IEPS versus Z/R for 26 oxides is shown in Fig. 3, where the Z/R values have been corrected for crystal field effects. While this plot can be regarded only as being in very approximate agreement with equation 24, it

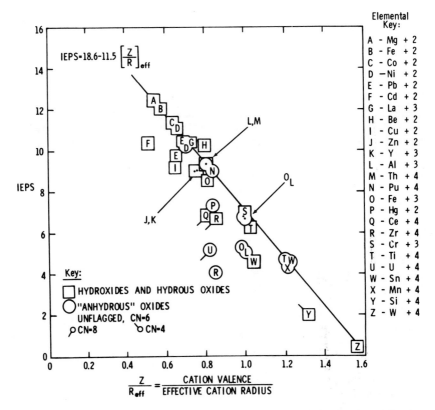

FIG. 3. Parks' equation for IEPS. Radius corrected for crystal field effects, coordination, and hydration (54).

does help to explain the fact that the IEPS is highest for the divalent metal (magnesium, calcium, ferrous, nickel, etc.) oxides, is generally in the medium pH range for most hydrated trivalent metal oxides, and is lowest (most acidic) for metals of valence 4 and higher.

The general conclusion, therefore, is that, as the valence of the surface metal atoms increases, the M—O bond strength increases, the O—H bond strength decreases, the electronegativity (ability of the oxygen atoms to attract electrons or to donate a proton) increases, and the overall basicity of the —MOH group decreases.

The hydroxyl-rich surface model described above has the advantage of providing a plausible explanation for many experimental facts—for example, the large number of oxides, silicates, and metals which have been found to give similar heats of adsorption with various classes of polar liquids. Moreover, the model does not appear to be limited to use with oxides of metals which form stable hydroxides in water. Neither —TiOH nor —CuOH groups are known, from independent analysis, to exist as stable soluble species, although both TiO_2 and CuO fit well with the other IEPS data of Fig. 3, and Parks concludes that both of these oxides ionize via the dissociative reactions of equations 19a and 20a. Parks further proposes the following general rules:

1. For hydrated or hydrous oxides, for which the coordination number of the metal atom is 6, or for which the coordination number is 4 *and* the cation radius is ≥ 1 Å, adsorption is dissociative (equations 19a and 20a), and the IEPS of the solid is equal to the IEP of dissolved hydroxo complexes.

2. For anhydrous oxides (for example, freshly calcined) and for hydrated oxides for which the coordination number is 4 and $r + > 1.0$ Å, adsorption is nondissociative, and (in the absence of extraneous potential determining ions) the IEPS is close to 7.

Applications and Consequences. The factors noted above help to explain the function of several surface cleaning techniques widely used at present for ferrous and aluminum alloys. Frequently, these involve etching the surface in oxidizing acids such as chromic, nitric, and sulfuric acids. This would be expected to:

1. Dissolve or otherwise remove particulate hydrated oxide particles (such as hydrated aluminum or ferric oxides) which might otherwise act as stress risers or weak points on the adherend surface.

2. Oxidize and/or dissolve hygroscopic surface impurities (either organic or inorganic hydroxides of the type noted above) which tend to increase the thickness of the absorbed water film.

3. Increase the valence of the surface metal ions, reducing the IEPS to a less basic and more reproducible range.

Many crystalline particulate materials, including the aluminum and magnesium silicates widely used as fillers in coatings and adhesives, consist of a mixed-oxide lattice structure. The IEPS of such particles depends on which oxide is exposed at the various crystalline planes. The present authors, for example, have studied (55) surface titration reactions corresponding to those shown in equations 19 and 20 for kaolin, and have observed an IEPS value of about 2 for the predominantly silica faces of the kaolin platelets and an IEPS of about 7 both for the aluminum hydroxide edges and for hydrated alumina particles found to be present as a surface impurity on commercial bleached kaolin. As a second example, primary fibrils of chrysotile asbestos normally consist of a hollow tube structure in which the tube wall consists of alternating silica and brucite layers. Commercial grades of chrysotile asbestos normally show a high IEPS (about 10 to 11) characteristic of the hydrated magnesium oxide outer surface. Washing in dilute monobasic acids can, however, dissolve and remove a major fraction of the outer magnesium oxide layer, reducing the IEPS to values of less than 4 and changing the adsorption properties of the fibrils to a surface approaching that characteristic of silica.

Interactions between Polar Organic Compounds and Oxide Surfaces

Previous sections of this paper have discussed the London (dispersion) force interactions which are operable across any adhesive–adherend interface, whatever the composition or degree of polarity of either surface. A detailed discussion of London forces has been reviewed by Debye (56) and by Good (14). A large body of experimental data, based primarily on heat of wetting measurements, however, shows that the magnitude of the interfacial forces attainable by London interactions is small relative to other forces which can be attained across a water–oxide, a water–polar solid, or an oxide–polar solid interface.

Typically, the magnitude of the interaction term ($\Phi\sqrt{\gamma_1\gamma_2}$ in equation 8) attributable solely to London force interactions is close to about 100 dynes/cm^2 for either water, polar, or nonpolar organic materials in contact with a large number of metals, oxides, and silicate surfaces (7, 47, 57). For such surfaces in contact with a nonpolar organic (for example, polyolefins), the London force interactions are the only forces operable, and the total interfacial interaction term is about 100 dynes/cm^2. The total free energy of interaction for such surfaces with water, however, is generally of the order of 500 dynes/cm^2 (45, 47), showing that the surface interactions due to hydrogen bonding and other dipole interactions are much larger than those attainable by simple London force interactions. Thus, while nonpolar polymers such as polyethylene, polypropylene, and polytetrafluoroethylene can provide strong mechanical bonds to high-energy surfaces, such nonpolar polymers are not generally able to retain adhesion after extended immersion in water.

Equation 3 predicts, for example, that a major free energy decrease should accompany the displacement of polyethylene from a silica surface by water. Combining equations 3 and 8 gives

$$\Delta F = 2\gamma_W + 2\Phi_{OS}\sqrt{\gamma_O \gamma_S} - 2\Phi_{OW}\sqrt{\gamma_O \gamma_W} - 2\Phi_{WS}\sqrt{\gamma_W \gamma_S} \quad (25)$$

Substituting Fowkes' dispersion force relationships for the O/W and O/S terms gives

$$\Delta F = 2\gamma_W + 2\sqrt{\gamma_O^d \gamma_S^d} - 2\sqrt{\gamma_O^d \gamma_W^d} - 2\Phi_{WS}\sqrt{\gamma_W \gamma_S} \quad (26)$$

Taking $\gamma_W = 72.8$, $\gamma_O = \gamma_O^d = 31$ for polyethylene, γ_S^d (for SiO_2) $= 100$, $\gamma_W^d = 21.8$, and the experimental value of 468 ergs/cm² used by Fowkes (22) for the last term in equation 26 gives

$$\Delta F = 145.6 + 111.2 - 93.4 - 468 = -305 \text{ ergs/cm}^2 \quad (27)$$

This large free energy decrease is seen to be due primarily to the large difference in the S/W and S/O interaction energies. To increase ΔF in equation 26, it is necessary to increase the second (O/S) interaction term by a substantial amount, by permitting the organic and solid phases to form dipole, ionic, or (possibly) covalent bonds across the interface. Such considerations have, of course, long motivated formulators to try an almost limitless variety of techniques to add "functionality" or "polarity" to commercial coatings and adhesives. But is polarity of an organic formulation *alone* sufficient to provide adhesion retention? Does any highly polar organic adhesive adhere to any polar substrate, or is there some rationale for predicting that some combinations of polar groups and oxide types serve the purpose better than others?

Thermoplastic polyamides, polyurethanes, and polyacrylonitriles are three examples of highly polar polymers which have a high cohesive energy density and an ability to form strong hydrogen bonds to oxide surfaces. Applied from solution or by powder, coating, or other fusion techniques, these polar polymers can wet, spread, and form a high degree of interfacial contact with steel or other metal surfaces. Such coatings are, however, known to be unsatisfactory in retention of adhesion after prolonged water immersion, and for this reason they are invariably applied over a primer (for example, baked phenolic-based) for use in commercial coatings. The reasons why such highly polar thermoplasts give unsatisfactory bond–water resistance include the following:

1. While the polar (amide, urethane, nitrile) groups can provide low organic–substrate interfacial tension, such polar groups also given low organic–water interfacial tensions, and the total ΔF term in equation 3 is generally negative at equilibrium in the polymer–oxide surface–water system of Fig. 1.

2. Highly polar thermoplasts also exhibit high permeability to water and a high-equilibrium water solubility at the interface.

3. Because of steric factors, the number of dipole bonds which can be formed between a polar polymer (for example, nylon 6,6) and an oxide surface is generally small relative to the number of hydrogen bonds which can be formed at the water–polar solid. Hence, such solids may adhere through a combination of a small number of dipole forces plus a larger number of London forces.

This section will, therefore, attempt to provide some qualitative guidelines for predicting which types of functional groups in an organic formulation can be most effective in providing and retaining adhesion to particular types of oxide surfaces.

Dipole and Ionic Interactions

Of the forces other than London forces listed in Table I, several can be eliminated from further consideration in this study. Metallic forces (with the possible exception of weak image forces which may enter into interactions between liquid or noble metals with polar organics) are assumed to be negligible between the surface proposed in Fig. 2 and either water or polar-organic solids. Dipole–induced dipole forces of the types discussed by Good (*14*) and Debye (*56*) are small in magnitude compared to the forces of present interest and will, therefore, also be omitted from further consideration. Thus, the following sections will consider:

1. Interactions between permanent dipoles, of which hydrogen bonds are the most important type of dipole–dipole interaction in the present case.

2. Ionic forces—bonds arising from Coulombic attraction of excess electric charges of oppositely charged ions.

3. Covalent bonds formable across an adhesive–adherend interface.

Hydrogen Bonding. Following the definitions of Pauling (*13*) and Noller (*58*), a hydrogen bond is formed by an atom of hydrogen strongly attracted to two atoms rather than to one such that it may be considered as acting as a bond between them. Since hydrogen, having only one (1s) orbital, can form only one covalent bond, the hydrogen bond is always partially ionic in character and can be formed only between the most electronegative elements.

Electronegativity is a term describing the ability of an atom to attract electrons to itself. Only fluorine, oxygen, nitrogen, and chlorine are sufficiently electronegative to form hydrogen bonds, although carbon atoms that are attached to several electron-withdrawing groups (as in chloroform or in hydrogen cyanide) can also form a polarized C—H bond. Apart from this polarized carbon case, the order of strength for hydrogen bonding is

$$F > O > N > Cl \tag{28}$$

Because of the small size of the proton, which permits a close approach to a negative charge site, dipole interactions involving hydrogen are generally considerably stronger than dipole interactions not involving hydrogen. One exception to this rule involves hydrogen bonds to chlorine, which are weak because of the large size of the chlorine atom relative to that of nitrogen or oxygen. Although H $\cdots$ F hydrogen bonds are very strong, fluorine will be excluded from further consideration in this paper because covalently bonded fluorine is not normally present in adhesive and coating formulations and also because many aspects of the interactions between fluorocarbons and oxide surfaces are not presently understood.

Following Pauling's definitions the general formula for a hydrogen bond between two electronegative atoms A and B may be written as

$$A\text{—}H \cdots B \tag{29}$$

where the dash symbolizes the shorter, stronger covalent bond between A and H and the dots represent the secondary valence hydrogen bond to B. In water, for example, where A = B = oxygen, the primary bond (HO—H) is about 1 Å in length and has a dissociation energy of about 120 kcal/mole, while the hydrogen bond H $\cdots$ O has a length of about 1.8 Å and a dissociation energy of about 5 kcal/mole.

Pauling illustrates the partially ionic character of the hydrogen bond in water by writing the following electron resonance formulas:

$$\text{—O—H:O—} \tag{30a}$$

$$\text{—O:H}^+\text{:}\bar{\text{O}}\text{—} \tag{30b}$$

$$\text{—}\bar{\text{O}}\text{:H}^+\text{—O—} \tag{30c}$$

Structures 30b and 30c represent the ionic contributions to the total bond energy. From the electronegativity difference (about 1.4 units) between hydrogen and oxygen and from the relative bond lengths, Pauling estimates that approximately 61% of the interaction energy is attributable to structure 30a and 39% to the ionic structures 30b and 30c.

The strength of this hydrogen bond increases either as the electron-donative capability of B increases, or as the electronegativity of A increases. Electron-withdrawing groups attached to A can decrease the electron density on A and thereby increase the ionic character of the A—H valence bond. For example, the strength of the internal hydrogen bonds decreases in the order carboxylic acids > phenols > alcohols because of the resonance structures which increase the degree of polarization of the O—H bond for phenols and carboxylic acids such as

$$\text{R—C}\begin{array}{c}\diagup\text{O}^-\\ \diagdown\overset{+}{\text{OH}}\end{array} \quad \text{and} \quad -:\!\!\bigcirc\!\!=\!\overset{+}{\text{OH}} \tag{31}$$

If the A—H valence bond on the proton donor group is not sufficiently polar, or if an unshared pair of electrons on the proton acceptor (B) group is not readily available, little H $\cdots$ B attraction results. But as the valence bond becomes more highly polarized and as the negative charge on the proton acceptor atom becomes more highly localized, the H $\cdots$ B attraction force increases, and ultimately it can become strong enough to transfer the proton completely via an acid–base reaction, yielding an essentially completely ionic bond:

$$\text{A—H} \cdots \text{B} \rightarrow \text{A}^- \text{H}^+ \text{—B} \tag{32}$$

The ability of ionic interfacial bonds to influence interfacial energetics is illustrated by the phenomena observed when a water-immiscible liquid organic acid (for example, oleic acid) is brought into contact with an aqueous caustic soda solution. As soon as a drop of the acid is introduced into the solution, the oil spontaneously emulsifies in the water phase. The reason is quite straightforward. Since the acidic carboxyl functions are initially confined to the oil phase, and the hydroxyl ions to the aqueous phase, the locus of the initial acid–base neutralization must be at the liquid–liquid interface. Since the acid–base reaction leading to the formation of a salt (sodium oleate in this case) is associated with a very large *negative* free energy change, the interfacial free energy change for this system must comprise the normal dispersion term characteristic of an oil–water interface (which is small and positive) and the neutralization energy term (which is large and negative). As a consequence, the interfacial free energy change at the oleic acid–aqueous caustic interface is large and *negative*, leading first to spontaneous emulsification, and (ultimately) to solution or micelle formation.

Essentially the same interfacial conditions should be obtained if the surface of a solid substance containing acidic surface groups is brought into intimate contact with a liquid containing basic groups; that is, the interfacial energy change on establishing contact will be large and negative (or the work of adhesion, large and positve). After solidification of the liquid phase, separation of these two solids would require either the separation of opposing electrostatic charges, or (in the presence of water) the regeneration of free acid and free base—both of which are high-energy-demand processes. Thus, the strength of this interfacial bond should be much greater than that expected if only nonspecific dispersion interactions were operative across the interface.

Interactions with Oxide Surfaces. From the above, it follows that an oxide surface, shown in Fig. 2, can interact with a polar organic group by a variety of mechanisms. Either the surface hydroxyl or the organic group can act as the acid (that is, as the proton donor in the Bronsted sense or the electron acceptor in the Lewis sense) or as the base (proton acceptor or electron donor). The bond between the surface group and the organic group can range anywhere from being wholly ionic to being very weakly ionic, as illustrated in equation 32. An oxide might interact, for example, with a ketone, with the ketone acting either as the base (electron donor):

$$-\text{MOH} + \text{R}-\overset{\overset{\text{O}}{\|}}{\text{C}}-\text{R} \longrightarrow -\text{MOH} \cdots \text{O}=\text{C}\overset{\text{R}}{\underset{\text{R}}{\diagup}} \qquad (33)$$

or as the acid (electron acceptor):

$$-\text{MOH} + \text{R}-\overset{\overset{\text{O}}{\|}}{\text{C}}-\text{R} \longrightarrow -\overset{\overset{\text{H}}{|}}{\text{MO}} \cdots \overset{\overset{\text{R}}{|}}{\underset{\underset{\text{R}}{|}}{\text{C}}} = \text{O} \qquad (34)$$

Similarly, an oxide may interact with a carboxylic acid to give either a dipole or an ionic bond:

$$-\text{MOH} + \text{R}\overset{\overset{\text{O}}{\|}}{\text{C}}-\text{OH} \longrightarrow -\overset{\overset{\text{H}}{|}}{\text{MO}} \cdots \text{HO}-\overset{\overset{\text{O}}{\|}}{\text{C}}\text{R}$$

or $\qquad\qquad\qquad\qquad\qquad\qquad\qquad\qquad\qquad\qquad\qquad\qquad (35)$

$$-\overset{+}{\text{MOH}}_2 \quad \overset{-}{\text{O}}-\overset{\overset{\text{O}}{\|}}{\text{C}}\text{R}$$

The oxide might also interact with an amine to give either dipole or ionic bonds:

$$\text{—MOH} + R_3N \rightarrow \text{—MOH} \cdots NR_3 \quad \text{or} \quad \text{—M}\overset{-}{\text{O}}\ \overset{+}{\text{H}}NR_3 \quad (36)$$

The surface may interact with strongly cationic groups to yield an ionic bond:

$$\text{—MOH} + R_4\overset{+}{\text{N}}\overset{-}{\text{Cl}} \rightarrow \text{—M}\overset{-}{\text{O}}N\overset{+}{R_4} + HCl \quad (37)$$

or with strongly anionic groups to yield

$$\text{—MOH} + NaO\overset{\overset{O}{\|}}{\underset{\underset{O}{\|}}{S}}\text{—R} + H_2O \rightarrow \text{—M}\overset{+}{O}H_2\overset{-}{O}\text{—}\overset{\overset{O}{\|}}{\underset{\underset{O}{\|}}{S}}\text{—R} + NaOH \quad (38)$$

To derive some general relationships that can predict which of the many possible interactions will predominate in any given surface–organic group combination, it is necessary to distinguish between two general interaction types. For *type A* interactions the surface provides the basic group which interacts with an organic *acid* (as in equations 34, 35, and 38). For *type B* interactions the surface provides the acidic group which interacts with an organic *base* (as in equations 33, 36, and 37). Within each type, we further distinguish two extreme conditions wherein the interaction is predominantly ionic or nonionic:

Type A-1: Dipole interaction with an organic acid:

[handwritten: Metal provides basic site]

$$\text{—MOH} + HXR \rightarrow \text{—M}\overset{\overset{H}{|}}{\text{O}} \cdots HXR \quad (39)$$

Type A-2: Ionic interaction with an organic acid:

$$\text{—MOH} + HXR \rightarrow \text{—M}\overset{+}{\text{O}}H_2\ \overset{-}{X}R \quad (40)$$

Type B-1: Dipole interaction with an organic base:

[handwritten: Metal provides acidic site]

$$\text{—MOH} + XR \rightarrow \text{—MOH} \cdots XR \quad (41)$$

Type B-2: Ionic interaction with an organic base:

$$\text{—MOH} + XR \rightarrow \text{—M}\overset{-}{\text{O}}H\overset{+}{X}R \quad (42)$$

where X is an electronegative atom, usually oxygen or nitrogen, although X may also be chlorine in equation 41. To predict the probability of the interaction for a given oxide surface–polar group combination, the equilibrium constants for type A or type B interactions may be expressed as

$$\kappa_A = \frac{[MOH_2^+][X^-]}{[MOH][HX]} \quad \text{ionic (base)} \quad (43)$$

$$\kappa_B = \frac{[MO^-][HX^+]}{[MOH][X]} \quad \text{ionic (acid)} \quad (44)$$

$$\Delta_A \equiv \log \kappa_A \tag{45}$$

$$\Delta_B \equiv \log \kappa_B \tag{46}$$

To derive Δ_A and Δ_B the metathetical surface reactions are written as

$$MO^- + H^+ \to MOH \qquad K_1 = (MOH)/(MO^-)(H^+) \tag{47}$$

$$MOH + H^+ \to MOH_2^+ \qquad K_2 = (MOH_2^+)/(MOH)(H^+) \tag{48}$$

and for

$$MO^- + 2H^+ \to MO^+H_2 \qquad K_1K_2 = (MOH_2^+)/(MO^-)(H^+)^2 \tag{49}$$

At the IEPS, $MOH_2^+ = MO^-$, and

$$K_1K_2 = \frac{1}{(10^{-\text{IEPS}})^2} \tag{50}$$

Assuming symmetry about the IEPS gives

$$K_1 = K_2 = \frac{1}{10^{-\text{IEPS}}} \tag{51}$$

From the definitions of the pK_A of an acid or base:
For an acid:

$$HX \to H^+ + X^- \tag{52}$$

$$K_{A(A)} \equiv \frac{(H^+)(X^-)}{HX} \tag{53}$$

For a base:

$$HX^+ + H_2O \to X + H_3O^+ \tag{54}$$

$$K_{A(B)} \equiv \frac{(X)(H^+)}{(HX^+)} \tag{55}$$

Commonly, these equilbrium constants are tabulated as pK_A values for an acid or for the conjugate base, where

$$pK_A \equiv -\log K_A \tag{56}$$

By appropriate substitution of equations 47 through 56 into equations 43 and 44, and rearrangement of terms, it follows that

$$\Delta_A = \text{IEPS} - pK_{A(A)} \tag{57}$$

$$\Delta_B = pK_{A(B)} - \text{IEPS} \tag{58}$$

Thus, if Δ_A or Δ_B is positive, the ionic reactions (type A-2 or B-2) should predominate, whereas if Δ_A or Δ_B is negative, the dipole type A-1 or B-1 interaction should predominate. Furthermore, it follows from the above discussion that the total interaction energy should increase as Δ_A or Δ_B increases—that is, as the ionic contributions to the interfacial bond increase. At very negative Δ values, ionic forces are negligible and dipole forces are weak. Nonpolar polymers (such as polyethylene) can be regarded both as very weak acids ($pK_{A(A)}$ is large and positive) and as very weak bases ($pK_{A(B)}$ is large and negative), giving large negative values of Δ_A and Δ_B with surfaces of any IEPS, and leading to the prediction that London forces (only) should be present across such interfaces.

As Δ_A or Δ_B becomes less negative, dipole–dipole forces should increase in strength, as shown schematically in Fig. 4. As Δ values increase to approach zero, type A-2 (or B-2) interactions begin to increase in importance and type A-1 (or B-1) interactions decrease. Ultimately, the most energetic ionic bonds should be formed at large positive values of Δ_A or Δ_B, as indicated in Fig. 4,

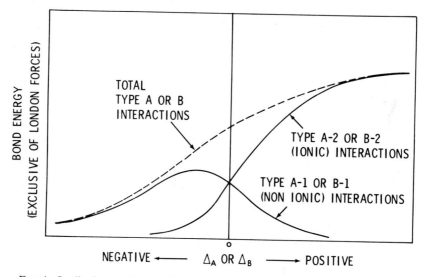

FIG. 4. Qualitative relationships between formation energy for electrostatic interactions and ionization parameters Δ_A or Δ_B.

reflecting the fact that ionic bonds are stronger and decrease in strength more slowly with distance than do hydrogen bonds or other dipole–dipole interactions in which ionic interactions play a minor role.

Tables IV and V show pK_A values, taken from Noller (58), for a large number of organic acids and bases, including water. These tables also show Δ_A or Δ_B values calculated for three oxide surfaces—SiO_2, Al_2O_3, and MgO,

TABLE IV
Polar Surface Interactions with an Organic Acid

		$\Delta_A \equiv IEPS - pK_{A(A)}$		
Organic Acid	$pK_{A(A)}$ (ref. 58)	SiO_2 (IEPS = 2)	Al_2O_3 (IEPS = 8)	MgO (IEPS = 12)
Dodecyl sulfonic acid	−1	3	9	13
Trichloroacetic acid	0.7	1.3	7.3	11.3
Chloroacetic acid	2.4	− 0.4	5.6	9.6
Phthalic acid	3.0	− 1	5.0	9.0
Benzoic acid	4.2	− 2.2	3.8	7.8
Adipic acid	4.4	− 2.4	3.6	7.6
Acetic acid	4.7	− 2.7	3.3	7.3
Hydrogen cyanide	6.7	− 4.7	1.3	5.3
Phenol	9.9	− 7.9	− 1.9	2.1
Ethyl mercaptan	10.6	− 8.6	− 2.6	1.4
Water	15.7	−13.7	− 7.7	− 3.7
Ethanol	16	−14	− 8	− 4
Acetone	20	−18	−12	− 8
Ethyl acetate	26	−24	−18	−14
Toluene	37	−35	−29	−25

having an IEPS of 2, 8, and 12, respectively. These tabulated values support the general conclusions that:

1. A relatively small number of oxide–organic combinations give positive values of Δ_A or Δ_B—that is, only those combinations in the upper right-hand corner of Table IV or in the upper left-hand corner of Table V represent *predominantly* ionic reactions.

2. Water interactions are always primarily nonionic, although water may interact either as a very weak acid or as a very weak base with surfaces of high or low IEPS. According to the values for water in Tables IV and V, for example, water should interact more strongly than phenol with acidic surfaces (for example, SiO_2) but less strongly with basic surfaces (for example, MgO).

TABLE V
Polar Surface Interactions with an Organic Base

Organic Base	$pK_{A(B)}$ (ref. 58)	$\Delta_B \equiv pK_{A(B)} - \text{IEPS}$		
		SiO_2 (IEPS = 2)	Al_2O_3 (IEPS = 8)	MgO (IEPS = 12)
Trimethyl dodecyl ammonium hydroxide	12.5	10.5	4.5	0.5
Piperidine	11.2	9.2	3.2	−0.8
Ethylamine	10.6	8.6	2.6	−1.4
Triethylamine	10.6	8.6	2.6	−1.4
Ethylenediamine	10	8	2.0	−2
Ethanolamine	9.5	7.5	1.5	−2.5
Benzylamine	9.4	7.4	1.4	−2.6
Pyridine	5.3	3.3	−2.7	−6.7
Aniline	4.6	2.6	−3.4	−7.4
Urea	1.0	−1.0	−7	−11
Acetamide	−1	−3	−9	−13
Water	−1.7	−3	−9	−13
Tetrahydrofuran	−2.2	−4.2	−10.2	−14.2
Ethyl ether	−3.6	−5.6	−11.6	−15.6
t-Butanol	−3.6	−5.6	−11.6	−15.6
n-Butanol	−4.1	−6.1	−12.1	−16.1
Acetic acid	−6.1	−8.1	−14.1	−18.1
Phenol	−6.7	−8.7	−14.7	−18.1
Acetone	−7.2	−9.2	−15.2	−19.2
Benzoic acid	−7.2	−9.2	−15.2	−19.2

TABLE VI
Summary of Surface Interfacial Interaction Mechanisms for Organic Compounds of Type R—O—H

Interactions with Compound ROH		Type	Mode	Requirements
—MOH + HOR →	—MOHOR with H above	A-1	Dipole	Basic surface + organic acid
	—M$\overset{+}{O}$H$_2\overset{-}{O}$R	A-2	Ionic	
	—MOR + H$_2$O	Covalent bond formation		
	—M$\overset{-}{O}$H$_2\overset{+}{O}$R	B-2	Ionic	Acidic surface + organic base
	—MOHOR with H below	B-1	Dipole	
where R = H, alkyl, aryl, or acyl group				

3. Whether a polar group interacts as an acid or as a base can be predicted from the IEPS of the oxide. Ketones should, for example, interact primarily via equation 33 with low IEPS surfaces and via equation 34 with high IEPS surfaces. Figure 5 illustrates the interaction types, and dipole orientations, for water, ketones, and other polar compounds predicted from the values of Δ_A and Δ_B shown in Tables IV, V, and VII.

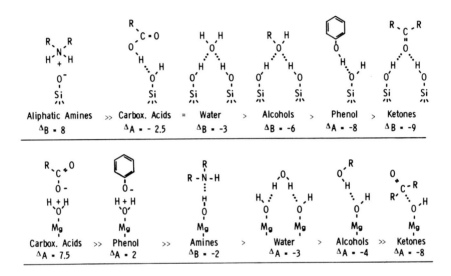

FIG. 5. Dipole orientations, dominant interaction modes, and relative bond strengths predicted for high IEPS surface (for example, MgO) and low IEPS surface (for example, SiO$_2$).

4. A surface and an organic group need not be of "opposite" charge to form an electrostatic interfacial attraction; for example, amines should bind to cationic surfaces by forces other than London forces and should adsorb on such surfaces (for example, MgO) from nonpolar media. Such bonds have, however, more negative values of Δ_A or Δ_B than would water with MgO and would have little tendency to resist displacement via reactions of the type

$$-\text{MgOH} \cdots \text{NR}_3 + 2\text{H}_2\text{O} \longrightarrow$$

$$-\text{MgO}\diagup^{\text{H}} \cdots \text{HOH} + \text{HOH} \cdots \text{NR}_3 \quad (59)$$

The above arguments are, admittedly, qualitative at this stage, and are based largely on relative values of pK_A and IEPS obtained from independent ionization measurements in water. In actual fact, neither the IEPS nor the pK_A value gives a direct quantitative measurement of the ionic character of a bond in a molecule. Rather, these express only the tendency of the molecule

TABLE VII

INTERACTION MECHANISMS PREDICTED FOR COMPOUND OF TYPE R—OH

(R = alkyl, aryl, or acyl group)

R	H		Alkyl		Aryl		Acyl	
	(Water)		(Alcohols)		(Phenols)		(Carboxylic, Acids)	
$pK_{A(A)}$	15.7		16		10		4.5	
$pK_{A(B)}$	−1.7		−4		−7		−6	
IEPS	Δ_A	Δ_B	Δ_A	Δ_B	Δ_A	Δ_B	Δ_A	Δ_B
2	−13.7	− 3.7	−14	− 6	−8	− 9	−2.5	− 8
8	− 7.7	− 9.7	− 8	−12	−2	−15	+3.5	−14
12	− 3.7	−13.7	− 4	−16	+2	−19	+7.5	−18

to ionize in one particular solvent (water). The ionic character of a bond is determined by the importance of the ionic structure when the nuclei are at their equilibrium distance in the quantum expression:

$$A:B \rightleftarrows A^+B^- \qquad (60)$$

$$\Psi = a\Psi_{A:B} + b\Psi_{A^+B^-} \qquad (61)$$

where Ψ is the wave function for a particular state, and a and b are the numerical coefficients, as defined by Pauling (*13*). The tendency to ionize in solution, on the other hand, is determined by the relative stability of the actual molecules in solution and the separated ions in solution. Nevertheless, Pauling points out that the tendency to ionize in solution accompanies large ionic character of bonds in general, since both result from great differences in electronegativity of the bonded atoms. Thus, for present purposes, it is assumed that the *relative* probabilities of interaction via either types A-1, A-2, B-1, or B-2 above can be predicted from the relative magnitudes of Δ_A and Δ_B for each interaction type.

Practical Consequences. The foregoing electrostatic theory leads to the prediction that water resistance should increase where ionic attractions for the organic–solid interface are large relative to ionic attractions for the water–solid interface in neutral systems—that is, with no additional potential-determining ions such as H^+ or OH^- present to shift the equilibria in equations 43 and 44. A corollary of this is that virtually every polar group should be thermodynamically displaceable by water if the H^+ or OH^- activity is sufficiently high. Addition of extraneous H^+ or OH^- ions should result either in development of mutual repulsion forces or in suppression of the degree of ionization of the organic group in either type A or type B interactions, as follows:

Type A: Basic surface with an organic acid:

$$—\overset{+}{\text{MOH}}_2 + \text{HXR} \underset{H_2O}{\overset{H^+}{\longleftarrow}} \left\{ \begin{array}{c} \text{MO}^{\diagup \text{H}} \cdots \text{HXR} \\ \text{or} \\ —\overset{+}{\text{MOH}}_2 \overset{-}{\text{XR}} \end{array} \right\} \underset{H_2O}{\overset{OH^-}{\longrightarrow}}$$

$$—\overline{\text{MO}} \cdots \text{HOH} + \text{HOH} \cdots \overline{\text{XR}} \quad (62)$$

Type B: Acidic surface with an organic base:

$$—\overset{+}{\text{MOH}}_2 + \text{H}\overset{+}{\text{X}}\text{R} \underset{H_2O}{\overset{H^+}{\longleftarrow}} \left\{ \begin{array}{c} —\text{MOH} \cdots \text{X}—\text{R} \\ \text{or} \\ —\overset{-}{\text{MOH}}\overset{+}{\text{X}}\text{R} \end{array} \right\} \underset{H_2O}{\overset{OH^-}{\longrightarrow}}$$

$$—\overline{\text{MO}} \cdots \text{HOH} + \text{XR} \quad (63)$$

These effects can be generalized by noting that, if the hydrogen ion activity in equations 43 and 44 is determined by the addition of extraneous electrolyte to the aqueous phase, then the regions wherein strong ionic attractions result are limited to the pH ranges:

For type A interactions:

$$pK_{A(A)} < pH < \text{IEPS} \quad (64)$$

For type B interactions:

$$\text{IEPS} < pH < pK_{A(B)} \quad (65)$$

where pH is that in the aqueous phase at the oil–water–solid contact line.

These equations can satisfactorily explain a considerable body of experimental observations. The present authors, for example, have studied the underwater adhesion of a large number of air-dried, solvent-based primer coatings containing mixtures of alkylated phenolic resins with either polyvinyl butyral or nitrile rubbers, of the type described in numerous military specifications. The adhesion of such coatings, applied to clean sandblasted steel surfaces, is generally excellent in water at pH ranges between neutral and about 12. At higher pH in the aqueous phase, however, thin coatings (about 2 to 3 mils) detach from the steel surface in very short periods. This is believed to be due to charge repulsion factors of the type shown in equation 62:

$$MO{\cdots}H{-}O{-}C_6H_4{-}\text{...} + 2OH^- \longrightarrow MO^- + {}^-O{-}C_6H_4{-}\text{...} + 2H_2O \tag{66}$$

Additional quantitative evidence of these effects can be obtained from the ore flotation literature, in which ionic surfactants are commonly added to aqueous ore slurries to cause partial dewetting of the particle surfaces by the aqueous phase. The success of a given flotation aid depends, therefore, on its ability to displace water at the mineral interface. Figure 6 summarizes the results of one series of experiments reported by Iwasaki et al. (59) for aqueous slurries of the mineral goethite (IEPS about 7). At a pH in the aqueous phase which is between the IEPS and the $pK_{A(B)}$ of the alkyl amine surfactant used, ore flotation and recovery are essentially 100%. Similarly, at a pH between the IEPS and the $pK_{A(A)}$ of the alkyl benzene sulfonate surfactant used in a second series of experiments, recovery is also essentially 100%. Outside these pH limits, with either surfactant, the ore recovery on froth-flotation drops essentially to zero.

Similar data have been reported in the flotation literature for other anionic and cationic surfactants. In general, it is observed that the pH range for efficient recovery is narrower for weakly ionized than for strongly ionized surfactants (carboxylic acids relative to sulfonates, or alkyl amines relative to quaternary salts), reflecting the pH range predicted by equation 64 or 65.

These considerations, together with the interactions shown in Tables VI and VIII, where equations 62 and 63 are illustrated for an organic compound of the type R—OH, indicate that nonionizable polar organic groups, such as the alcohols, should not give bonds which are as strong as the maximum attainable with ionizable groups (for example, phenolics and carboxylic acids) but should give bonds which are more resistant to displacement in aqueous media of high or low pH. Thus, the ability of aliphatic hydroxyl groups to form hydrogen bonds to a wide variety of oxides (differing widely in IEPS) through *combinations* of type A-1 and B-1 interactions, together with the relative stability of such bonds in alkaline or acidic media, helps to explain the important role of hydroxyl-containing polymers in the technology of adhesives and coatings.

It would be incorrect to infer, from the preceding discussion, that a polymer must contain a very high proportion of ionizable (or hydroxyl) groups to provide a durable bond. If this were true, then polyols such as polyvinyl alcohol or polyelectrolytes such as polyacrylic or polyethyleneimine could represent the ultimate choice for coating or adhesive applications.

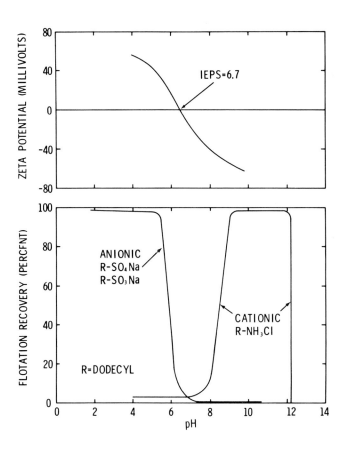

FIG. 6. Flotation data for mineral goethite using anionic and cationic flotation aids (59).

TABLE VIII

MECHANISMS FOR DISPLACEMENT OF ORGANIC GROUPS BY WATER AT HIGH OR LOW pH
(Compound of type ROH interacting with surface of $6 < \text{IEPS} < 8$)

R = alkyl group

$$-\overset{+}{\text{M}}\text{OH}_2\cdots\text{OH}_2 + \text{HOCH}_2- \xrightarrow[\text{H}_2\text{O}]{\text{H}^+}$$

$$\left.\begin{array}{l} -\text{M}-\overset{\text{H}}{\text{O}}\cdots\text{HOCH}_2- \\ \text{or} \\ -\text{MOH}\cdots\overset{\text{H}}{\text{O}}-\text{CH}_2- \end{array}\right\} \xrightarrow[\text{H}_2\text{O}]{\text{OH}^-} \left\{\begin{array}{l} -\text{M}\bar{\text{O}}\cdots\text{HOCH}_2- + \text{H}_2\text{O} \\ \text{or} \\ -\text{M}\bar{\text{O}}\cdots\text{HOH} + \text{HOCH}_2- \end{array}\right.$$

Displacement of alcohol by stronger base (water)

Type A-1 or B-1 hydrogen bonds $\Delta_A \approx \Delta_B \approx -10$

Hydrogen bonding by both water and alcohol

R = acyl group

$$-\overset{+}{\text{M}}\text{OH}_2\cdots\text{OH}_2 + \text{HO}-\overset{\overset{\text{O}}{\|}}{\text{C}}-\text{CH}_2- \xrightarrow[\text{H}_2\text{O}]{\text{H}^+} -\text{MOH}_2\cdots\bar{\text{O}}-\overset{\overset{\text{O}}{\|}}{\overset{+}{\text{C}}}-\text{CH}_2- \xrightarrow[\text{H}_2\text{O}]{\text{OH}^-} -\text{M}\bar{\text{O}}\cdots\text{HOH} + \text{HOH} + \bar{\text{O}}\overset{\overset{\text{O}}{\|}}{\text{C}}\text{CH}_2-$$

Displacement by stronger base (water)

Primarily A-2 interactions $\Delta_A \approx 3$

Charge repulsion between ionized acid and anionic surface

Order of decreasing base strength: $\text{H}_2\text{O} >$ alcohol $>$ carboxylic acid
Order of decreasing acid strength: Carboxylic acid $\gg$ water $\approx$ alcohol

Apart from the obvious objections to such hydrophilic polymers on water sorption, swelling, and transport grounds, experimental evidence indicates that a surprisingly small proportion of such groups are usually found to be optimum for bonding purposes. Maleic anhydride, for example, is frequently copolymerized with polyolefins at total levels of about 1 to 4% maleic acid by weight, and similar levels of vinyl pyridine are usually incorporated in elastomer terpolymers for bonding applications. In addition, the hydroxyl content in a variety of widely used coating and adhesive resins (including most bisphenol-A epoxies, polyvinyl acetals, alkyds, and hydroxy-acrylate copolymers) rarely exceeds 2 to 4% by weight.

A consideration of the geometric factors involved in contacting a polymer with an oxide surface, which is highly irregular on a molecular scale and which contains about one ionic hydroxyl group per 50 to 100 $Å^2$, leads to the conclusion that only a relatively small fraction of the ionic groups in a polyelectrolyte can enter into ionic bond formation at the interface, with the remainder contributing to the undesirable water sorption effects noted above. Hence, the small amount of ionizable groups normally used to improve bonding characteristics of low-polarity polymers should be regarded as well-spaced *anchoring points*, rather than as the desired major component at a resin–oxide interface.

Covalent Interfacial Bonding

As the above discussion has illustrated, the electronic structures of most oxide surfaces and the functional moieties of polar organic polymers favor the formation of dipolar or Coulombic adhesive–adherend interactions. In principle, the establishment of a preponderance of covalent bonds across a resin–oxide interface should (because of the far greater force-atom-separation distance degeneracy of covalent linkages) provide a stronger adhesive joint and (because of the electronic charge symmetry about a covalent bond) provide a bond insensitive to rupture or displacement by polar species such as water.

It is instructive to compare the success of efforts reported in the literature on water-repellent coatings for textiles involving the formation of covalent bonds between cellulose fibers and organic compounds with the paucity of success reported when similar reaction techniques were used to form covalent bonds to metal oxide surfaces. Baird, for example, in a review article (*60*) lists almost 300 references on reactions between organic compounds and the surface of cellulose fibers. Generally, such reactions take advantage of the fact that the hydroxyl and methylol groups on the cellulose surface have essentially the same reactivity as do aliphatic alcohols, and hence undergo etherification via Williamson-type reactions with alcohols, esterification reactions with organic and mineral acids, and addition and condensation reactions with reagents such as isocyanates, epoxy resins, and aldehydes.

The hydroxyl groups on metal oxide surfaces, however, apparently either do not react with or do not form stable covalent linkages with organic compounds of the type used to react with cellulosic hydroxyls. This is largely due to the fact that the electropositivity of virtually all the common metals and metalloids found in the mineral oxides is considerably greater than the electropositivity of carbon atoms in cellulose or other organic compounds. As a result, the electron density on the oxygen atom in the oxide surface group (—MOH) is considerably higher than that on the oxygen in water or in most organic compounds. To illustrate the effect of the oxygen electron density on reaction rate and hydrolytic stability, consider first the esterification of aliphatic hydroxyl compounds, which is believed to initiate through donation of electrons by the alcoholic oxygen to the carbonyl carbon of the acid group, followed by elimination of water:

$$R_1OH + R_2\overset{\overset{O}{\|}}{C}OH \rightleftarrows R_1-\underset{\underset{H}{|}}{O}\cdots \overset{\overset{O}{\diagup}}{C}-\underset{\underset{\underset{H}{|}}{O}}{|}R_2 \rightleftarrows R_1O\overset{\overset{O}{\|}}{C}R_2 + H_2O \qquad (67)$$

If the electron density on the hydroxyl oxygen is low, as in phenolic hydroxyls wherein the oxygen atom contributes electrons to the benzene ring, the corresponding reaction proceeds with much more difficulty; phenol does not esterify when boiled in most organic acids.

Reaction 67 is, of course, a reversible reaction. If the electron density on the oxygen atom is high, as in the surface group —MOH, the formation of an ester bond which is stable in the presence of water becomes very improbable. Reaction between alumina and a carboxylic acid, for example, should not proceed by water elimination to yield an Al—O—COR bond, but rather should form the ionic linkage shown above in reaction type A-2:

$$-AlOH + HO\overset{\overset{O}{\|}}{C}R \rightleftarrows -Al\overset{+}{O}H_2\overset{-\overset{O}{\|}}{O}CR \qquad (68)$$

In general, therefore, covalently bonded adhesive–adherend interfaces rarely exist in oxide–hydrocarbon systems. One important possible exception to this conclusion involves surface reactions with alkoxy derivatives of tetravalent metals (silicon, zirconium, titanium) discussed in the following section.

A second possible exception, based on speculation rather than on experimental grounds, may involve reaction between metal oxides in the low to medium IEPS range and certain resins derived from formaldehyde—for example, phenol, cresol, melamine, urea, and resorcinol–formaldehyde resins. Such resins play a dominant role in the technology of structural adhesives. Heat-cured coatings, adhesives, and primers based on such resins are capable, as a generic class, of providing exceptional underwater adhesion retention to iron, steel, and aluminum alloys. Such resins are almost never used alone for such applications, but rather are combined either with high-molecular-weight elastomers, such as nitrile or Neoprene rubbers, or with polymers having recurring aliphatic hydroxyl groups along the backbone, such as high-molecular-weight bisphenol-A epoxy resins, alkyd resins, and polyvinyl acetals. All rubber-phenolic, vinyl-phenolic, and epoxy-phenolic adhesives and virtually all high-performance baking enamels and primers fall into these categories. The long-term, high-temperature cure conditions used with such resin combinations (frequently in the presence of strong acid catalysts) lead to the evolution of water via methylol hydroxyl condensation reactions of the type:

$$\text{Ar(OH)}-CH_2OR_1 + HOR_2 \xrightarrow{H^+, \Delta} \text{Ar(OH)}-CH_2OR_2 + R_1OH \quad (69)$$

where R_1 is usually H but may also be an alkyl group as in methylated or butylated phenolics and melamines, and where R_2 is the epoxy, alkyd, acetal, or other polyhydroxy resin coreactant. The reaction conditions required for reaction 69 are also precisely the conditions that give the maximum probability of forming ether linkages between oxide surfaces and formaldehyde resins, as, for example:

$$-MOH + HOCH_2-\text{Ar(OH)}- \xrightarrow{H^+, \Delta} -MOCH_2-\text{Ar(OH)}- + H_2O \quad (70)$$

Whether or not such ether linkages can be formed at a metal oxide interface, the question still remains whether the surface bond will be susceptible to hydrolysis in water because of the strongly ionic character of the —M—O—C bond. Hydrolysis, however, cannot occur more rapidly than the rate at which water molecules can reach the bonds. The highly crosslinked polymer network formed at the interface between an oxide and a phenolic or melamine resin-based formulation, via the condensation reactions shown in equations 69 and 70, can provide an extremely effective barrier to water penetration, after which disruption of the bond by liquid water or water vapor can become a very slow process.

Thus, an attempt has been made to categorize the large number of possible surface–organic interaction mechanisms into five general types: one type in which a covalent bond is formed via a condensation reaction between the surface (—MOH) group and an organic group, plus four types of dipole–dipole or ionic interactions. Although the foregoing treatments apply to any organic compound containing oxygen, nitrogen, or chlorine atoms, the results of the preceding sections can be summarized by illustrating, in Tables VI and VII, these five mechanisms for a surface interacting with a compound of the formula H—O—R, where R might be any saturated or unsaturated, substituted or unsubstituted, alkyl, aryl, or acyl ($R-\overset{\overset{O}{\parallel}}{C}-$) group. Similarly, Table VIII shows possible mechanisms for loss in bond strength, for an adsorbed R—OH group, in the presence of water and extraneous H^+ or OH^- ions.

Silane and Titanate Coupling Agents

During the past decade, a large number of "sizing" or "coupling" agents have become commercially available, which consist of mixtures of hydrolyzable and nonhydrolyzable organic groups covalently bonded to Group IV, tetravalent, metal atoms—chiefly silicon, titanium, or zirconium. The success of such compounds used either as primers to pretreat oxide surfaces or as integral blend additives in coatings and adhesives, in improving bond resistance to water, is frequently cited as evidence of covalent adhesive–adherend bonding.

As noted above, the hydrolytic stability of a covalent bond between a metal atom and an organic group varies inversely with the degree of polarization of the covalent bond. If a silicon or other Group IV metal atom is bonded to a strongly electronegative atom, such as chlorine, the bond is highly polarized and hydrolysis proceeds rapidly and exothermically, to yield (initially) a silanol group via

$$\equiv\overset{\delta+}{\text{Si}}-\overset{\delta-}{\text{Cl}} + H_2O \longrightarrow \equiv\underset{\underset{\text{Cl}\bar{\delta}}{|}}{\overset{\delta+}{\text{Si}}} \quad \underset{\overset{|}{\delta H}}{\overset{-\delta}{\text{O}-H}} \longrightarrow \equiv\text{Si}-\text{OH} + \text{HCl} \quad (71)$$

As the electronegativity of the group attached to the silicon atom decreases, hydrolytic stability increases. Thus, alkoxy compounds (Si–OR, Ti–OR, where R is an alkyl group) hydrolyze more slowly than the halides, via

$$\equiv\text{Si}-\text{OR} + H_2O \rightarrow \equiv\text{SiOH} + \text{ROH} \qquad (72)$$

Silicon bonded to still weaker electronegative groups (Si—C—, or Si—O—Ar, where Ar is an aromatic group) are relatively stable to hydrolysis. Following the accepted conventions, Si—R denotes a group bonded to silicon by a nonhydrolyzable covalent bond (for example, Si—C—), and Si—X denotes silicon bonded to a hydrolyzable group. The most widely used "coupling agents" of this type are either of the class $R-Si-(X)_3$, as illustrated by the compounds listed in Table IX, or of the class containing four hydrolyzable groups—for example, tetraethyl silicate or tetrabutyl titanate.

Although the rate is slow at neutral pH and ambient temperature, the silanol groups formed in equations 71 and 72 can thereafter condense to yield network structures of the type

$$R-Si-(OH)_3 \rightleftarrows \left[\begin{matrix} R & R \\ | & | \\ -Si-O-Si- \\ | & | \\ O & O \\ | & | \end{matrix} \right]_{n/2} + 1.5\, n\text{-}H_2O \qquad (73)$$

Alkoxy silanes and titanates are generally preferred over the corresponding halogen derivatives because of their higher boiling point, greater storage stability, and greater compatibility in resin solutions, and because they liberate an alcohol, rather than HCl, as the condensation product.

Although not yet proved by independent analysis, the silanol groups in equations 71, 72, and 73 are also believed (61) to be capable of condensation with the hydroxyl groups on oxide surfaces to yield an M—O—Si covalent bond via the (reversible) reaction

$$-\text{MOH} + \text{HO}-\text{Si}\equiv \rightleftarrows -\text{MO}-\text{Si}\equiv + H_2O \qquad (74)$$

TABLE IX
COMMERCIAL SILANE AND TITANATE MONOMERS

Monomer	Empirical Formula	Rate of Hydrolysis	Boiling Point (C)	Commercial Supplier
Ethyl orthosilicate	$(C_2H_5O)_4Si$	Slow	166	(1)
Tetrabutyl titanate	$(C_4H_9O)_4Si$	Rapid	Decomposes	(2)
Tetra(triethanolamine) titanate	$[(HOC_2H_4)_2NC_2H_4O]_4Ti$	Slow	Decomposes	(2)
Vinyltrichlorosilane	$C_2H_3SiCl_3$	Rapid	91	(1, 3)
Vinyltriethoxysilane	$C_2H_3Si(OC_2H_5)_3$	Slow	160	(1, 3)
γ-Methacryloxypropyl-trimethoxysilane	$CH_2CCH_3COO(CH_2)_3Si(OCH_3)_3$	Slow	255	(1)
β-(3,4-Epoxycyclohexyl)-ethyltrimethoxysilane	$(CH_2)_2Si(OCH_3)_3$ attached to epoxycyclohexyl	Slow	310	(1)
γ-Glycidoxypropyltrimethoxysilane	$CH_2\text{—}CHCH_2O(CH_2)_3Si(OCH_3)_3$	Slow	290	(1, 3)
γ-Aminopropyltriethoxysilane	$NH_2(CH_2)_3Si(OC_2H_5)_3$	Slow	217	(1, 3)
N-β-(Aminoethyl)-γ-aminopropyltrimethoxysilane	$NH_2(CH_2)_2NH(CH_2)_3Si(OCH_3)_3$	Slow	259	(1, 3)

1. Union Carbide Corporation, Silicones Division.
2. DuPont, Organic Chemical Division.
3. Dow Corning Corporation.

The overall (although admittedly idealized) reaction for the formation of a cross-linked surface zone can therefore be visualized as

$$\begin{array}{c}\diagdown\\ \diagup\end{array}\!\!M\text{—OH} \qquad\qquad \begin{array}{c}\diagdown\\ \diagup\end{array}\!\!M\text{—O—}\underset{\underset{O}{|}}{\overset{\overset{|}{}}{Si}}\text{—R}$$

$$\begin{array}{c}\diagdown\\ \diagup\end{array}\!\!M\text{—OH} + 3R\text{—Si}(X)_3 + 3H_2O \longrightarrow \begin{array}{c}\diagdown\\ \diagup\end{array}\!\!M\text{—O—}\underset{\underset{O}{|}}{\overset{\overset{|}{}}{Si}}\text{—R} + 9HX \qquad (75)$$

$$\begin{array}{c}\diagdown\\ \diagup\end{array}\!\!M\text{—OH} \qquad\qquad \begin{array}{c}\diagdown\\ \diagup\end{array}\!\!M\text{—O—}\underset{\underset{|}{O}}{\overset{|}{Si}}\text{—R}$$

If the nonhydrolyzable group (R) contains a functional organic moiety such as a methacryloxy, vinyl, amine, or oxirane group, these may then form covalent bonds with corresponding reactive groups in the adhesive phase, leading to a bond which should, in theory, provide a high degree of water resistance.

Whether or not the details of the above reaction are correct, the experimental fact is that small amounts of such additives incorporated in primer formulations or used as integral blend additives do, under certain circumstances, provide significant improvements in the retention of adhesion or in the retention of bulk physical properties for a variety of organic–inorganic composites after prolonged water immersion. Sterman and Toogood (62), for example, have published a variety of data showing that less than 1% of γ-aminopropyltriethoxysilane, added to the resin component of heat-cleaned, glass-cloth-reinforced epoxy and polyester laminates, gave a dramatic improvement in retention of flexural strength after a two-hour boil. Bass and Porter (63) reported that alkoxy silanes impart permanent water repellency to glass, textile, and masonry surfaces. Similarly, the present authors have noted that small additions of amino functional silanes—for example, γ-aminopropyltriethoxysilanes—added to the B component of aliphatic amine-cured, bisphenol-A epoxy resin coatings, can provide major improvements in underwater adhesion. Significantly, the benefits of such additives in adhesive or coating formulations generally lie in adhesion retention after water immersion, rather than in providing an appreciable increase in initial physical properties.

It is also an established fact that such additives are not a universal cure-all for water resistance, but rather have been beneficial only in certain specific formulations. Such results can be interpreted by noting that, when small amounts (about 1% or less) of the reactive silane or titanate are added to an organic formulation, the following events must occur in proper sequence:

1. The additive must first diffuse to and selectively adsorb in high concentration at the interface.

2. The additive must, thereafter, condense with itself and with the surface hydroxyl groups, and (lastly) react with the bulk organic phase.

Clearly, if the rate of diffusion to the interface is slow relative either to the rate of cure (solidification) of the bulk resin phase or to the rate of reaction with the bulk resin, or if either the surface activity of the additive or the rate of silanol condensation at the interface is slow, the additive will remain in, or preferentially react with, the bulk phase rather than at the interface. For this reason, the most effective additives are low-molecular-weight monomers having an amphipathic structure, which, therefore, tend to migrate rapidly to an organic–polar solid interface.

As noted above, the rates of hydrolysis of the alkoxy compounds and the silanol condensation reactions are relatively sluggish at normal pH and ambient temperature. These reactions are, however, catalyzed by both acid and base (64, 65). At low pH, the overall rate is proportional to the concentration of acid, water, and alkoxy groups. At high pH, the rate is proportional to the concentration of base and alkoxy groups. Under all conditions, sufficient water must be present at the interface to provide the stoichiometric requirements of equation 75. Normally, this water is provided by the hydrated layer present on oxide surfaces as shown in Fig. 2, although for highly filled composite materials or systems containing freshly desiccated inorganic components, a small additional quantity of water may have to be added to complete the condensation reactions.

In summary, it is clear that small additions of hydrolyzable titanium or silicon-based monomers, particularly those that *also* contain nonhydrolyzable, functional organic groups, can offer major improvements in resisting adhesion loss at an oxide–organic interface, after prolonged water immersion, if the additive can diffuse to the interface, react with bulk organic phase, and condense with itself and with surface hydroxyl groups in proper sequence. It is not clear at present, however, whether the covalent surface bonding mechanism shown in equations 74 and 75 is responsible for these improvements. That is, the enhanced water resistance might also be attributable to kinetic effects (such as reduced permeation rates for water to the interface through a highly cross-linked adjacent polymer zone) or to wettability effects—for example, to a higher surface–water interfacial free energy for a surface covered with condensed silanol groups relative to the original hydroxyl-rich metal oxide surface.

Coordinate Interfacial Bonding

An entirely new—and regrettably little understood—dimension is added to the problem of characterizing adhesive–adherend bonding when the possibilities of *coordinate* bonds are considered. By coordinate bonds are meant the strong localized bonds which are formed between metal ions containing incompletely filled *d*-orbitals and electronegative atoms containing one or more pairs of unshared electrons in the outermost valence shells. Virtually all the common metallic elements found in most inorganic metallic and nonmetallic substances exhibit this coordinative tendency to some degree. The most noteworthy are iron, nickel, and cobalt; aluminum and chromium; zinc and copper; and magnesium and calcium. The primary electron-donor atoms which participate in coordination are fluorine, oxygen, and nitrogen.

Most of the common metals tend to form highly symmetrical coordination complexes, with each electron donor atom equidistant from the metal cation and from other donors; the most frequently encountered symmetry is sixfold (octahedral)—MX_6. If the donor atoms are anions (for example OH^-), the resulting complex, if isolated, will carry a negative charge equal to the difference between the cation charge and the coordination number (for example, $[M^{++}X_6^-]^{4-}$). If the donor atoms are electrostatically neutral (for example, NH_3), the complex (if isolated) will carry a positive charge corresponding to the cation valence. Any intermediate structure (including uncharged complexes) is possible.

This tendency toward symmetrical coordination is so strong that it, in large measure, dictates the crystal structure of most hydrous metal oxides; for example, gibbsite ($Al(OH)_3$), brucite ($Mg(OH)_2$), and goethite ($Fe(OH)_3$) all exist as close-packed, face-centered cubic lattices of hydroxyl ions with the metal cations positioned interstitially with octahedral symmetry relative to neighboring hydroxyls. Each hydroxyl is bonded ionically to one, and coordinately to a second, metal ion.

On the surface of a perfect crystal face of such an oxide, all the metal valence and coordinate bonds are satisfied; interactions between such a surface and another liquid or solid phase would thus be expected to be limited to those involving the surface hydroxyls alone, without participation of the metal. If, however, the oxide is disorganized (as is undoubtedly true of oxide films on most metals), the surface may contain a disarray of hydroxyl groups, and the metal ions (particularly those present in protrusions on the surface) may be incompletely coordinated. In this event, the vacant coordination sites may be occupied by adventitious impurities such as water molecules, or they may be filled by electronegative functions of the adhesive. Which will occur preferentially will depend on (1) the relative electronegativities of the function and of water, (2) the size of the electronegative function, and (3) the *steric* relationship of one electronegative function of the adhesive to its nearest neighbor.

For example, primary alkyl amino and phenolic hydroxyl groups usually coordinate more readily with cations than with water or hydroxyl ions, while alcoholic hydroxyl, ether, and keto-oxygen groups coordinate more weakly. Secondary or tertiary alkyl amines coordinate with greater difficulty than primary amines, ostensibly because the alkyl substitutions make their entry into the coordination sphere difficult. If the steric or geometric disposition of two electronegative functions on an organic compound is such as to make it possible for the two functions to occupy adjacent coordination sites about a cation, a very stable, water-insensitive coordination complex or chelate may be formed. For most (hexacoordinate) metals such a condition is satisfied when two such functions are attached to *two adjacent alkyl* carbons (as in ethylenediamine) or *alternate* adjacent aromatic carbons (as in resorcinol). One may thus infer that, to maximize bond strength and minimize water deterioration at oxide–polymer interfaces, one should design or select adhesives that contain significant amounts of coordinating electronegative groups, juxtaposed in "pairs" to increase the probability of metal ion chelation. It is provocative that many common adhesives, such as the amine-cured bisphenol-A epoxies, the resorcinol formaldehydes, the high nitrile rubbers, the urethanes, and the aminoacrylics, appear to satisfy these criteria. The dispositions of electronegative functions in such polymers are shown schematically in Fig. 7.

Thus, the possibilities for establishment of coordinate bonds and formation of chelate complexes between polar adhesives and metal oxide surfaces are real, and the role of such bonding in the development of superior adhesion may be of greater importance than has heretofore been realized. More detailed study of interface structure and composition will be necessary before the ultimate importance of these bonding mechanisms can be assessed.

Summary and Conclusions

This discussion has focused on possible ionic, dipole–dipole, covalent, or coordinate interactions between metal, metal oxide, or silicate surfaces and (1) water, (2) polar organic polymers (containing electronegative atoms such as nitrogen, oxygen, or chlorine), or (3) organic derivatives of tetravalent metal atoms.

The major objective has been to develop a qualitative technique for estimating the relative strengths of the various interfacial bonds (other than London forces) which can be formed between organic compounds and oxide surfaces, and to predict which types of bonds should be most able to resist displacement by, or hydrolytic attack in, water.

The general procedure has been to propose a model for the hydrated oxide surface which is present on clean metal oxide surfaces at normal ambient

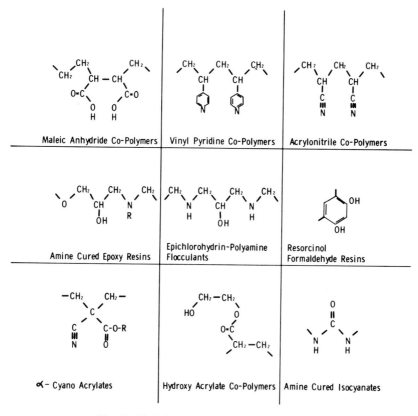

FIG. 7. Chelating moieties in polymeric adhesives.

temperatures and humidities, and then to interpret the stability and probability of forming electrostatic bonds in terms of electronegativity differences between the metal atoms in the surface (—M—O—H) group and the carbon, nitrogen, oxygen, or chlorine atoms interacting with these surface oxide groups.

Interactions with Water. In the absence of extraneous potential-determining ions in the aqueous phase, the interactions between water and oxide surfaces are predominantly via dipole interactions (hydrogen bonds) wherein the surface acts as either the acid (proton donor) or the base (proton acceptor):

$$-\text{MOH} \cdots \text{OH}_2 \rightleftarrows -\text{MOH} + \text{H}_2\text{O} \rightleftarrows -\text{MO} \cdots \text{HOH}\diagup\text{H}$$

As the pH in the aqueous phase is altered by electrolyte addition, the surface acquires an ionic charge via

$$-\overset{+}{\text{MOH}_2} \cdots \text{OH}_2 \underset{}{\overset{\text{H}^+}{\rightleftarrows}} -\text{MOH} + \text{H}_2\text{O} \underset{}{\overset{\text{OH}^-}{\rightleftarrows}} -\overset{-}{\text{MO}} \cdots \text{HOH}$$

On the basis of prior work by Parks (54) and others, the IEPS (isoelectric point of the surface) can be interpreted in terms of an electrostatic model which predicts that the IEPS decreases either as the valence of the metal atom (M) increases or as the radius of the oxolated surface cation decreases.

Interactions with Polar Organic Compounds. In contact with polar organics containing nitrogen, oxygen, or chlorine atoms, the large number of potential electrostatic interaction mechanisms have been categorized into four general types:

Type A-1: Dipole interaction with an organic acid:

$$-\text{MOH} + \text{HXR} \longrightarrow -\overset{\text{H}}{\underset{|}{\text{MO}}} \cdots \text{HXR}$$

Type A-2: Ionic interaction with an organic acid:

$$-\text{MOH} + \text{HXR} \longrightarrow -\overset{+}{\text{MOH}_2} \cdots \overset{-}{\text{XR}}$$

Type B-1: Dipole interaction with an organic base:

$$-\text{MOH} + \text{XR} \longrightarrow -\text{MOH} \cdots \text{XR}$$

Type B-2: Ionic interaction with an organic base:

$$-\text{MOH} + \text{XR} \longrightarrow -\overset{-}{\text{MOH}}\overset{+}{\text{XR}}$$

Also considered was a fifth type of interaction wherein a covalent linkage is formed via a condensation reaction between the surface and the organic group:

$$-\text{MOH} + \text{HOR} \rightleftarrows -\text{MOR} + \text{H}_2\text{O}$$

In the absence of extraneous ions, the relative probability for each of these five interactions depends on the IEPS and the pK_A of the organic acid or organic base. For type A-1 or A-2 interactions,

$$\Delta_A \equiv \text{IEPS} - pK_A \text{(organic acid)} = \log^{-1}\frac{[\text{MOH}_2^+][\text{X}\bar{\text{R}}]}{[\text{MOH}][\text{HXR}]}$$

Thus, for $\Delta_A < 0$, interaction type A-1 predominates over type A-2 and the interfacial bond is primarily nonionic. As Δ_A continues to decrease to more negative values, the strength of the dipole–dipole interactions decreases and the bond becomes progressively more susceptible to displacement by water via

$$-\text{MO}\cdots\overset{\mid}{\underset{}{\text{H}}}\text{XR} + 2\text{H}_2\text{O} \rightleftarrows -\text{MO}\cdots\overset{\mid}{\underset{}{\text{H}}}\text{OH} + \text{H}_2\text{O}\cdots\text{HXR}$$

As Δ_A increases, an individual interfacial bond becomes progressively more ionic and stronger, and it should be more capable of resisting displacement by water. Although only a small number of metal oxide–organic combinations give positive values of Δ_A, such bonds are essentially entirely ionic and, for energetic reasons, should resist water displacement.

Similarly, it was shown that

$$\Delta_B \equiv \text{p}K_A \text{ (organic base)} - \text{IEPS} = \log^{-1}\frac{[\text{MO}^-][\text{H}\overset{+}{\text{X}}\text{R}]}{[\text{MOH}][\text{XR}]}$$

Analogously, for the combinations wherein Δ_B is positive, the interfacial bonds are predominantly ionic (type B-2) and resistant to water displacement, while as Δ_B decreases, to more negative values, the interfacial bond involves the progressively weaker dipole interaction (type B-1) and is more easily displaceable via

$$-\text{MOH}\cdots\text{XR} + 2\text{H}_2\text{O} \rightleftarrows -\text{MOH}\cdots\text{OH}_2 + \text{HOH}\cdots\text{XR}$$

For cases where an organic compound can interact either as a weak acid or as a weak base, the dominant mode (dipole orientation) is predicted by the larger numerical value of Δ_A or Δ_B.

The ability of all such bonds to resist water displacement depends strongly on the pH in the aqueous phase. If the pH at the interface is altered by the addition of extraneous potential-determining ions (such as H^+ or OH^-), the theory predicts that maximum resistance to displacement by water should occur only over the pH ranges:

For type A interactions:

$$\text{IEPS} > \text{pH} > \text{p}K_{A(A)}$$

For type B interactions:

$$\text{p}K_{A(B)} > \text{pH} > \text{IEPS}$$

Under either strongly acidic or strongly alkaline conditions in the aqueous phase, all bond types described above should become weak relative to water, owing either to the development of mutual repulsion forces or to suppression of ionization in the surface or organic groups (equations 62 and 63).

Covalent bonds to carbon atoms, of the type —M—O—C—, can be formed, the present theory predicts, most readily under anhydrous conditions, or in heat-cured systems *and* in surface–organic combinations where Δ_A or Δ_B is close to zero. Because of the electronegativity difference between most metal atoms and carbon atoms in organic compounds, the resulting M—O—C bond is generally highly polarized and therefore is susceptible to hydrolytic attack, reverting in the presence of water to dipole or ionic bonds of type A or B.

Surface Interactions with Organic Compounds of Tetravalent Metals. The electrostatic models proposed can be applied to interactions with organic compounds of silicon, titanium, or zirconium and other tetravalent metal compounds of the formula X_{4-n}—Si—R_n, where X is an electronegative (hydrolyzable) group and R is a nonhydrolyzable group—for example, bonded via an Si—C covalent bond. Although it is clear that such compounds can provide major improvements in bond resistance to water if allowed to diffuse to the interface, condense with surface water to form a cross-linked interfacial zone, and then react with the bulk resin phase in proper sequence, it is not clear whether such improvements are evidence for the formation of hydrolytically stable covalent bonds at the interface. Rather, such improved water resistance may be attributable to other factors, which include reduced permeability of water through the cross-linked interfacial resin zone or to differences in the wettability of a silanol-rich versus the original, hydrated metal oxide surface.

Coordination Complexes. Certain steric configurations in organic compounds, particularly those in which two electronegative groups are separated by two or three carbon atoms, should, in theory, permit the two electronegative functions to occupy adjacent coordination sites about a surface cation and thereby provide a stable, water-insensitive, coordination complex or chelate structure at a resin–oxide interface.

References

1. J. J. Bikerman, "The Science of Adhesive Joints," Academic Press, New York, 1961, p. 133.
2. J. J. Bikerman, in "Adhesion and Cohesion," ed. by P. Weiss, Elsevier, Amsterdam, 1962, p. 36.
3. R. L. Patrick, C. M. Doede, and W. A. Vaughan, Jr., *J. Phys. Chem.*, **61**, 1036 (1957).
4. L. H. Sharpe and H. Schonhorn, *Advan. Chem. Ser.*, **43**, 189 (1964).
5. W. A. Zisman, in "Adhesion and Cohesion," ed. by P. Weiss, Elsevier, Amsterdam, 1962, p. 176.

6. W. A. Zisman, *Ind. Eng. Chem.*, **55**, 18 (1963).
7. W. A. Zisman, *Advan. Chem. Ser.*, **43**, 1 (1964).
8. H. Schonhorn, *J. Appl. Polymer Sci.*, **8**, 355 (1964).
9. J. R. Huntsberger, *Advan. Chem. Ser.*, **43**, 180 (1964).
10. J. R. Huntsberger, in "Treatise on Adhesion and Adhesives," Vol. 1, ed. by R. L. Patrick, Dekker, New York, 1967, Chapter IV.
11. J. H. Hildebrand and R. L. Scott, "The Solubility of Non-Electrolytes," Reinhold, New York, 3rd ed., 1950.
12. J. H. Hildebrand and R. L. Scott, "Regular Solutions," Prentice-Hall, Englewood Cliffs, New Jersey, 1962.
13. L. Pauling, "The Nature of the Chemical Bond," Cornell University Press, Ithaca, New York, 3rd ed., 1960.
14. R. J. Good, in "Treatise on Adhesion and Adhesives," Vol. 1, ed. by R. L. Patrick, Dekker, New York, 1967, Chapter II.
15. L. A. Girifalco and R. J. Good, *J. Phys. Chem.*, **61**, 904 (1957).
16. J. L. Gardon, in "Treatise on Adhesion and Adhesives," Vol. 1, ed. by R. L. Patrick, Dekker, New York, 1967, Chapter VIII.
17. R. J. Good, L. A. Girifalco, and G. Kraus, *J. Phys. Chem.*, **62**, 1418 (1958).
18. R. J. Good and L. A. Girifalco, *J. Phys. Chem.*, **64**, 561 (1960).
19. R. J. Good, *Advan. Chem. Ser.*, **43**, 74 (1964).
20. F. M. Fowkes, *J. Phys. Chem.*, **66**, 382 (1962).
21. F. M. Fowkes, "Chemistry and Physics of Interfaces," ed. by D. E. Gushee, American Chemical Society, Washington, D.C., 1965, Chapter I.
22. F. M. Fowkes, in "Fundamental Phenomena in the Materials Sciences," Vol. 2, "Surface Phenomena," ed. by L. J. Bonis and H. H. Hausner, Plenum Press, New York, 1966, p. 139.
23. H. R. Baker, C. R. Singleterry, and W. A. Zisman, "Factors Effecting Surface Chemical Displacement of Bulk Water from Solid Surfaces," presented at the Division of Colloid and Surface Chemistry, National American Chemical Society Meeting, Atlantic City, 1967.
24. M. K. Bernett and W. A. Zisman, *J. Phys. Chem.*, **70**, 1064 (1966).
25. W. A. Zisman, *Ind. Eng. Chem.*, **57**, 27 (1965).
26. S. S. Voyutski, *Adhesives Age*, **5**, 30 (1962).
27. R. R. Stromberg, in "Treatise on Adhesion and Adhesives," Vol. 1, ed. by R. L. Patrick, Dekker, New York, 1967, Chapter III.
28. J. H. Anderson and K. A. Wickerscheim, "Solid Surfaces," (Proc. Intern. Congr. Phys. Chem. Solid Surfaces), North-Holland, Amsterdam, 1964, p. 252.
29. J. J. Bikerman, "Surface Chemistry, Theory and Applications, "Academic Press, New York, 1958, Chapter III.
30. P. C. Carman, *Trans. Faraday Soc.*, **36**, 964 (1940).
31. R. K. Iler, "Colloid Chemistry of Silica and Silicates," Cornell University Press, Ithaca, New York, 1955.
32. G. Salomon, in "Adhesion and Adhesives," Vol. 1, ed. by R. Houwink and G. Salomon, Elsevier, Amsterdam, 1965, p. 18.
33. G. J. Young, *J. Colloid Sci.*, **13**, 67 (1958).
34. C. S. Brooks, *J. Colloid Sci.*, **13**, 522 (1958).
35. C. S. Brooks, *J. Phys. Chem.*, **64**, 532 (1960).
36. A. C. Zettlemoyer, "Chemistry and Physics of Interfaces," ed. by D. E. Gushee, American Chemical Society, Washington, D.C., 1965, Chapter XII.

37. A. C. Zettlemoyer, "The Physical Chemistry of Surfaces and Surface Heterogeneities," this volume.
38. L. Ter-Minassian-Saraga, *Advan. Chem. Ser.*, **43**, 232 (1964).
39. P. J. W. Debye and L. K. H. Van Beek, *J. Chem. Phys.*, **31**, 1595 (1959).
40. F. P. Bowden and W. R. Throssell, *Nature*, **167**, 601 (1957).
41. F. P. Bowden and D. Tabor, "Structure and Properties of Solid Surfaces," ed. by R. Gomer and C. H. Smith, University of Chicago Press, Chicago, 1963, Chapter VI.
42. F. P. Bowden, in "Adhesion and Cohesion," ed. by P. Weiss, Elsevier, Amsterdam, 1962, p. 121.
43. E. G. Shafran and W. A. Zisman, NRL Report No. 6496 (March 24, 1967).
44. D. V. Keller, *Wear*, **6**, 353 (1963).
45. G. E. Boyd and H. K. Livingston, *J. Am. Chem. Soc.*, **64**, 2383 (1964).
46. F. M. Fowkes, in "Treatise on Adhesion and Adhesives," Vol. 1, ed. by R. L. Patrick, Dekker, New York, 1967, Chapter IX.
47. W. H. Wade and N. Hackerman, *Advan. Chem. Ser.*, **43**, 222 (1964).
48. G. D. Cheever, "Wetting of Phosphate Interfaces by Polymer Liquids," this volume; *J. Paint Technol.*, **39**, 1 (1967).
49. G. R. Irwin, in "Treatise on Adhesion and Adhesives," Vol. 1, ed. by R. L. Patrick, Dekker, New York, 1967, Chapter VII.
50. E. Orowan, *Z. Physik*, **79**, 580 (1932).
51. F. Smith and D. J. Kidd, *Am. Mineralogist*, **34**, 403 (1949).
52. H. B. Weiser, "Inorganic Colloid Chemistry, II. The Hydrous Oxides and Hydroxides," Wiley, New York, 1935, p. 142.
53. A. C. Zettlemoyer and J. J. Chessick, *Advan. Chem. Ser.*, **43**, 88 (1964).
54. G. A. Parks, *Chem. Rev.*, **65**, 177 (1965).
55. A. S. Michaels and J. C. Bolger, *Ind. Eng. Chem., Fundamentals*, **3**, 14 (1964).
56. P. J. W. Debye, in "Adhesion and Cohesion," ed. by P. Weiss, Elsevier, Amsterdam, 1962, p. 1.
57. H. F. Mark, in "Adhesion and Cohesion," ed. by P. Weiss, Elsevier, Amsterdam, 1962, p. 240.
58. C. R. Noller, "Chemistry of Organic Compounds," Saunders, Philadelphia, 1965.
59. I. Iwasaki, S. R. B. Cooke, and A. F. Columbo, *U.S. Bur. Mines Rept. Invest.*, **5593** (1960).
60. W. Baird, in "Waterproofing and Water-Repellency," ed. by J. L. Moilliet, Elsevier, Amsterdam, 1963, Chapter IV.
61. O. K. Johannson, F. O. Stark, G. E. Vogel, and R. M. Fleishmann, *J. Composite Materials*, **1**, 278 (1967).
62. S. Sterman and J. C. Toogood, *Adhesives Age*, **8**, No. 7, 34 (1965).
63. R. L. Bass and M. R. Porter, in "Waterproofing and Water-Repellency," ed. by J. L. Moilliet, Elsevier, Amsterdam, 1963, Chapter V.
64. J. F. Hobden and H. H. G. Jellinek, *J. Polymer Sci.*, **11**, 365 (1963).
65. J. A. C. Watt, *J. Textile Inst.*, **51**, T1 (1960).

Discussion

MATIJEVIĆ: I would like to project what I think is the future of this type of approach. To treat a metal oxide interface in the same way as the interface of silica, for example, is an obvious oversimplification. The acid–base interaction at the interface is just one of the possibilities which may be considered in the explanation of the surface properties. Any metal oxide in solution is in "equilibrium" with a variety of species, particularly hydrolyzed species, which originate from the dissolution of some of these metal ions. It is known that such hydrolyzed species adsorb preferentially. Thus, if we take the isoelectric point as a criterion, we should consider in each case the causes for the existence of the isoelectric point. I believe that the adsorption of complex hydrolyzed metal ions is in effect responsible for the surface properties of metal hydrous oxides. One could visualize the following process taking place on suspending a metal oxide, M_pO_q, in water:

$$M_pO_q \text{ (solid)} \xrightarrow[\text{partial dissolution, hydrolysis}]{H_2O} M_x(OH)_y^{(zx-y)+} \text{ (soluble)} \xrightarrow{\text{adsorption}} M_pO_q/M_x(OH)_y^{(zx-y)+} \text{ (charged particle)}$$

The data by Parks could probably be reinterpreted along the same line of thought. One could find the relationship between the size of the ion and hydrolyzability. Then adsorptivities of these ions could be correlated to isoelectric points. Following Dr. Michaels' example for flotation, if one adds 10 micromoles of a ferric salt to beryl, the mineral can be floated much more efficiently because it adsorbs the hydrolyzed ferric ions which, in turn, cause charge reversal. It would seem, then, that the next logical step would be to establish in each case the composition of the solution in equilibrium with the solid hydrous metal oxide and correlate the adsorptivity of all species with the interfacial properties of the solid. Presently, only a few cases could be handled in the proposed way, because the necessary information on the composition and the thermodynamic properties of solutions of polyvalent metal ions is seldom known with certainty.

With regard to Dr. Michaels' reference to chelation, I feel that he pointed his finger at one of the most important phenomena. We have recently shown that metal chelates show striking coagulation and reversal of charge effects. For example, trisphenanthroline nickel ion coagulates a silver bromide sol in concentrations more than four orders of magnitude lower than the nickel ion alone. The chelate also reverses the charge most efficiently, while nickel ion cannot produce charge reversal at all. It is particularly interesting that these strong effects by chelates are observed at low values at which, according to the thermodynamic information available, the chelate species should be completely decomposed. The obvious explanation seems to be that chelation may take place at the interface even though the corresponding process in solution does not take place. The study of these phenomena are of great importance, and I fully support all of Dr. Michaels' comments on this subject.

H. MARK: Dr. Michaels made an extremely interesting suggestion that one could design the resinous components so that they would fit in the certain pre-existing spacings in the lattice. There is actually experimental evidence for such a phenomenon in the so-called substantivity of certain dyestuffs. There, the matrix is usually a crystalline fiber, such as cellulose or one of the known isotactic polymers. The fitting of the polar groups of the dyestuff molecule on the chains of the polymer must be very close and is usually a matter of a tenth of an angstrom. If the respective intramolecular distances are very similar, one finds remarkably strong bonding by polar bonds, hydrogen bridges, or coordination complex formation.

FOWKES: The use of acid–base interaction to promote adhesion is certainly very old in the art and particularly in the case of petroleum products to metal, such as in rolling and cutting oils. The new contribution here is the idea of using isoelectric points as a means of characterizing the acidity or basicity of a metal oxide. The real question is, How good is this, and we haven't the data yet to understand it. But I could point out at least a few cases where, with some evidence, you might be right. For example, in the case of alumina versus silica, in the absence of water, alumina is highly acidic and not basic, and indicator dyes show this very clearly. However, if you try to improve the adhesion of petroleum products to an aluminum metal surface covered with an oxide, you cannot give it any increased adhesion by using a basic additive. This was the direction predicted by the isoelectric point for the alumina, in this case, so I can think of at least one example where the isoelectric point is a better prediction for the adhesion by acid–base interaction than an indicator dye. This whole idea needs quite a bit more experimental check to know whether we should accept it generally.

MICHAELS: I agree. There is a lot of experimental confirmation yet required.

HERMANS: I happen to know that the oxidation of some metals can be very strongly affected by traces of materials. For example, liquid magnesium, in the presence of oxygen, oxidizes very rapidly, but this is slowed down tremendously by traces of sulfur dioxide gas or by traces of beryllium in the metal. If such cases exist, would they also have a large effect on your considerations?

MICHAELS: Yes. There is strong evidence that the oxidation state of a metal in a metal oxide controls its surface-charge characteristics. The higher valency oxides are, as a rule, more acidic than the lower ones. It is only logical to expect that, if an adhesive or organic phase which is brought in contact with a metal oxide surface contains either an oxidant or a reductant, it will affect the oxidation state of the metal. Iron, for example, can be reduced from the plus 3 to plus 2 valence state with hydrazine or other organic soluble reducing agents; this would be expected to alter the surface bonding characteristics materially. In general, commonly used surface pretreatments of metals tend to elevate the oxides to the highest valency state. Acidic treatments and oxidizing agent treatments are examples of efforts to eliminate bonding variability that might result from differences in the oxidation state of the metal in the oxide layer.

HERMANS: Yes, I can see that, but would you say that a trace of beryllium could have this kind of effect? You are almost implying that most of the beryllium is on the surface.

MICHAELS: This would be the interpretation I would offer, with the information at hand. I do not know what kind of general conclusion to draw about the situation at the solid–gas interface and at the solid–water interface.

ZETTLEMOYER: In the case of nickel in iron, the nickel at the iron surface is more concentrated than in the bulk. The inhibiting characteristics of nickel for iron are seemingly because of the surface activity. In the case of chromium in iron, it is not that way; the chromium concentration at the surface is the same as in the bulk. So here the characteristics of the mixed oxides are important in inhibiting and making Fe–Cr stainless.

Are we not talking about the formation of the interface in adhesion and not adhesion itself? By that I mean, I thought that adhesive failure was a large majority of the time in the adherend—that is, in one of the phases and not at the interface. Along with this, I would like to hear comments on the fate of the water, which seems to be rather important, since it is almost always present on the oxide.

MICHAELS: In arriving at these Δ_A and Δ_B values, we concerned ourselves solely with the interactions: metal oxide–water and adhesive–water; but it is really surface hydroxyls and the reactive moieties in the adhesive which are involved in bonding. These two react together to form the bond. The quantity that was made use of in the interpretation of the bond durability under wet conditions was the displacement capability of water in the bond. The difference between the interaction parameter oxide–adhesive and the interaction parameter of the oxide with water determines if the reaction adhesive–adherend plus water goes to adhesive plus water and adherend plus water. This difference-calculation is a measure of the free energy change accompanying the displacement of the bond by a water layer. Nowhere can we draw any inferences as to what physically happens when we peel an adherend and adhesive apart in the presence of water. There is no meaningful inference you can draw from these calculations as to what processes occur during the bond destruction. All these numbers give is an indication as to whether water displacement is the more probable state under the circumstances.

ZETTLEMOYER: That is the most you can say.

EIRICH: I would like to continue some of Dr. Hermans' comments and refer to the effect of impurities on the surface structure and acidity. Surfaces accumulate inhomogeneities, impurities, and dislocations and present a state very different from impurities in the interior of the material. This must affect the local electrical properties, the pH or pK values, and make it doubtful to what extent one can correlate bulk properties with the surface properties which are responsible for ionic or polar bonds. The most interesting result of this lecture to me is the correlation between the tendency of a material toward surface dislocation and disturbances and its ability to undergo a variety of surface bondings. Have you, for instance, ever looked into the difference between the IEPS of pure and impure magnesium oxide, and if so, is there any correlation between the degree and type of added impurities and the change of surface properties?

MICHAELS: With regard to the latter question, there is much experimental evidence in Parks' data, for example, on the effects of the impurity population density in oxides on isoelectric points. There is no question but that there is a definite (and predictable, in many cases) shift in the IEPS, with impurity concentration within the oxide, as well as with the hydration state of the oxide. There is no question but that the impurity factor is a very important one characterizing the oxidized surface. You suggested that we have been using bulk properties to account for surface properties, but really, when the chips are down, the isoelectric point measurement is a surface measurement and not a bulk measurement. We are talking about electrokinetic measurements, and we are inferring ion distribution characteristics of surfaces from these electrokinetic measurements. So, I would say that it really is a specific surface characteristic that we are determining when we measure IEPS. On the other hand, the acid–base ionization constants that are used in the other calculation are bulk properties, there is no question about that; the only uncertainty here is that, in solution, the reactive moiety or the ionizing moiety is completely surrounded by water, while at the surface it is not completely surrounded by water. When it is not completely surrounded by water, are the proton donor–acceptor characteristics altered? I am sure the answer is yes, and it may be substantial. The fact that certain indicator dyes, when adsorbed on alumina, show color suggesting much lower pH values than are expected from IEPS measurements may very well be due to the fact that the environment around the chromophore in the dye is different on the surface than it is in solution at the corresponding pH. I have often wondered about this in trying to interpret surface properties by dye adsorption. How really meaningful are such measurements, because of the change in morphology of these dye molecules in the adsorbed state.

JELLINEK: I would like to make a comment on the water layer which you have on this oxide. There is more and more evidence that this layer is structured beyond the first boundary layer which permits an ice-like structure, for instance, where you have capillaries. The properties of surface water are quite different from those of bulk water. I do not know what difference this would have on the adhesion concentration.

MICHAELS: Even allowing for structuring of water, at a substantial distance away from the interface or surface, it is the energy associated with that structure, large or small, compared with the energy associated with the proton transfer reaction that gives rise to the ionic bonds, which is important. If it is small, then it probably does not make any difference in bonding. If, on the other hand, there is a big free energy decrease associated with water-structuring, some sort of an ice-like surface is formed, and then the arguments about water displacement are no longer valid. But I have no way of knowing it.

JELLINEK: The heat of fusion must come in somehow.

MICHAELS: That is probably not a very important contribution—that is, only 80 calories per gram, as compared with 50 or 60 kilocalories per mole for a neutralization reaction. But if there is something more to it than the heat of fusion, then this is not applicable.

ONSAGER: When the difference in energies between the modifications of salt and water is involved, the heat of fusion is a good deal bigger, but the free energy of fusion is zero at the melting point. I am strongly inclined to go along with Dr. Michaels; it cannot amount to a whole lot.

ROSOFF: The interfacial tensions under discussion are largely negative. Therefore, the interface would be unstable, and I am wondering just how much of the stability of the bond is being affected by kinetic effects.

MICHAELS: I am not quite sure what you mean by unstable. A negative surface tension or a negative free surface energy merely means that there is a strong driving force for establishment of that interface.

ROSOFF: There is a mutual penetration of two phases.

MICHAELS: When there is an extraordinarily large negative interfacial free energy, the rate-determining steps in establishing the bond have nothing to do with the free energy considerations. They are controlled by flow, diffusion, and orientation effects. Resin rheology and capillary penetration into surface irregularities probably dominate the establishment of the bond. The factors that govern the durability of the bond may very well be diffusional.

LEE: I would like you to elaborate on the relationship between Coulombic forces and interfacial tension. Also, what is the significance of spontaneous emulsification between a liquid adhesive and a solid substrate?

MICHAELS: The relationship between spontaneous emulsification, and the interaction of liquid adhesive with a solid substrate, is merely that both phenomena involve specific interactions that give rise to negative interfacial tensions. The observed spreading rates will be much faster than the predicted spreading rates, based strictly on dispersion-force interactions. There are large spreading stresses generated at that interface during the process of spreading which can create large shear stresses in the fluid and make the fluid boundaries advance faster. Some evidence of this was observed in working with very basic adhesives on very acidic surfaces. They spread like wildfire on surfaces—far faster than predicted on the basis of simple wetting measurements. Your first question was, What is the significance of Coulombic interactions across the interface in producing negative interfacial tensions? Calculation of the free energy change as a result of the formation of an ion pair by proton transfer results in free energy changes of the order of 75 to 100 kilocalories per mole pair. For any reasonable assumption as to the population density of such ion pairs on the surface, even if there is one for every 1000 Å or one for every 10000 Å, the total energy per unit area, liberated by the formation of these ions pairs, is measured in thousands of ergs per square centimeter. This is so much greater than the dispersion interaction energies that you can forget the former as being contributory to the formation of this interface. So, it does not take very many ion-pair interactions to make an enormous change in the free energy change on wetting. This is the most striking conclusion to be drawn from this type of analysis. You do not need very many of these groups on the surface to make a big difference in the wetting characteristics, and presumably in the adhesive bond strength.

ZIMMT: Metal surfaces are not always homogeneous. In an iron surface, there are cathodic and anodic areas. How does this affect the bulk property of adhesion?

MICHAELS: If the cathodic and anodic areas on an iron surface are represented by different valence states in the metal, the more basic or lower valence sites may form much weaker bonds than the more acidic areas. When applying corrosion protective coatings on metals, application of cathodic potentials to these metals to protect them against corrosion may actually ruin the bond. The reason you may ruin the bond is that you deprotonate the metal in the process of applying the negative potential, and you may raise the pH at the interface to the point where the adhesive just comes off because there is no longer any opportunity for Coulombic interaction. Thus, it is possible that local potential differences, of only a few millivolts on a metal surface, may cause marked changes in the degree of adhesion. I think the practice of oxidizing pretreatment to raise all metal ions to a constant valence state bears on this same question.

SCHWARTZ: The dispersion forces and also the Coulombic forces are supposedly not highly directional. On the other hand, hydrogen bonds and coordination bonds are directional. In making your calculations, you commented that even a small population in the surface is sufficient to account for a high bonding energy on the basis of ergs per square centimeter. Do you also have to apply a correction factor to account for surface hydroxyl groups that might not be oriented in the right direction to form hydrogen bonds?

MICHAELS: In this estimation, what we have assumed is that there really is no variability in the hydrogen bond structure, and that, orientation notwithstanding, hydrogen bond energies are all roughly in the range of 2 to 5 kilocalories per mole. When you are talking about 100 kilocalories for ionic bonds, then you could not care less about the remainder of the hydrogen bonds. So, we have sidestepped the issue about directionality of hydrogen bonds by focusing our attention on the much stronger ionic ones.

SCHWARTZ: If the Coulombic bonds are present, and if they dominate, they will dominate all the other effects.

MICHAELS: If they are present to the extent of maybe 0.1% of the total bonding possibilities or more, they probably dominate.

LAUKONIS: The isoelectric point values for the various iron oxides are, for the most part, between pH values of 5 and 7. The phosphating solutions I have used have a pH of about 2.8. Can the considerations which you have outlined be used to predict the nature of the bonding between a phosphate coating and an iron oxide surface?

MICHAELS: With phosphates, for example, the pH determines the state of ionization of the phosphate ion, and this, in turn, determines the crystal structure of the resulting iron phosphate that forms. Whether it is monobasic, dibasic, or tribasic phosphate ion that ends up on the lattice, is governed largely by the pH of the reaction. I am not sure, but the bonding that you get between the phosphate layer and the substrate may be governed more by the crystal structure of the phosphate as influenced by the pH, than by the pH itself. But, I think that there should be evidence in the literature regarding, for example, the metal cation population density in the various phosphates, relative to the oxides, that will tell you when you get a good "fit." If you get a good fit, you might get a good bond.

MATIJEVIĆ: If one takes an aluminum salt in the presence of hydroxyl ions at different ratios of hydroxyl to aluminum and then adds phosphates, there is no precipitation as long as there is a deficiency of anions. When enough hydroxyl and phosphate ions are added to reach an isoequivalent point, precipitation does occur with all the aluminum in the precipitate. I think that there is much more involved in such a process than a simple interaction of surfaces. It is known that with some of these ions interpenetrating complexes are formed. This means that in a hydroxylated species, a certain number of coordinated hydroxyl ligands are substituted by, for example, phosphate or sulfate ions. The iron–phosphate system in solution represents such a case. The extent of such a reaction would then also affect the properties of the surfaces. This is again a case which people have studied very little, because the processes appear to be too involved.

LAUKONIS: Am I correct in assuming that if, in the process of bond formation, ion exchange of some sort takes place between the hydrated oxide surface and the phosphate molecule, this assures that the bond is ionic? There is still the problem of where and when the waters of hydration enter into the attachment of the phosphates.

MATIJEVIĆ: Apparently a covalent bond is formed by internal ion exchange. Some of these complex inorganic species retain the same charge because the exchange of "penetrating" species is equivalent. Such is the case for basic aluminum sulfate.

Also, further to your remarks, I would very strictly distinguish between the zero point of charge and the isoelectric point. I think silica exhibits an isoelectric point. Metal oxides have a zero point of charge. I feel that there is a big difference between the two.

FOWKES: In the semiconductor industry, amines are used to promote adhesion to silica surfaces, and this is a very fine way of getting stronger bonds, as measured initially in the laboratory. After a week or two weeks or a few months, this bond deteriorates completely. The silica bond with an amine just gets weaker and weaker, with time, in a system that was kept chemically clean, and since this is the kind of bond we are considering today, it is worth remembering this. I am beginning to think I understand it now. We are studying the electrical layers within solids and particularly oxides, and these electrical double layers in oxides really determine the charge of the surface. The relaxation time for such electrical double layers at room temperature is established in some weeks or months, and consequently the time effects matches very nicely. So, in the surface of SiO_2 we have oxide ions, sodium ions as counter ions, 20 to 100 Å deep in the oxide, and this determines the surface oxide charge. Now, if we bring up a positive ion in the surface, instead of a negative, we are going to have a relaxation of this positive counter ion in the solvent, and this is where some solid-state physics and solid-state chemistry come into understanding what determines the electrical properties of these solids. So, this is just a warning that the initial results are not the whole thing. You want this bond to last.

Diffusion and Reaction of Polymers in a Capillary Matrix

H. F. MARK

Polytechnic Institute of Brooklyn
Brooklyn, New York

Introduction

Transport of matter and energy through porous systems has been and still is of great theoretical and practical interest, and it becomes even more important if it is accompanied by chemical reactions. First, it should be realized that all processes of life in plants and animals take place in capillary fibers, strands, membranes, and tissues through diffusion of aqueous solutions of polymeric, oligomeric, and organic and inorganic substances which react with each other in the most complex and efficient way. Respiration, digestion, vision, and, in fact, any muscle action are controlled by diffusion and reaction of organic molecules in a porous polymeric matrix the structure of which is of preponderant influence on the smooth and regular course of all their functions.

This report does not intend to include systems and processes of this type but rather to limit itself to another class of phenomena which, although restricted to nonliving matter, still embrace numerous processes of considerable theoretical and practical interest. Processes like the seeping and draining of water or oil through soil and sand, the drying of laundry in air, or writing with ink on paper are all characterized by the cooperation of a capillary system—soil, sand, fabric, paper—with a diffusing fluid such as water, oil, ink, or air. Similarly the hardening of glue, concrete, or plaster of Paris involves chemical reactions inside porous systems. This widespread occurrence of diffusion and reaction in capillary beds has led to a large body of experimental and theoretical studies which are reported in some fifty scientific and technical journals and have been condensed in reference books (*1*). It is the purpose of this introductory report to present a brief review of the present state of the art in such a manner that it serves to clarify the main theme of this symposium–namely, interface conversion for polymer coatings.

Important Processes Involving Porous Systems

The natural movement of water through soil and sand greatly influences forestry, agriculture, harbors, highways, bridges, and buildings. In view of its obvious importance this discipline has received a modern and thorough fundamental treatment in the science of soil mechanics (2). Equally important is the seepage of oil and gas alone or of oil, gas, and water through all kinds of geological formations: It is the first basic concern of the oil and gas industry, which has been and still is devoting many intense and systematic investigations to this subject matter (3).

Less global, but still of considerable importance, are all processes of technical filtration beginning with the purification of water by diffusion through beds of sand and clay to the separation of chemical precipitates from mother liquors. In many cases the porous systems are natural polymers such as wood, flour, cotton, paper, or screens made of synthetic polymers and are fabricated with the aid of processes which control the character of the porosity or capillarity of the filter. Practical interest concentrates on the rate of flow through the bed and on the efficiency of separation. These two requirements are, in general, of opposing character, and in each case it requires a clear theoretical understanding and a skillful experimental operation to arrive at an optimal compromise which permits the maintenance of favorable conditions over longer periods of time (4).

Modern, more important improvements of the old principle of filtration are the various chromatographics, some of which use natural polymers (for example, cellulose, rubber, and dextran) and synthetic polymers as porous beds. Advanced mathematical analyses, intricate designs, and ingenious techniques have led to almost miraculous features of separation, purification, and identification of complicated substances on a submicro scale. For all this a well-founded understanding of the laws of flow and permeation and a reliable characterization of the capillary systems were the first requirements for perfection (5).

Equally indispensable for fundamental progress in physics and chemistry are the processes of gas and vapor diffusion through porous matrices, such as sintered metal powders, foamed ceramics, and molecular sieves which have led to new industrial processes for gas purification and isotope separation (6).

More practical, but not less important, are all techniques which lead to the drying of gases, solvents, and industrial polymeric goods such as paper, films, bulk fibers, and fabrics. Each of these operations requires, for controlled speed and degree, a reliable and advanced understanding of the motion of fluids through a permeable solid matrix and its quantitative characterization on the basis of reproducible measurements (7).

Another wide and old field which belongs to the scope of this report includes all processes of printing, staining, and dyeing in the course of which dissolved or suspended molecules or colloidal particles have to be distributed and deposited in the cavities and interstices of organic networks such as wood and paper and such as films, sheets, and fabrics of many natural and synthetic polymers. Extremely severe conditions are presently demanded concerning the speed and precision of the techniques and concerning the esthetics, durability, and low cost of the products. All this can be optimized only by a thorough understanding of each step which contributes to the ultimate pattern of process and product (8).

All activities enumerated until now involve the transport of fluids with dissolved or dispersed solutes through capillary beds without any chemical changes in the system. There are however, other phenomena in the course of which the motion of one or more ingredients in a porous matrix is accompanied by chemical processes. These are controlled by the local concentrations of the individual reagents and are, therefore ultimately connected with and strongly influenced by the relative rates of physical diffusion and chemical reaction. A large group of such processes are all those catalyzed reactions that occur in the fluid phase inside of solid, heterogeneous catalysts where the reagents diffuse to the active centers and undergo in their immediate neighborhood some chemical change; the reaction products must then diffuse out in order to liberate the active center for the next elementary reaction step. Since, in general, no polymeric materials are involved in these procedures, they will be mentioned only briefly here without any more detailed discussion (9).

In other areas, however, there exists considerable knowledge and experience on phenomena where polymeric reactions take place in capillary beds. One important, well-known, and thoroughly studied case is the setting of various types of concrete and plaster compositions. It comprises the reaction of the individual components of a solid permeable bed with carbon dioxide and water and the subsequent crystallization of calcium aluminates and silicates in the cavities of the matrix. The rates and extent of the various steps are controlled by the local concentrations of the reagents, which, in turn, depend on the laws of diffusion and on the character of the porous bed. There are rather elaborate experimental and theoretical treatments of these phenomena (10), some of which have a close similarity to the formation of the various iron and zinc phosphate compositions which are formed in the phosphating of steel.

Other polymerization processes in capillary matrices take place in the resin treatment of paper and textiles. The porous matrix, sheet, web, or fabric is impregnated with a polymerizable composition, and the resin is formed *in situ* by condensation or addition polymerization of its diffusible components. In some cases the imbibition of the goods with the resin formed is completed under such conditions that polymerization does not occur until later when the resin is cured by heat or by the action of radiation. Under such conditions the two contributing factors, diffusion and reaction, are substantially separated from each other and can be individually controlled. In other cases, however, impregnation and setting are carried out in one operation, sometimes because of simplicity and speed, sometimes because the results are satisfactory and equal to those of a two-step process (*11*).

A review of theoretical understanding and experimental techniques in all enumerated cases leads to the general conclusion that the following steps should be developed in the establishment of a polymeric coating on a smooth steel surface:

1. Thorough cleaning of the surface.
2. Deposition of a strongly adhering porous layer of an iron or zinc phosphate composition of known porosity and known specific surface.
3. Uniform imbibition of this capillary bed with a thickened mixture of monomers which are capable of undergoing addition polymerization.
4. Initiating this polymerization at ambient or only slightly elevated temperatures either with irradiation or with an appropriate catalytic system in such a manner that all organic components remain in or on the bed.

Porous Systems

The number, variety, and application of porous systems are almost unlimited.

Distillation columns are filled with Raschig rings or Berl saddles with dimensions of several inches. Water purification beds contain pebbles and sand which produce pores, capillaries, and slits in the range of centimeters and millimeters. Pumice and industrial fillers made of limestone, glass filter, felt, or cotton possess pore size diameters between 10^{-2} and 10^{-3} cm. Filter paper and heterogeneous solid catalysts go down to 10^{-6} cm, and molecular sieves together with colloid filters contain even voids and cavities of molecular size 10^{-7} to 10^{-8} cm.

Two quantities, which can be measured without too much difficulty, are commonly used to characterize porous systems in a preliminary way but often to a sufficient degree: the porosity, P, and the specific surface, S (*12*). The porosity is the ratio of the volume of voids to the total volume of the sample:

$$P = \frac{\text{Void volume}}{\text{Total volume}} = \frac{V_v}{V_t} \qquad (1)$$

Being a ratio of two volumes, it is a simple number and does not refer to any given weight or volume of the material as long as the porous system is uniform. There are many ways to measure directly the porosity of a given material.

1. If the specific gravity, d_s, of the solid component is known, one measures the specific gravity, d_p, of the porous material and obtains P as

$$P = \frac{d_s - d_p}{d_s} = 1 - \frac{d_p}{d_s} \qquad (2)$$

2. If the sample is soft and compressible like a loaf of bread, a sponge, or a filter pad, one compresses it to the complete (or substantially complete) disappearance of all voids and compares the original volume, V_o, with the compressed volume, V_c; the porosity, P, is then

$$P = \frac{V_o - V_c}{V_o} = 1 - \frac{V_c}{V_o} \qquad (3)$$

3. If the material is rigid and a sample of irregular shape is given which should not be subdivided or damaged, one places the specimen in a vessel of known volume, evacuates it first, connects it then with an argon atmosphere of known pressure and volume, and determines the amount of argon needed to fill the voids of the sample. In the case of hard materials (rock, coke, catalysts) with fine pores, one can also inject mercury under pressure and compare the weight before and after injection.

4. It is also possible to get an approximate estimate of the porous character of a sample by microscopic inspection of cross sections.

Table I lists several common porous materials and indicates the range of their porosity.

The specific surface, S, of a material is the solid–gas or solid–liquid interface which 1 g of the material exhibits if it is submersed in a fluid; it is expressed in square meters (or square centimeters) per gram and has the dimension of l^2 per gram. Whereas P provides only information on the total void space, S also tells something about the fineness of the pores. There are several ways to measure S directly; they are, however, not as clear-cut and as precise as those to determine P:

1. An estimate of average pore size and even an impression of pore shape and size distribution can be obtained from the microscopic observation of several cross sections of a sample; it is, in general, qualitative, and a quantitative evaluation requires a rather time-consuming operation.

2. Surfaces are known to adsorb molecules from the gas or vapor phase. Under the influence of the free surface energy a monomolecular layer is first adsorbed which permits the quantitative determination of the surface area if

the absolute dimensions of the adsorbed molecules are known (in angstrom units) and if the number of moles of material adsorbed in the monolayer is known. Langmuir has demonstrated how, under these conditions, surface areas can be quantitatively determined from monolayers. If the surfaces are plain and simple, it is, in general, possible to establish the existence of a

TABLE I
LIST OF THE POROSITY, P, OF A FEW IMPORTANT SYSTEMS

System	Porosity Range (%)
Berl saddles	70–85
Raschig rings	55–65
Glass fiber webs	85–95
Steel wool	70–75
Sand	40–50
Soil	45–55
Brick	15–35
Sandstone (oil sand)	10–35
Limestone	5–10
Leather	55–60
Filter paper	15–25
Cigarette filters	20–50

monomolecular layer and arrive at a reliable figure for the surface area. With systems of more complicated character it is difficult to prevent the formation of multiple layers, and it was, therefore, of great importance that Brunauer, Emmett, and Teller (13) developed a method to measure surface areas of complex systems such as silica gel, carbon black, cotton fibers, and cracking catalysts even if thicker layers are formed. Brunauer et al. (13) expanded Langmuir's theory of monolayers in a very elegant fashion for the case where certain parts of the surface may already be covered with double or triple layers and arrived at an equation which, by extrapolation to very low relative gas or vapor pressures, allows the quantitative determination of the area which is accessible to the gas for surface adsorption. The so-called BET relation can be expressed as

$$\frac{\beta}{V(1-\beta)} = \frac{1}{V_i G} + \beta \frac{(G-1)}{V_i G} \qquad (4)$$

where β = the relative pressure = p/p_0.
V = the volume of the adsorbed gas or vapor.
V_i = the volume of gas or vapor which is necessary to produce a monolayer at the accessible surface.
G = an empirical constant.

In the use of equation 4 one measures p and V at a given temperature, T (p_0, the vapor pressure of the adsorbate, is known for that temperature), and plots $\beta/V(1 - \beta)$ versus β. This gives a straight line with the intercept

$$i = \frac{1}{V_i G}$$

and the slope

$$S = \frac{G - 1}{V_i G}$$

Since i and S can be measured, one obtains G and V_i as

$$G = \frac{S + i}{i}$$

and

$$V = \frac{1}{S + i}$$

This method is very reliable and relatively simple, and it is widely used in practical surface measurements.

Table II gives a list of characteristic values for the accessible surfaces of

TABLE II
SPECIFIC SURFACE, S, OF A FEW IMPORTANT SYTEMS

System	Specific Surface Range (cm^{-1})
Berl saddles and Raschig rings	3–7
Glass fiber webs	500–700
Steel wool	30–40
Sand	150–250
Silica powder	7000–9000
Leather	1×10^4–2×10^4
Cracking catalysts	5×10^5–6×10^5
Carbon black	Up to 4×10^6

several materials, and Fig. 1 shows how the graphic evaluation of equation 4 is actually carried out (*13*).

Once the values of P and S are known, it is possible to arrive at the average diameter, $\bar{d}$, of the voids if one assumes that the pores are essentially spherical:

$$\bar{d} = \frac{GV}{S} \tag{5}$$

if V is the total pore volume in 1 g of the material. If P is known, the value of V can readily be obtained.

In many cases, however, the voids and cavities are far from being spherical and vary over a wide range in shape and size. Under such conditions the meaning of an average value for these two quantities loses its physical significance, and interest must be focused on the pore size and shape distribution.

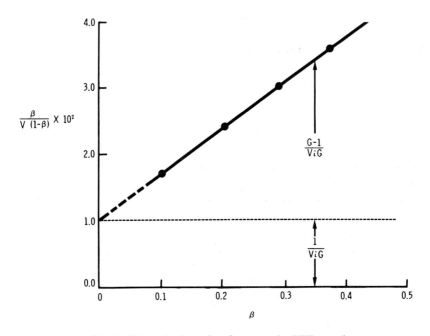

FIG. 1. Determination of surface areas by BET equation.

A clear concept of such distribution functions is difficult to define, and any reliable measurement is even more complicated.

Microscopic inspection of a sufficient number of cross sections gives a certain impression of the size and shape distribution and can also lead to somewhat quantitative formulations, but, since the method is essentially statistical, it requires a rather large number of individual data to give meaningful results.

Pore or capillary size distribution is a controlling factor in the functioning of gel permeation chromatography where dissolved polymer molecules flow and diffuse through a gelatinous network which possesses pores of different volume; the smaller molecules can penetrate into more cavities than the larger ones, and their efflux time is longer. This effect leads to a very efficient way to establish the polymolecularity of a polymer solution. Conversely, if one measures the efflux times of dilute solutions which contain uniform polymeric solutes, one can hope to arrive at a rather detailed pore size distribution function (*14*). In any event, it is relatively easy to arrive at a clear concept and at a precise and reliable determination of the average void size of porous systems, whereas the clear definition and the measurement of a cavity size distribution function are rather complicated and difficult.

As a consequence only very few attempts have been made to attack the numerous problems of permeation phenomena through porous beds in a straightforward manner by measuring P and S and predicting from them the rates of seeping, filtration, diffusion, drying, or setting in a quantitative manner. Instead, one has measured those quantities which have been of practical interest and importance and has established empirical relationships which describe to a certain degree the behavior of the system under investigation. Some of these formal laws proved to cover a rather wide range of conditions and not only were of immediate practical value but evidently contained also a nucleus of fundamental significance. The general approach was therefore to establish a few such relationships over a wide range by appropriate experimental work and then to interpret them in terms of P and S and, eventually, also of the pore size distribution function. However, before demonstrating the results of these procedures it will be opportune to interpose a short discussion on the fundamental properties of fluids (gases and liquids).

Fluids

Fluids range from materials of low viscosity and high compressibility (gases) to systems of medium or high viscosity and low compressibility (liquids and solutions); in the context of this Symposium we shall be interested mainly in liquids and solutions, including particularly polymer solutions.

The motion of liquids under the influence of external forces (gravity, pressure, shear) can be described in phenomenological terms by the general differential equations of hydrodynamics such as the equation of Stokes and Navier which connects the density and velocity of each volume element of the liquid with its position (x, y, z) and with the external forces to which the system is exposed. This equation represents (as the corresponding general equations of mechanics—Lagrange—and of electrodynamics—Maxwell) an all-embracing frame and obtains direct physical significance only if certain geometrical and physical specifications are added to it. These additional facts give the special character of the liquid (viscosity, Newtonian or non-Newtonian, compressibility, etc.), of the forces (pressure, shear, etc.), and the boundary conditions of the entire process. Once such specifications are available, a solution of the basic equation can be attempted, which in simple cases may lead to representation in the form of a closed algebraic equation, but which in somewhat more complicated cases would be given by a computer-programmed treatment.

Developments along these lines have led to another basic conclusion—namely, that turbulence can be excluded with good approximation if the motion of the fluid is slow and if its viscosity is high (15). For the movement of liquids in capillary beds, water and oil seepage, filtration, or coating of fabrics or paper with polymeric systems, the conditions are usually such that one can confine all considerations essentially to the laminar flow of substantially incompressible liquids or solutions through a porous bed under the influence either of gravity or of pressure differentials which are maintained in the system.

A well-known and particularly useful solution of the Stokes-Navier equation is related to the slow laminar flow of an incompressible Newtonian liquid through a straight capillary of given dimensions under the influence of a constant pressure differential. It was independently derived by Hagen and Poiseuille (16) and states that the amount of liquid, Q, which passes per unit time through a capillary of length l and radius r under the influence of Δp is given by

$$Q = \frac{\pi}{8} \cdot \frac{r^4}{l} \cdot \frac{\Delta p}{\eta} \tag{6}$$

where η = Newtonian viscosity of the incompressible liquid.

This relation has been fully verified by numerous experiments and is widely used to determine the viscosity of liquids and solutions in a routine manner. Being a result of formal hydrodynamics, it is valid only as long as the dimensions of the capillary are large in comparison to the molecules of the liquid and the system can be treated as a continuum.

This is true in most of the cases enumerated under Porous Systems, but it may not always be fulfilled in vapor-phase chromatography and in the flow of reacting gas mixtures through porous catalysts. Although such cases do not fall directly within the scope or main interest of this Symposium, an equation may be added here which was derived by Knudsen (17), which deals with the diffusion of gases through capillaries and voids the dimensions of which are comparable with the mean free path of the molecules of the gas. It states that the amount of material, Q, passing at pressure p under a pressure differential of Δp through a capillary of given dimensions, a (radius) and h (length), in unit time is expressed by

$$Q = \frac{4}{3} \frac{a^3}{h} \bar{v} \frac{\Delta p}{p} \tag{7}$$

In this equation $\bar{v}$ is the mean velocity of the gas molecules, which according to the kinetic theory of gases is given by

$$\bar{v} = \sqrt{\frac{2\pi RT}{M}}$$

where R = gas constant.
T = absolute temperature.
M = molecular weight of the gas.

The physical properties of gases (heat capacity, compressibility, heat conductivity, etc.) are well understood on the basis of the classical kinetic theory together with certain refinements (van der Waals' forces, etc.) which become necessary in the high-pressure range. This is achieved by a statistical treatment of independently moving particles, atoms, and molecules, with no or only little interaction. The corresponding physical properties of crystalline solids are equally well represented by the dynamic theory of rigid three-dimensional lattices with complete interdependence of the individual particles (atoms, ions, or molecules). As a consequence of strong particle interaction, the system is forced into the highly ordered crystalline state against the randomizing influence of the thermal motion. Liquids are in between and cannot be treated by either of the two available methods—statistics or dynamics.

A successful and ingenious way to arrive, nevertheless, at a remarkably satisfactory picture of liquid structures was conceived by Eyring (18). He considers a liquid as a densely packed mixture of particles, atoms, or molecules, and of "holes," which are small in comparison to the particles and permit their relatively free motion in respect to each other. This concept, together with certain additional refinements, permits a rather satisfactory quantitative understanding of the most important experimental data such as density, compressibility, heat capacity, viscosity, and x-ray structure (19).

Interaction of Fluids with Surfaces

When a fluid comes in contact with a surface, in all cases a certain interaction exists which leads to a more or less pronounced condensation of it in the immediate neighborhood of the interface. In gases one observes the phenomenon of adsorption, which has already been mentioned, and also, in the presence of very small pores and interstices, the phenomenon of capillary condensation. In liquids there occurs, in many cases, the formation of a film along the wall which may have flow characteristics different from those of the bulk liquid. If a liquid is in contact with a solid, the work, W, which is necessary to remove a unit area of the liquid from the solid surface is given by

$$W = J_s + J_l - J_{sl} \tag{8}$$

where J_s, J_l, and J_{sl} are the surface tensions of the solid and the liquid and the interfacial tension between both phases. If a droplet of a liquid (l) is placed on a solid (s) in air (a), the liquid forms a definite angle, θ, with the surface. The magnitude of the "contact angle" is given by the relation

$$\cos\theta = \frac{J_{sa} - J_{sl}}{J_{la}} \tag{9}$$

which contains the three pertinent interfacial tensions J_{sa}, J_{sl}, and J_{la}. When $\cos\theta$ is greater than zero, the contact angle is less than $90°$. In such cases it is said that the liquid has a special wettability to the solid and will cover its surface readily and/or penetrate into pores and voids of its structure.

The formation of surface film and the wettability of solids by different materials plays an important role in the flow of liquids through capillary beds and particularly in the replacement of one liquid by another in a porous system. Such effects are significant in the flushing of oil wells with water, in the washing of filters, and in the separation of substances in chromatographic columns.

Empirical Approach to Permeation

The necessity of getting useful working equations for certain particularly important practical problems of permeation through capillary layers led to early formulation of empirical rules and laws, some of which covered a remarkably wide range of beds (for example, rock, sand, cotton, pads, paper, chromatographic fillers) and of fluids (for example, water, oil, acids, bases, organic solvents, vapors) and found extensive practical application in many fields. After their simplest forms had been established, they were subsequently refined and improved, and, eventually, attempts were made to rationalize them in terms of the fundamental description of the porous bed (compare Porous Systems) and of the basic properties of the permeating fluid (compare Fluids).

An early and quite general approach to such an empirical formulation was taken by Darcy (20), who carried out systematic tests on the permeation of fluids through filter beds of many kinds and summarized his results in the equation

$$Q = C \cdot \frac{A}{l} \cdot \Delta p \tag{10}$$

where Q is the total amount of fluid percolating in unit time, A is the area, l is the length of the filter bed, Δp is the pressure differential which causes the flow, and C is an empirical constant which depends on the character of the bed and the nature of the fluid. If the flow is laminar and the fluid Newtonian, Darcy's law can be written as

$$Q = \frac{k}{\eta} \cdot \frac{A}{l} \cdot \Delta p \qquad (11)$$

where the specific permeability, k, now depends only on the character of the bed and, eventually, on the interaction between the bed and the percolating fluid. This quantity can be determined by a series of experiments in the course of which A, l, Δp, and η are varied and can serve to describe a given porous system quantitatively in respect to its specific permeability. The dimension of k is square centimeters, and, in the oil industry, specific permeabilities are measured in Darcy units, where

$$1 \text{ Darcy} = 9.87 \times 10^{-9} \text{ cm}^2$$

In groundwater hydrology it is customary to express permeability in terms of velocity of the percolating water per unit drop of the pressure differential. Since most porous beds (filters of sand, cotton, and paper) are somewhat compressible, k will depend on external stresses to which the bed is exposed during the permeation. Examples include the seeping of groundwater through soil on which a large building has been erected, the flow of a solution through filter cloth under pressure, or in a very important special case the drying of a wet sheet of paper on the cylinders of a paper machine.

For many purposes it is practical to consider the outer forces on the percolating fluid in terms of a gravity action, dg, where d is the specific gravity of the fluid and g is the gravitational force, and of an external pressure action grad p, which is a vector and a function of x, y, and z. Using this concept and putting Darcy's law in a differential form, one obtains the vector, q, of the local seepage or filter velocity as

$$q = \frac{k}{\eta} (\text{grad } p + dg) \qquad (12)$$

or, if the permeability and viscosity are also functions of x, y, and z,

$$q = \text{grad } \frac{kp}{\eta} + \frac{kdg}{\eta} \qquad (13)$$

A particularly well-known and important application of equation 13 concerns the theory of filtration, where the process is usually carried out under such conditions that a suspension of small solid particles in a liquid is passed through a "filter cake" which increases in thickness or height, h, as the filtration proceeds. The amount of solid material deposited is proportional to the throughput of suspension.

Under constant filter pressure Δp, the throughput becomes

$$q = \frac{k}{\eta} \cdot \frac{\Delta p}{h}$$

Since dh/dt is proportional to q, one arrives at

$$\frac{dh}{dt} = c \cdot q = -\frac{k}{\eta} \cdot \frac{\Delta p}{q^2} \cdot \frac{dg}{dt} \tag{14}$$

This is a differential equation for q as a function of time with the solution

$$q = \sqrt{\frac{k \Delta p}{2\eta(ct + c')}} \tag{15}$$

where c' is the integration constant and depends on the boundary conditions of the process. If k, η, and Δp are known, equation 15 may be used to compute the throughput at any given time.

There are many similar applications of Darcy's law to such processes as the steady-state seepage of groundwater through soil, rock, and sand, and of oil through the porous material surrounding the bores of oil wells, and to other operations such as the drying of textiles and the control of chromatographic separations. Table III contains the permeability ranges of several porous beds which are of importance for many practical processes.

TABLE III
Permeability, k, for Various Substances

System	Permeability (cm²)
Berl saddles and Raschig rings	$1–4 \times 10^{-3}$
Glass fiber webs	$2–5 \times 10^{-7}$
Steel wool	$0.4–1.0 \times 10^{-4}$
Sand	$0.2–2.0 \times 10^{-6}$
Soil	$0.3–1.5 \times 10^{-7}$
Brick	$5 \times 10^{-11}–2 \times 10^{-9}$
Sandstone	$5 \times 10^{-12}–3 \times 10^{-8}$
Limestone	$2 \times 10^{-11}–5 \times 10^{-10}$
Leather	$1 \times 10^{-10}–1 \times 10^{-9}$
Filter paper	$2 \times 10^{-7}–1 \times 10^{-6}$
Cigarette filters	About 1×10^{-5}

Besides these practical applications of Darcy's law in its simple empirical form, it was obviously of interest to express the permeability, k, in terms of quantities which can be measured for the porous bed without referring to the special process under consideration. These quantities will be the porosity, P, and the internal surface, S. Several attempts have been made to correlate for a given bed the permeability which the bed exhibits for a given fluid with its P and S values which were independently determined by one of the absolute methods mentioned above.

Physical Models for Permeation

The empirical approach to percolatives and permeatives provided for a fairly satisfactory description of the motion of fluids in and through capillary beds in terms of the dependence of throughput on the macroscopic variables such as filter area, filter depth, and pressure gradient and permitted the definition and the measurement of a specific permeability constant, k. It was obviously desirable to correlate this constant with the porosity, P, and the specific surface, S, of the bed with the aid of appropriately chosen models. In general, two types of models have been used for this step of rationalization: the capillary model and the sphere model.

In both cases one makes allowance for the statistical character of the voids and capillaries in the bed by assuming that it can be replaced either by a random arrangement of capillaries of various length and bore through which the fluid moves or by a random arrangement of solid spheres of various diameters around which the fluid flows. Each approach culminates in an equation which combines k with P and S and which, eventually, also contains quantities which refer to the model itself. There are numerous attempts to arrive at such a rationalization or explanation of permeability, but it will suffice for the purpose of this report to select two of them, one for the capillary model and one for the sphere model, to illustrate the procedures and to show the essential results (*21*).

Kozeny (*22*) assumes that the porous system can be considered as an assembly of randomly arranged capillaries which are embedded in an impermeable matrix so that the fluid has to move through these capillaries while it passes through the capillary bed. Using the Hagen-Poiseuille equation for the flow through each capillary and averaging over length and width, Kozeny finds that the permeability, k, of such an idealized bed can be expressed in terms of P and S:

$$k = c \cdot \frac{P^3}{S^2} \qquad (16)$$

The proportionality factor, c, is known as the "Kozeny constant" and depends on the shape of the capillary cross sections; it is 0.50 for a circular shape, 0.56 for a square shape, 0.60 for a unilateral triangular shape, and 0.66 for the shape of a slit.

Equation 16 does, in fact, allow one to compute the actual permeability, k, of many beds from the independently measured porosity and specific surface in reasonable agreement with the experimental data. Discrepancies and inconsistencies have been largely removed by certain refinements of the theory, one of which introduces an additional quantity for the characterization of the bed—namely, the tortuosity, T, of the path along which the fluid moves. The permeability is then given by

$$k = c \cdot \frac{P^3}{S^2} \cdot \frac{1}{T} \tag{17}$$

and it is clear that the introduction of such an empirical (or semiempirical) constant permits a closer agreement with the experimental data.

Brinkman (23) considers a porous bed as an assembly of packed spheres with radius r and calculates the permeability of such a system by considering the viscous flow around the individual spheres. If a certain type of packing is assumed, hexagonal or cubic dense packing, P and S are simple functions of r, and k can be expressed in terms of r and of the number of spheres, which is given by the dimensions of the bed. In reality, the packing of spheres cannot be considered as dense packing. This complication can be taken care of by introducing an independently variable porosity, P, in the calculation. In these terms the permeability, k, of the sphere model is given by

$$k = \frac{r^2}{18}\left(3 + \frac{4}{1-P}\left(\frac{8}{1-P} - 3\right)^{1/3}\right) \tag{18}$$

Also, this model permits a first-order degree of agreement with the experiments but needs further improvements and refinements to achieve a truly satisfactory fit (24).

The model approach for the rationalization of permeability makes use of some type of statistical averaging over the elements of the porous bed and over the individual steps of the path of the percolating fluid. This links the treatment of flow phenomena in porous systems to the more general field of random motion of particles including molecules (Brownian movement) and the resulting effect of diffusion (25). In this domain it is well known that the distance, x, at which a randomly moving particle arrives after time t is not proportional to t but is proportional to $\sqrt{t}$. This is a direct consequence of the differential equation which describes diffusion phenomena; its solution can be expressed by

$$\overline{x^2(t)} = KDt \tag{19}$$

where D is the diffusion coefficient and K is a numerical constant which depends on the geometrical conditions under which the diffusion takes place.

Using considerations of this type, Rideal and Washburn (26) derived the relation

$$x^2 = \frac{\gamma \cos \theta}{2\eta} \bar{r} t \tag{20}$$

for the distance x over which a liquid travels in time t if its viscosity is η and if it moves through capillaries of the average radius $\bar{r}$ under the influence of a force which is given by the contact angle, θ, and the interfacial tension, γ. This equation is very useful to compute the average radius of a given capillary system. One has to know η, γ, and $\cos \theta$, all of which can be determined by independent experiments, and one has to measure the distance, x, over which the liquid moves in time t. This equation has been used with success before and has now been adopted by Cheever (27) to determine the porous character of phosphated steel surfaces.

Chemical Reactions in Capillary Beds

If chemical reactions occur between the substances which percolate in a capillary bed or between the diffusing material and the walls of the porous medium, the controlling factors are the rate of diffusion and the rate of chemical conversion. As long as the diffusion is rapid and can provide for a substantially steady-state concentration of the reagents at any point, the process follows essentially the kinetics dictated by the chemistry of the reaction itself. Many examples of this behavior are known, but it will suffice here to mention only a few for which rather elaborate studies exist and in which typical polymeric porous materials are involved.

One is the pulping of wood, in the course of which the reaction liquor—acidic or alkaline—diffuses into the interior of the porous wood chips, reacts with the lignin, solubilizes it, and is ultimately washed out of the remaining cellulose with an excess of water or steam. In terms of its kinetics the sulfite pulping process is subdivided into penetration, sulfonation, hydrolysis, and washing, and it is customary to separate the individual steps by permitting adequate time for the diffusion to establish a substantially steady-state concentration of the acid in the chips. This can be done by allowing the acidic solution to penetrate the chips at a temperature which is so low that no reaction occurs, and by raising the temperature as soon as the distribution of the acid in the porous system is substantially uniform. Under such conditions the sulfonation of lignin is eventually a reaction of the second order with an activation energy around 20 kcal, which leads to a more or less complete solubilization and ultimately to a dissolution of the lignin. Elaborate studies have been made to establish the porosity and specific surface of chips made from a variety of wood, and to establish the rate of penetration and of sulfonation as a function of temperature and of the removal rate of the solubilized lignin sulfonate (28). If once the two controlling constants, the diffusion constant, D, and the reaction constant, k, have been determined by independent tests, the total process can be described by a differential equation combining diffusion and reaction in an additive manner

$$\frac{\partial c}{\partial t} = D \frac{\partial^2 c}{\partial x^2} - kf(c) \qquad (21)$$

where c is the concentration of the percolating reagent, and $f(c)$ depends on the order of the chemical reaction. In many cases it was found that D is not a constant but depends on c, so that equation 21 assumes the somewhat more complicated form

$$\frac{\partial c}{\partial t} = \frac{\partial}{\partial x}\left(D \frac{\partial c}{\partial x}\right) - kf(x) \qquad (22)$$

which has also been treated in the literature for several cases (29).

Other important processes in which diffusion in a capillary bed and chemical reaction of the resulting system cooperate are the different resin treatments of paper and fabrics. In the case of paper the principal objective is to increase the wet strength, whereas in the case of fabrics such properties as crease and wrinkle resistance have to be improved.

The wet strengthening of paper is usually carried out in such a manner that a reinforcing solution—urea or melamine formaldehyde, alkylene imines, and others—is allowed to penetrate the fibers or the wet sheet at a temperature at which only diffusion but no polymerization occurs until the porous medium is uniformly penetrated with the resin former. Then the temperature is raised, and the polymer forms with or without actual chemical combination with the solid phase of the porous bed. Many studies exist which intend to clarify the organic, physical, and colloid chemistry of this important process; they all consider it to be a superposition of diffusion and chemical reaction (30).

Equally important and interesting are the various resin treatments of fabrics in the course of which polymerizable solutions are allowed to penetrate the porous system at a temperature which is below that at which resin formation occurs. Once the penetration is complete and the distribution of the resin former is substantially uniform, the polymerization is initiated by raising the temperature, by irradiation, or by the addition of a suitable rapidly diffusing catalyst.

References

1. R. M. Barrer, "Flow and Diffusion of Gases through Solids," Cambridge University Press, London, 1941; M. Muskat, "Flow through Porous Media," McGraw-Hill, New York, 1946; P. C. Carman, "Flow of Gases through Porous Media," Academic Press, New York, 1956; A. E. Scheidegger, "Flow through Porous Media," Macmillan, New York, 1960; M. E. Harr, "Ground Water and Seepage," McGraw-Hill, New York, 1962.
2. K. Terzaghi, "Soil Mechanics," Chapman and Hall, London, 1951.
3. M. Muskat, "Principles of Oil Production," McGraw-Hill, New York, 1949.
4. B. F. Ruth, G. H. Montillon, and R. E. Montonna, *Ind. Eng. Chem.*, **25**, 153 (1933).
5. L. Zechmeister and L. Cholnoky, "Chromatography," Wiley, New York, 1943; H. H. Strain, "Chromatography," Interscience, New York, 1945.
6. C. G. Maier, *J. Chem. Phys.*, **7**, 854 (1939); M. Benedict, in "Encyclopedia of Chemical Technology," Vol. 5, ed. by R. E. Kirk and D. F. Othmer, Wiley, New York, 1st ed., 1950.
7. W. R. Marshall, Jr., in "Chemical Engineers' Handbook," ed. by J. H. Perry, McGraw-Hill, New York, 1950; and in "Encyclopedia of Chemical Technology," Vol. 7, ed. by R. E. Kirk and D. F. Othmer, Wiley, New York, 2nd ed., 1965.
8. L. Diserens, "The Chemical Technology of Dyeing and Printing," Reinhold, New York, 2nd ed., 1948; T. Vickerstaff, "The Physical Chemistry of Dyeing," Interscience New York, 1950.
9. W. Broetz and H. Spengler, *Brennstoff-Chem.*, **31**, 97 (1950); H. S. Taylor, G. M. Schwab, and R. Spence, "Catalysis," Van Nostrand, New York, 1937; G. M. Schwab, "Katalyst," Springer, Vienna, 1941.
10. R. H. Bogen, "Chemistry of Portland Cement," Reinhold, New York, 1947.
11. S. M. Atlas and H. Mark, "Resin Treatment of Textile Materials," Wiley, New York, in press.
12. A. E. Scheidegger, "Flow through Porous Media," Macmillan, New York, 1962, pp. 5–28.
13. S. Brunauer, P. Emmett, and E. Teller, *J. Am. Chem. Soc.*, **60**, 309 (1938); S. Brunauer, "Adsorption of Gases and Vapors," Princeton University Press, Princeton, New Jersey, 1943.
14. W. G. Lloyd and T. Alfrey, *J. Polymer Sci.*, **62**, 159, 301 (1962); J. C. Moore, *ibid.*, **A2**, 835 (1964).
15. A. E. Scheidegger, "Flow through Porous Media," Macmillan, New York, 1960, pp. 29–44.
16. F. R. Eirich, "Rheology," Vol. I, Academic Press, New York, 1956, Chapter 2.

17. M. Knudsen, *Ann. Phys.*, **28**, (4), 75 (1909).
18. K. J. Laidler and H. Eyring, "Theory of Rate Processes," McGraw-Hill, New York, 1941.
19. A. E. Scheidegger, "Flow through Porous Media," Macmillan, New York, 1960, pp. 158–193.
20. H. Darcy, "Fontaines," Dalmont, Paris, 1856.
21. A. E. Scheidegger, "Flow through Porous Media," Macmillan, New York, 1960, pp. 112–157.
22. J. Kozeny, *Vienna Acad. Trans.*, **136**, 271 (1927).
23. H. C. Brinkman, *Appl. Sci. Res. J.*, **A1**, 27, 81 (1948).
24. J. Happel and B. J. Byrne, *Ind. Eng. Chem.*, **46**, 1181 (1954); J. Happel and N. Epstein, *ibid.*, **46**, 1187 (1954).
25. A. Einstein, *Ann. Phys.*, **33**, 1275 (1910).
26. E. K. Rideal, *Phil. Mag.*, **44**, 1152 (1922); G. F. N. Calderwood and E. W. J. Mardles, *J. Textile Inst.*, **46**, T161 (1955).
27. G. D. Cheever, "Wetting of Phosphate Interfaces by Polymer Liquids," this volume.
28. J. P. Casey, "Pulp and Paper," Vol. I, Interscience, New York, 1960, pp. 174 *et seq.*
29. R. M. Barrer, "Diffusion in and through Solids," Cambridge University Press, London, 1951, pp. 47 *et seq.*
30. J. P. Casey, "Pulp and Paper," Vol. II, Interscience, New York, 1960, pp. 1163 *et seq.*

Discussion

JELLINEK: Brunauer recently published the BET method for measuring pore-size distributions.

H. MARK: As long as the pores are not too small, one can operate with some modification of Poiseuille's law, but in the case of very small cavities and capillaries one has to use Knudsen's law in addition or instead. In view of gel permeation chromatography, it would be important to have a complete knowledge of the pore-size distribution, but most existing methods for measuring it are complicated and not quite satisfactory.

GRISKEY: All the equations that you showed were from materials that were considered, the flow equations at least, to be incompressible. In certain situations—for example, in filtration—cellulose acetate dope (polymer in acetone or methyl chloride) is compressible when put through a filter press or filter unit. In some systems, the material itself is either viscoelastic or compressible and can change shape. Are there any equations that you know of or any other data that will handle situations of this type?

H. MARK: In many cases of percolation—water, oil solvents, vapors—the systems are Newtonian, but there exist equations for permeation which take into account the viscoelasticity of the fluid. One case which was studied with some care is the flow of polyamide, polyester, and polyolefin melts through a sand pack which occurs during the melt spinning of these materials.

EIRICH: Nylon happens ordinarily to exhibit rather Newtonian flow, but high-molecular-weight polymers cause trouble with their non-Newtonian character and melt elasticity. Depending on the average rate of shear, to avoid some of the complications just mentioned, one may have to adjust the entrance conditions, lengthen or shorten the capillaries, apply the Philippoff-Bagley or Rabinowitz corrections, etc.

BOLGER: Michaels and Bolger [*Ind. Eng. Chem., Fund.*, **1**, 24 (1962)] applied a Kozeny-Carman model to explain the filtration and sedimentation rates for flocculated mineral particle suspensions. Consequently, one could predict, rather nicely, the settled bed volumes and sedimentation rates from porosity and specific surface areas measured independently from rheology data. Recently, Fitch [*Ind. Eng. Chem.*, **58**, No. 10, 18 (1966)] reported the use of the equations derived from this model to design commercial filtration and thickening equipment. Our work represents an attempt to measure the geometric parameters you cite as determined independently from viscosity data for non-Newtonian systems.

BULT: If you have increased porosity, let us say through sandblasting the surface or phosphating, you have already proved that you deal with the roughness of the surface, or the porosity, as the case may be. In phosphating, you have altered the characteristics of the surface. There are no single experiments where only the porosity or only the roughness has been altered without affecting other properties. So, in effect, does the roughness or porosity of the surface have any influence on adhesion of the coating?

H. MARK: The sandblasted surface of a metal is rough—that is, it contains many grooves and holes—but it is not a capillary bed with a large number of crevasses, cracks, and cavities. The phosphated surface, on the other hand, is a porous system which can be controlled in terms of thickness, porosity, and specific surface. Once these essential quantities are known, one can design a primer composition for optimum adherence to the walls of this capillary bed. On top of the primer one then puts a finish—in one or more layers—which has good adhesion to the primer and the desired properties of hardness, toughness, and resistance against atmospheric and environmental conditions.

KUMINS: My colleagues and I have shown that the initial stages of adsorption preinduce an organization of the polymer structure which may extend as much as 8000 Å from the surface. It is very possible that in a capillary the organization of the polymer structure markedly changes the physical properties of the macromolecule, particularly if the capillary diameters are of the order of magnitude of molecular dimensions or even 16,000 Å across. In addition to having a mechanical keying in the asperities, this organization of matter will help in securing a better bond.

VALENTINE: I think that possibly there is a little misunderstanding with the question that Dr. Bult raised, and I was ready to raise the same point, because Dr. Mark suggested that blast cleaning produces a roughened surface which, perhaps, could be treated in the same way as our capillary matrices. Our work has shown that physical effects of altering the surface are quite secondary in connection with the chemical changes of altering the cleanliness of the surface. I feel it is unjustified to assume that we get mechanical keying and so on from a blast-cleaned surface. I am not talking about phosphating. We are quite sure from our work that these physical factors are very, very small, indeed, and the important thing is to get a chemically clean surface. When you do this, you are also altering the roughness at the same time. When we have tried to separate the two factors, it has always come out that the chemical factors count many times more than roughness factors.

H. MARK: I agree with you and will strike the word sandblasting from the list.

Interfacial Turbulence: Spontaneous Emulsification and Evaporative Convection
A Contributed Discussion

D. T. WASAN

Department of Chemical Engineering
Illinois Institute of Technology
Chicago, Illinois

Fill a beaker with water, and then sprinkle talc particles on the surface. When enough talc accumulates at the surface, one observes an erratic motion of the particles due to a peculiar twitching of the surface. This phenomenon bears the name Marangoni effect (1) and results from unbalanced surface tension forces.

Now consider what happens at a liquid–liquid interface when a similar phenomenon occurs when unequilibrated liquids are brought into contact. If a solution of 40% methanol in toluene is placed quietly upon water, the toluene remains clear, while a turbid emulsion appears in water. However, if pure toluene is placed upon water containing methanol, no spontaneous emulsification occurs.

Nature has hidden from us many types of convective flows which we cannot observe with a naked eye. Evaporating liquids, even though great pains may be taken not to disturb them, generally become alive with motion of one kind or another as they evaporate. Varley (2), in 1836, first noticed "with the aid of a microscope, motions of extremely curious and wonderful characters in fluids undergoing evaporation." Thompson (3), in 1855, gave the first explanation of the phenomenon in a note, "On Certain Curious Motions Observable at the Surface of Wine and Other Alcoholic Liquors." As the wine evaporated, the surface would twitch and writhe. He explained the motion in terms of local variations of surface tension.

Experiments have shown that drying paint films often display steady cellular convection cells which are induced by surface tension (4). Floating is the term used to describe the mottled, splotchy, or streaked appearance exhibited by a paint film. Patton (5) has described the phenomena as follows:

"When solvent volatilizes at the surfaces of a wet film, solvent further down in the body of the film commences to diffuse or migrate to the surface to replenish the surface supply. To expedite this movement, escape channels are often formed. Such escape routes terminate by erupting at the surface and frequently lead to the formation of geometrical patterns that are observable at the surface of the film. When observed under a microscope, and especially in the case of fast-evaporating lacquers of low viscosity, the surface of a wet film is seen to be made up of miniature seething volcanoes that spew forth paint. This paint in turn spreads away from each volcanic apex to the outside of this erupting center."

In another experiment with the casting of cellulose acetate film, when a polymeric film is cast on a hot plate, microscopic examination reveals that, at the phase interface, highly localized convective movement occurs as acetone volatizes.

These are only a few of the many examples of the interfacial turbulence phenomenon which is generally triggered by local variation of surface tension. This change in surface tension may be caused by two mechanisms, either by local differences in concentration of the solute as it occurs in the case of the toluene–methanol–water system, or by gradients in temperature as it occurs in evaporation systems. The latter phenomenon has been well summarized by Berg *et al.* (*6*).

Ward and Brooks (*7*) were the first to report the existence of spontaneous, highly localized, interfacial agitation accompanying mass transfer. Enhanced mass transfer due to interfacial turbulence has been reported by Sherwood and Wei (*8*), who investigated mass transfer with interfacial chemical reaction, and by Lewis (*9*).

Sternling and Scriven (*10*), in their pioneering paper, made an attempt to analyze theoretically the phenomena of interfacial turbulence. A schematic diagram of the flow disturbance as viewed by them is shown in Fig. 1. If solute diffuses from phase A to phase B, the roll cell is viewed to convey liquid rich in solute from phase A and liquid lean in solute from phase B toward the interface at point 1 as shown in the figure. According to these investigators, "if viscosity is higher in phase A and in addition if the diffusivity is lower in phase A, the flow-induced concentration upset there is less affected by diffusion than in phase B, hence the effect of the disturbance is greater on the side of phase A and the interfacial solute concentration is increased at 1. Because of symmetry and the necessary conservation of solute, the change in solute concentration is in the opposite direction of 2. Thus variations in concentration, hence also in interfacial tension, are induced along the interface. The interface is no longer in mechanical equilibrium and seeks a state of lower free energy through expansions of regions of higher tension (the Marangoni effect).

Because there can be no discontinuity in velocity at the interface, motion in it induces flows in adjoining fluids."

From their analysis, Sternling and Scriven (10–12) predict several conditions favorable to the inception of surface agitation driven by interfacial

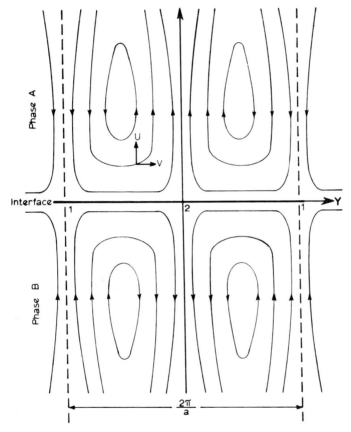

FIG. 1. Schematic diagram of flow disturbance showing circulation pattern of two-dimensional roll cells in cross section.

tension: (1) solute transfer out of the phase of higher viscosity; (2) solute transfer out of the phase in which its diffusivity is lower; (3) large differences in kinematic viscosity and solute diffusivity between the two phases; (4) interfacial tension highly sensitive to solute concentration; and (6) low viscosity and diffusivity in both phases. Interfacial turbulence theoretically results when any of these conditions occur.

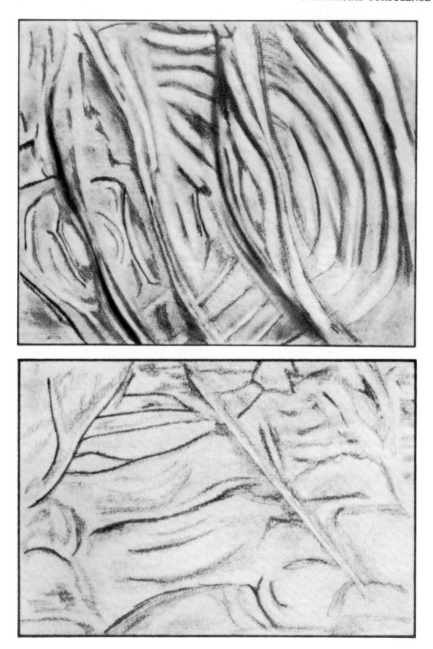

FIG. 2. Two kinds of ripples. Initial acid concentration of 10%. (Top) Ripple confined in stripes, 22-hours, height of view 7.76 mm. (Bottom) Ripple confined in cells, 2 hours, height of view 6.50 mm.

Orell and Westwater (*13*), using a Schlieren technique, studied spontaneous interfacial cellular convection accompanying mass transfer in an ethylene glycol–acetic acid–ethyl acetate system and verified in part the Sternling-Scriven theory. These investigators varied acetic acid concentration between 0.1 and 10.0 wt % as a solute and studied the diffusion of acetic acid from ethylene glycol into ethyl acetate. Schlieren photography revealed the finer structure of the interfacial convection as made up of ripples; these are sketched in Fig. 2.

The enhancement of the rate of mass transfer due to concentration driving force-induced interfacial turbulence has been investigated in several typical liquid–liquid systems by Olander and Reddy (*14*). The overall mass transfer coefficient was observed to increase sharply by as much as a factor of 4 with an increase in the driving force.

Bakker *et al.* (*15*) carried out experiments on interfacial movement induced by differences in interfacial tension and identified "macro" scale and "micro" scale movements. The occurrence of the first depends on the geometry of the interface and on flow conditions, the latter on the physical properties of the phase. These investigators suggested a criterion for classifying solutes according to their ability to impede or promote movement of a free surface.

Recently Linde *et al.* (*16*) examined the oscillatory regime of Marangoni instability for mass transfer over the liquid–gas phase boundary and further verified the main predictions of the Sternling-Scriven theory for several two-phase systems under the conditions when simultaneous conversion of latent heat is negligible.

In a most recent study conducted at our laboratory, Kintner (*17*) photographed through a microscope the convection patterns at the interface of liquid–liquid drops and found them to be circular in shape.

In the past, observations on interfacial turbulence phenomena have been confined to nonflow systems. But recently Clark (*18*) observed the surface tension-driven convective flows in two-phase laminar vaporization systems at high concentration levels of the volatile components.

Brian *et al.* (*19*) recently re-examined the wetted wall column data of carbon dioxide absorption in monoethanolamine and argued that the physical mass transfer coefficient is increased by the carbon dioxide–monoethanolamine chemical absorption process owing to interfacial turbulence driven by surface tension gradient.

Finally, it may be said that interfacial turbulence is probably present in many of the transport processes, though its degree may vary considerably. The interface itself may be visualized not as a quiet and still affair, but generally undergoing constant twitching and movement.

References

1. C. Marangoni, *Nuovo Cimento*, **16,** 239 (1871).
2. C. Varley, *Trans. Soc. Arts*, **50,** 190 (1836).
3. J. Thompson, *Phil. Mag.*, **10,** 330 (1855).
4. J. R. A. Pearson, *J. Fluid Mech.*, **4,** 489 (1958).
5. T. C. Patton, "Paint Flow and Pigment Dispersion," Interscience, New York, 1964.
6. J. C. Berg, A. Acrivos, and M. Boudart, *Advan. Chem. Eng.*, **6,** 61 (1966).
7. F. H. Ward and L. H. Brooks, *Trans. Faraday Soc.*, **48,** 1124 (1952).
8. T. K. Sherwood and J. C. Wei, *Ind. Eng. Chem.*, **49,** 1030 (1957).
9. J. B. Lewis, *Chem. Eng. Sci.*, **3,** 248, 260 (1954); **8,** 295 (1958).
10. C. V. Sternling and L. E. Scriven, *Am. Inst. Chem. Engrs. J.*, **5,** 514 (1959).
11. L. E. Scriven and C. V. Sternling, *Nature*, **187,** 186 (1960).
12. L. E. Scriven and C. V. Sternling, *J. Fluid Mech.*, **11,** 321 (1964).
13. A. Orell and J. W. Westwater, *Am. Inst. Chem. Engrs. J.*, **8,** 350 (1962).
14. D. R. Olander and L. B. Reddy, *Chem. Eng. Sci.*, **19,** 67 (1964).
15. C. A. P. Bakker, P. M. van Buytenen, and W. J. Beek, *Chem. Eng. Sci.*, **21,** 1039 (1966).
16. H. Linde, E. Schwarz, and K. Groger, *Chem. Eng. Sci.*, **22,** 823 (1967).
17. R. C. Kintner, private communication, 1967.
18. M. W. Clark, "Mass and Heat Transfer Processes in Laminar Two-Phase Flow," Ph.D. Thesis, University of California, 1967.
19. P. L. T. Brian, J. E. Vivian, and D. C. Matiatos, *Am. Inst. Chem. Engrs. J.*, **13,** 28 (1967).

SESSION II

Chairman

F. M. FOWKES

Epitaxy and Corrosion Resistance of Inorganic Protective Layers on Metals

A Contributed Discussion

A. Neuhaus and M. Gebhardt

Mineralogical-Petrological Institute
Bonn University
Bonn, West Germany

Introduction

Concerning the Concept of Epitaxy

The corrosion protection given by a reactive covering layer on a metal depends essentially on the phase nature of the layer crystallites, their texture, their growth, and the adhesive bonding between the carrier metal crystallites and the cover layer crystallites (corrosion-protective layer). Depending on the kind of protection required, the phases must be inert to special chemical influences—for example, O_2, H_2O, N_2, N_2O_5, CO_2, and Cl_2—at different pressure and temperature conditions (normal corrosion, aqueous and hydrothermal corrosion, and high-temperature corrosion). The cover layer must be microcrystalline. The protective layer therefore must consist of as many guest crystallites as possible, distributed over the carrier surface uniformly, and of about the same size; and they should be densely packed.

The crystallites of the layer must show (chemisorption) main valence bonds as strong as possible and as many as possible on the carrying surface—that is to say, in the "interface." The latter occurs especially when carrier crystal and guest crystal have grown together epitaxially (*1, 2*). Good protective coatings must have matured—that is, they must be reactive with the carrier–metal substrate; and further they must have been formed epitaxially. Intergrown layers are any layer formations, nuclei formations, or growth processes on solid body surfaces, regardless of whether the "layer phases" have entered into reactive exchange of carrier with the solid body surface (carrier) or are only adsorbed on them. Such layer formations are therefore always crystalline, with a very changeable degree of dispersion. If we assume the carrier surface to be crystalline also, for a random degree of dispersion of the crystallites (micro-macro-monocrystalline), three orientation means of carrier–guest growths result (Fig. 1). Carrier and layer crystallites are completely nonoriented with one another, according to Fig. 1*a*. According to Fig. 1*b*, an orientation of texture may exist; that is, the guest crystallites all separate together with the equally low-numbered stable lattice planes on the carrier surface. They are not oriented with respect to one another. Finally, in Fig. 1*c* layer crystallites and the crystallites of the carrier show both textural and azimuthal orientation toward one another. We shall designate as epitaxial orientations only these latter orientation relations.

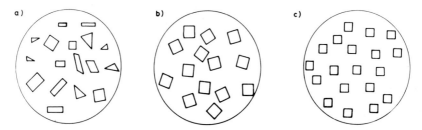

FIG. 1. Growths. (*a*) Fully nonoriented. (*b*) Texture orientation: a guest plane parallel to the host surface. (*c*) Epitaxy: texture and azimuthal orientation.

Further, we designate the growth zone of carrier surface and layer crystallites, which is connected and mostly is also materially heterogeneous, as the "contact zone" or as the "interface" and verify that the "interface" comprises only the directly intergrowing boundary lattice planes of carrier and guest crystallites, or a "transition zone" of many lattice planes. Since most of the cover layers discussed here form with the layers near the surface of the carrier reactive transition zones (many-layered interface), the concept "epitaxy" is to be suitably supplemented with the concept of "topotaxy," the successively repeated epitaxial displacement of the upper carrier layers through the layer phases. This is explained by using the example of the epitaxial–topotaxial formation of surface layers of iron (or nickel or cobalt) with oxygen and water in the temperature range of normal atmospheric corrosion up to high-temperature corrosion. Subsequently, there are treated examples and groups of examples of protective layers, again from the point of view of epitaxy and topotaxy.

Epitaxial Corrosion Layers on Iron, Cobalt, and Nickel

The corrosion of metals takes place, in general, spontaneously, with the release of free energy. In the primary contact of a "pure" metal surface with the atmosphere, phases are naturally formed which result from the high atmospheric oxygen and water content. Pronounced oxidation phases are formed. The coating thicknesses of these generally very thin (≤ 0.1 micron) primary corrosion films may be very different, depending on the growth velocity and the nature of the reaction species. Here, we are interested only in the ideal case of an essentially impervious coating. This is generally feasible only in a precise experiment, but it is occasionally also observed in corrosion, which takes place freely. For the case of the impervious crystallite coating of the initial corrosion layer, it follows that further growth in thickness is caused almost exclusively through epitaxially directed diffusion transport (topotaxial transport) of the carrier metal through the initial corrosion film up to the particular "coating-atmosphere" phase boundary. As corrosion progresses, the reaction film thickens and the path of diffusion of the carrier metal to the phase boundary coating-atmosphere is lengthened, more and more, until after achievement of a certain thickness of the layer, growth kinetics and growth law change their character (2).

With progressive thickening of the primary layer, the oxygen potential drops more and more at the phase boundary "metal–oxide film." With increase of temperature (hydrothermal or high-temperature corrosion) there is then formed, after the redox reactions have occurred in the solid state between the metal and the primary oxide, a varying, thick layer of less highly oxidized phases which is designated as the secondary layer. The growth mechanism of the primary layer changes almost simultaneously owing to the complete exhaustion of the iron supply to the phase boundary "layer-atmosphere." Also, the growth mechanism of the primary layer changes because of the atmosphere which results in the formation of higher oxidized and eventually hydrated terminal phases.

In the oxidation corrosion of iron–copper–nickel metal surfaces, there result collectively very complex-structured corrosion layers with a rigid sequence of phases with respect to both time and space. These phases are not always built one after the other with an increasing degree of oxidation from the carrier to the atmosphere. Also, when enough water is present, we have to consider the degree of hydration (Table I). In the special case of the slow oxidation of iron at moderately increased temperatures (for example, 400°C) and strongly reduced O_2 pressure (for example, 10^{-4} torr), a very thin film of Fe_3O_4 (primary oxide layer) forms first. With progressive growth in thickness of this film, a metastable FeO layer is formed. While redox reactions in solid phase proceed, γ-Fe_2O_3 is formed at the contact of the unchanged carrier lattice and the atmosphere. In the case of layer thicknesses of about 2000 Å, there is finally a change of the reaction mechanism whereby γ-Fe_2O_3 is converted into α-Fe_2O_3. Accordingly, the oxidation of iron at increased temperatures leads to a layer-and-phase sequence with increasing oxygen content from the

TABLE I

Oxide and Hydroxide Corrosion Products of Pure Iron, Nickel, and Cobalt Produced under Various Environmental Conditions

Increasing Degree of Oxidation →	Corrosion Products on Iron			Corrosion Products on Nickel	Corrosion Products on Cobalt
	Atmospheric Corrosion $-50°C < T < +50°C$	Hydrothermal Corrosion $T > 50°C$	High-Temperature Corrosion $T > 600°C$, $P \leq 1$ atm (dry oxides)		
	FeO wüstite (metastable)	Fe(OH)$_2$·2H$_2$O	FeO (stable)	Ni(OH)$_2$	Co(OH)$_2$
	Fe(OH)$_2$	Fe(OH)$_2$ hexagonal	—	NiO	CoO
	Fe$_3$(O, OH)$_4$ hydromagnetite	Fe$_3$(O, OH)$_4$ cubic	Fe$_3$O$_4$	Ni$_3$O$_4$	Co$_3$O$_4$
	Fe$_3$O$_4$ magnetite	Fe$_3$O$_4$ cubic			
	γ-Fe$_2$O$_3$ maghemite	γ-Fe$_2$O$_3$ tetragonal	γ-Fe$_2$O$_3$	Ni(OH)$_3$	Co$_2$O$_3$
	α-FeOOH goethite	α-FeOOH orthorhombic-rhombic	α-Fe$_2$O$_3$	Ni$_2$O$_3$	
	β-FeOOH	β-FeOOH tetragonal		NiO$_2$	
	—	γ-FeOOH orthorhombic-rhombic			
	γ-FeOOH lepidocrocite	δ-FeOOH hexagonal			

inside to the outside in steps and therefore leads to optimum stability, with respect both to the free atmosphere and to the iron substrate. One then has the following sequence of structures (written as stoichiometric phases): Fe–FeO–Fe$_3$O$_4$–γ-Fe$_2$O$_3$–α-Fe$_2$O$_3$.

All these phases are connected with one another through orientation relations, and they form reaction zones which are more or less deep-seated between carrier crystallites and guest phases. Therefore, in a rigid sense they are not only eptiaxial (rigidly two-dimensional relationships), but also, through

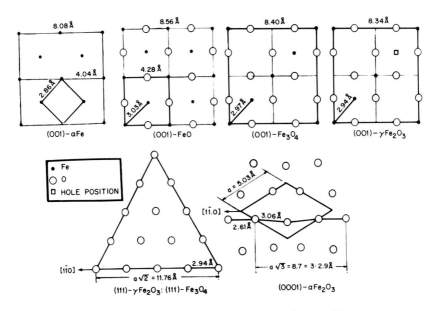

FIG. 2. Affinitive network planes of iron and iron oxides.

successively repeated epitaxial reactions and through topotaxial events, as Fig. 2 shows, they form whole chains of epitaxial–topotaxial intergrowing guest phases (orientation chains). In Table II are assembled the known epitaxies and topotaxies between iron metal, iron oxides, and iron hydroxides (2).

Protective Layers and Epitaxy

Burnishing Layers on Iron and Iron Alloys

Burnishing is defined as the production of thin, brown-black to deep-black oxide films on iron and iron alloys through the oxidation of carefully cleaned metal surfaces. The three following process variables are used in their production: (1) reaction with high-pressure steam, (2) oxidation in hot electrolyte solutions, and (3) reaction with oxidizing molten salts. In our own experiments (2), we used chiefly the third method of production. If in this case one selects a melt of sodium hydroxide + sodium nitrate + sodium nitrite as the polishing bath, then the following polishing phases are formed

TABLE II
Orientation Relationships of Iron Metal, Iron Oxide, and Iron Hydroxide

Epitaxies	Fe–FeO	Fe–Fe(OH)$_2$	Fe–Fe$_3$(O, OH)$_4$
	Fe–Fe$_3$O$_4$	Fe–γ-Fe$_2$O$_3$	Fe$_3$O$_4$–FeO
	Fe$_3$O$_4$–γ-Fe$_2$O$_3$	Fe$_3$O$_4$–α-Fe$_2$O$_3$	
Topotaxies	Fe$_3$O$_4$–γ-Fe$_2$O$_3$	Fe$_3$O$_4$–α-Fe$_2$O$_3$	Fe$_3$O$_4$–FeO
	γ-Fe$_2$O$_3$–α-Fe$_2$O$_3$	γ-FeOOH–γ-Fe$_2$O$_3$	α-FeOOH–α-Fe$_2$O$_3$
	δ-FeOOH–α-FeOOH	δ-FeOOH–α-Fe$_2$O$_3$	β-FeOOH–γ-Fe$_2$O$_3$

Orientation and nuclei chains:

$$\text{Fe} \begin{cases} \text{FeO–Fe}_3\text{O}_4\text{–}\gamma\text{-Fe}_2\text{O}_3\text{–}\alpha\text{-Fe}_2\text{O}_3 \\ \text{Fe(OH)}_2\text{–Fe}_3(\text{O, OH})_4\text{–Fe}_3\text{O}_4\text{–}\gamma\text{-Fe}_2\text{O}_3\text{–}\alpha\text{-Fe}_2\text{O}_3 \end{cases}$$

γ-FeOOH $\qquad\qquad\qquad$ δ-FeOOH
$\qquad\searrow\qquad\qquad\swarrow$
$\qquad -\gamma\text{-Fe}_2\text{O}_3\text{–}\alpha\text{-Fe}_2\text{O}_3-$
$\qquad\swarrow\qquad\qquad\searrow$
β-FeOOH $\qquad\qquad\qquad$ α-FeOOH

Hereditary orientation: α-Fe–(Fe$_3$O$_4$)–α-Fe$_2$O$_3$
$\qquad\qquad\qquad$ (001) $\qquad$ (11$\bar{2}$4)

on α-Fe: γ-Fe$_2$O$_3$, Fe$_3$O$_4$, Fe$_4$N, Fe$_3$N, and Fe$_2$N. In addition, the following other phases in the burnishing layers have been detected on several steels: (a) on V2A steel: NiFe$_2$O$_4$, Fe(Cr,Fe)$_2$O$_4$, and NiCr$_2$O$_4$; (b) on Co–W steel of the approximate composition 50% Fe, 40% Co, 7% W, and 2.5% Cr: Co$_3$O$_4$, CoFe$_2$O$_4$, and Co$_2$N; (c) on an Fe–Mn steel: MnO and MnFe$_2$O$_4$. All these layer phases are oxides or nitrides throughout at the selected reaction temperatures and reaction conditions. The systematic wearing away of these burnishing layers of different formation periods and thickness through ionic-radiation etching gave the following sequence of phases: α-Fe–Fe(N)–Fe$_4$N–Fe$_3$N–Fe$_2$N–Fe$_3$O$_4$–γ-Fe$_2$O$_3$. The following layer-formation mechanism results.

For short-time action of the burnishing melt on the pure iron surface, there is formed first, in analogy to the oxide corrosion layers (3), a very thin Fe$_3$O$_4$ film. However, in deviation from the simple oxide corrosion, active nitrogen (N*) is always present as a reaction species (formed by the reciprocal effect of iron with NO$_3^-$ and NO$_2^-$). This N* diffuses through the already formed primary film of magnetite and is first absorbed in the regions of the iron lattice adjacent to the surface in the form of a solid solution, Fe(N). In the further course of the burnishing reaction, the oxide layer thickens outward, thus in the direction of the melt, where then topotaxial γ-Fe$_2$O$_3$ is formed in the outer parts. At the same time, however, the diffusion of N* occurs in the direction of the remaining pure iron lattice with formation of other nitride phases.

The former layer formations correspond to a great extent to oxide corrosion and therefore do not require any further explanation. However, the epitaxial growth of nitride secondary layers at the phase boundary of the oxide primary layer with the pure iron lattice which results in the thickening of the oxide corrosion layer is novel. The nitride formation originally consists solely in a statistical absorption of N* in the octahedral holes of the metal lattice. It leads rather rapidly, with increasing N* supply, to the formation of genuine nitride phases with a nitrogen content increasing from the inside outward. With the advancing growth of the thickness of this nitride layer and simultaneously advancing growth in thickness of the oxide layer, the diffusion path for N* increases successively and rapidly. This finally does not advance up to the phase boundary of the unchanged carrier lattice. This gives the sequence of nitride phases: Fe(N)–Fe$_4$N–Fe$_3$N–Fe$_2$N–. All these secondary phase formations naturally are based on the reactions in the solid state. Therefore, with the burnishing layers on α-Fe, there results a sequence of phases with a rising degree of nitride content from the inside to the outside, or degree of oxidation: Fe–Fe(N)–Fe$_4$N–Fe$_3$N–Fe$_2$N–(FeO)–Fe$_3$O$_4$–γ-Fe$_2$O$_3$–(α-Fe$_2$O$_3$).

The epitaxies between the particular adjacent phases of this sequence of phases are listed in Table III. Figure 3 gives in addition a representation of the lattice geometric relationships between α-Fe and the nitride phases. As can be seen from both, the host–guest lattice comparisons are excellent throughout.

In the replacement of the pure iron carrier, through iron-, cobalt-, and nitrogen- containing hosts, there are formed additional oxides such as CoFe$_2$O$_4$

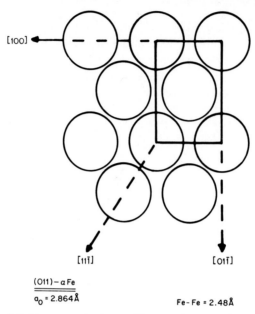

FIG. 3a. Affinitive network planes of iron and iron nitrides: (011)-α-Fe.

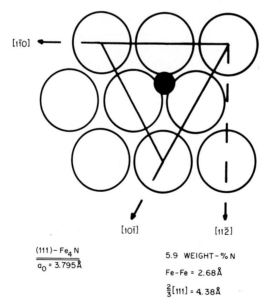

FIG. 3b. Affinitive network planes of iron and iron nitrides: (111)-Fe$_4$N.

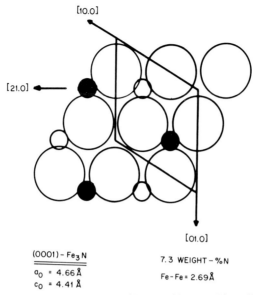

FIG. 3c. Affinitive network planes of iron and iron nitrides: (0001)-Fe_3N.

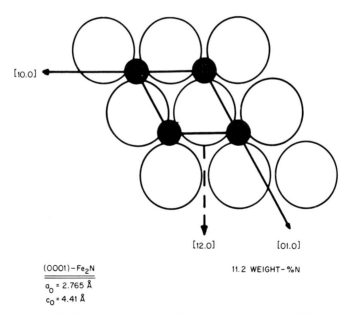

FIG. 3d. Affinitive network planes of iron and iron nitrides: (0001)-Fe_2N.

and Co_3O_4 as well as nitrides—for example, Co_2N—which on the basis of their lattice parameters can be arranged readily in the epitaxial chains.

Carbide, Nitride, and Boride Layers on Metal Substrates

In many applications of metals, their bulk properties are less important than their surface properties. For many applications, a surface improvement is sufficient on some materials through appropriate coating layers. Thus,

TABLE III

TOPOTAXY CHAIN IN BURNISHING LAYERS ON IRON

Phases		Epitaxies		Periods (Å)		Misfit (%)
Host	Guest	Host	Guest	Host	Guest	
α-Fe	Fe_4N	(011) [100]	(111) [1$\bar{1}$0]	2.86	2.68	−6
Fe_4N	Fe_3N	(111) [1$\bar{1}$0]	(0001) [21.0]	2.68	2.69	<1
Fe_3N	Fe_2N	(0001) [21.0]	(0001) [10.0]	2.69	2.77	+3
Fe_2N	Fe_3O_4	(0001) [10.0]	(111) [1$\bar{1}$0]	2.77	2.97	+7
Fe_3O_4	γ-Fe_2O_3	(111) [1$\bar{1}$0]	(111) [1$\bar{1}$0]	2.97	2.94	−1

iron or steel surfaces are improved with the best results through thin chromium, vanadium, niobium, tungsten, or molybdenum layers, which subsequently can be improved further through formation of carbides, nitrides, or borides.

Formation of Carbides on Chromed Steel. As a first example for an improvement of a steel surface, we use galvanic chroming with subsequent gas carburetion (process for carbide formation). The galvanic chroming of iron or steel leads to a chromed layer alloyed metallically with the substrate metal, which is not connected directly with the pure iron lattice but via an iron–chromium mixed crystal layer. In the subsequent carburetion, for example, with CCl_4 and H_2 at higher temperatures, a heterogeneous one-phase layer is formed in the previously described burnishing layers which consists of a sequence of layers built up from phases with an increasing carbon content from the inside to the outside (4). This sequence of layers is made up of (1) a thin outer layer consisting of Cr_3C_2, (2) a thicker middle layer of Cr_7C_3, and, finally, (3) in contact with the metallic chrome, a thin layer which is built up from phase $Cr_{23}C_6$. The orientation relations to be expected, thus epitaxies, topotaxies, and nuclei chains, are recorded in Table IV.

TABLE IV
EPITAXIES IN CARBIDE–CHROME LAYERS

Phase:	Cr	$Cr_{23}C_6$	Cr_7C_3	Cr_3C_2
Symmetry:	Cubic-Body Center	Cubic	Hexagonal	Orthorhombic-Rhombic
Intergrowth plane	(011)	(001)	(0001)	(001)
Direction	[1$\bar{1}$1]	[110]	[10.0]	[010]
Periods (Å)	15.02	15.07	13.98	11.46
Misfit (%)		+0.3	−6.9	−10.9

Formation of Nitrides of Titanium. In the preparation of titanium nitrides with pure nitrogen at temperatures of about 950°C, there is formed in 24 hours a nitride layer about 80 microns thick and of a golden color, which in itself is again built up of several zones (*4*). The outermost film, only 0.1 micron thick, consists of the cubic δ-phase TiN. It is preceded by an 8- to 10-micron-thick layer of the hexagonal ε-phase Ti_3N, which gives the high microhardness of 1500 kg/mm². After this follows the innermost layer with a solid solution of nitrogen in titanium with a decreasing lattice period and microhardness (1300–700 kg/mm²) toward the pure carrier lattice.

The formation of the nitride layer is diffusion-controlled with transportation of N* toward the pure titanium lattice. The correlation between Ti, Ti(N), Ti_3N, and TiN compounds (Table V) indicates once again that topotaxial phases are formed.

Formation of Borides on Tungsten. The preparation of tungsten borides takes place in the temperature range of 1100°C to 1500°C by means of the gas reaction with BCl_3 and H_2. A layer about 100 microns thick is formed, for example, after a reaction period of two hours, which again consists of a phase sequence with the metal lattice having a decreasing degree of boride content (*4*): W_2B_5–WB–W_2B–W. The corresponding microhardnesses in this connection decrease from 2800 (WB + W_2B_5) to 2000 (W_2B) to 300 (W-metal) kg/mm². The boride formation on tungsten therefore effects an enormous increase in surface hardness, abrasion, and corrosion resistance.

For the quality of the surface layer acting as an improvement layer, the adhesive bond between metal carrier and layer phases is naturally essential.

TABLE V
Epitaxies in Nitride Layers on Titanium

Phase: Symmetry:	α-Ti Hexagonal	Ti(+N) Hexagonal	Ti$_3$N Hexagonal	TiN Cubic
Intergrowth plane	(0001)	(0001)	(0001)	(111)
Direction Periods (Å)	[10.0] 2.95	[10.0] 2.96	[12.0] 2.84	[1$\bar{1}$0] 3.00
Misfit (%)		← +0.3 → ←	−3.6 → ←	+5.6 →
			(+1.7)	

After the excellent correlation (Table VI), one can expect that the adhesive bond between tungsten and the innermost boride layer (W$_2$B) will be excellent, while the outermost layer zone in this respect is far less favorable.

Gas Reactions with High-Melting Metals. For high-melting metals such as niobium, tantalum, tungsten, molybdenum, and uranium, studies have been made mainly of the reaction phases formed at higher temperatures with oxygen, nitrogen, and carbon and of their epitaxial relations to the carrier metal. Table VII lists the epitaxy possibilities derivable from the structural

TABLE VI
Epitaxies in Boride Layers on Tungsten

Phase: Symmetry:	W Cubic-Body Center	W$_2$B Cubic	WB Orthorhombic–Rhombic	W$_2$B$_5$ Hexagonal
Intergrowth plane	(011)	(011)	(010)	(0001)
First direction Periods (Å)	[100] 3.16	[1$\bar{1}$1] 3.23	[100] 3.19	[10.0] 3.00
Misfit (%)	← +2.2 →	← −1.2 →	← −6.0 →	
Second direction Periods (Å)	[1$\bar{1}$1] 5.48	[100] 5.58	[101] 4.43	[12.0] 5.19
Misfit (%)	← +1.8 →	← −20.6 →	← +17.1 →	

→ good adhesive bond
→ poor adhesive bond

TABLE VII

Epitaxies in Reaction Layers on High-Melting Metals

Phases		Epitaxies		Periods (Å)		Misfit
Host	Guest	Host	Guest	Host	Guest	(%)
Nb (cubic-body center)	NbO_x	(001) [100]	(001) [100]	3.30	3.38	+3
Nb (cubic-body center)	NbO_z	(001) [100]	(001) [100]	6.60	6.64	+1
Nb (cubic-body center)	NbO	(011) [111]	(011) [110]	2.86	2.97	+4
Nb (cubic-body center)	$Nb_2N(\beta)$	(010) [100]	(0001) [10.0]	3.30	3.06	−7
Nb (cubic-body center)	$Nb_2N(\beta)$	(011) [100]	(01$\bar{1}$0) [10.0]	3.30	3.06	−7
Nb (cubic-body center)	NbN(δ')	(011) [100]	(0001) [10.0]	3.30	2.97	−9
Nb (cubic-body center)	NbN(δ')	(011) [$\bar{1}\bar{1}$1]	(01$\bar{1}$0) [00.1]	5.72	5.55	−3
Nb (cubic-body center)	NbN(ε)	(011) [$\bar{1}\bar{1}$1]	(01$\bar{1}$0) [00.1]	5.72	5.64	−2
Nb (cubic-body center)	NbN(ε)	(011) [100]	(0001) [10.0]	3.30	2.96	−9
Nb (cubic-body center)	Nb_2C	(011) [100]	(0001) [10.0]	3.30	3.13	−5
Nb (cubic-body center)	Nb_2C	(011) [100]	(01$\bar{1}$0) [10.0]	3.30	3.13	−5
Nb (cubic-body center)	NbC	(011) [100]	(111) [$\bar{1}$10]	3.30	3.15	−4
W (cubic-body center)	WO_2	(011) [01$\bar{1}$]	(100) [010]	4.47	4.84	+8
W (cubic-body center)	W_2N	(011) [100]	(0001) [10.0]	3.16	2.89	−9
W (cubic-body center)	WN	(011) [100]	(0001) [10.0]	3.16	2.89	−9
W (cubic-body center)	W_2C	(110) [001]	(0001) [10.0]	3.16	2.99	−6
W (cubic-body center)	W_2C	(110) [001] [1$\bar{1}$0]	(10$\bar{1}$0) [01.0] [00.1]	3.16 4.47	2.99 4.71	−6 +6

TABLE VII (Continued)

Phases		Epitaxies		Periods (A)		Misfit (%)
Host	Guest	Host	Guest	Host	Guest	
Mo (cubic-body center)	Mo$_2$O	(011) [01$\bar{1}$]	(001) [100]	4.45	4.86	+9
Mo (cubic-body center)	Mo$_2$N	(011) [01$\bar{1}$]·	(001) [100]	4.45	4.17	−6
Mo (cubic-body center)	MoN	(011) [100]	(0001) [12.0]	9.45	9.91	+5
Mo (cubic-body center)	Mo$_2$C(β)	(011) [100]	(0001) [10.0]	3.15	3.00	−5
Mo (cubic-body center)	Mo$_2$C(β)	(011) [01$\bar{1}$]	(01$\bar{1}$0) [00.1]	4.45	4.73	+3
Mo (cubic-body center)	MoC(γ')	(011) [100]	(0001) [10.0]	3.15	2.90	−8
Mo (cubic-body center)	MoC(γ')	(011) [100]	(0001) [10.0]	3.15	2.93	−7
γ-U (cubic-body center)	UN	(011) [01$\bar{1}$]	(001) [100]	4.91	4.89	−1
γ-U (cubic-body center)	UN$_2$	(011) [01$\bar{1}$]	(001) [100]	4.91	5.32	+8
γ-U (cubic-body center)	UC	(011) [100]	(111) [1$\bar{1}$0]	3.47	3.50	+1
γ-U (cubic-body center)	UC$_2$ (tetragonal)	(011) [100]	(001) [100]	3.47	3.51	+1

data known to date and with them the nuclei relations of this group of metals with their oxide, nitride, and carbide phases. Tantalum has been omitted from Table VII because of its analogy with niobium.

Table VII confirms once more that oxide, nitride, and carbide reaction layers on metals obtained by a suitable reaction process are essentially of epitaxial or topotaxial nature.

Oxalation Layers on Iron, Zinc, Manganese, Magnesium, and Aluminum

Oxalation is defined as the production of crystalline protective layers from baths which consist mainly of oxalic acid or alkali oxalate with varying additions of heavy metal oxalates.

Phase Stability of the Oxalation Layers. In the normal pickling operation by an acid bath of alkali oxalate on a metal substrate A (A = Fe, Mn, Zn, Mg), oxalation layers are formed which are built up from a homogeneous phase of A-oxalate dihydrate. If the initial bath contains heavy metal ions

(B), such as Fe^{++} and Mn^{++}, there is formed once more only a monophase layer built up of mixed crystals of A-oxalate and B-oxalate, written as $(B,A)C_2O_4 \cdot 2H_2O$. In this case, the portion of metal A (metal substrate) is about 10 atom %. If a very strong pickling reaction occurs which results in the enrichment of the ions of the substrate metal in the immediate vicinity of the metal surface (A) to be oxalated, primary oxalation layers are formed each time. These thin layers almost always consist of the pure carrier metal–oxalate, which gradually is transformed into an oxalate layer with a predominant portion of primary bath metal ion (B) (Table VIII).

TABLE VIII

PHASES IN OXALATION LAYERS ON IRON, MANGANESE, ZINC, MAGNESIUM, AND ALUMINUM

Metal Substrate	Baths		
	Alkali Oxalate	Iron Oxalate	Manganese Oxalate
Fe	$FeC_2O_4 \cdot 2H_2O$	$FeC_2O_4 \cdot 2H_2O$	$FeC_2O_4 \cdot 2H_2O$ $(Mn, Fe)C_2O_4 \cdot 2H_2O$
Mn	$MnC_2O_4 \cdot 2H_2O$	$FeC_2O_4 \cdot 2H_2O$ $(Fe, Mn)C_2O_4 \cdot 2H_2O$	$MnC_2O_4 \cdot 2H_2O$
Zn	$ZnC_2O_4 \cdot 2H_2O$	$ZnC_2O_4 \cdot 2H_2O$ $(Fe, Zn)C_2O_4 \cdot 2H_2O$	$ZnC_2O_4 \cdot 2H_2O$ $(Mn, Zn)C_2O_4 \cdot 2H_2O$
Mg	$MgC_2O_4 \cdot 2H_2O$	$MgC_2O_4 \cdot 2H_2O$ $(Fe, Mg)C_2O_4 \cdot 2H_2O$	$MgC_2O_4 \cdot 2H_2O$ $(Mn, Mg)C_2O_4 \cdot 2H_2O$
Al	—	$FeC_2O_4 \cdot 2H_2O$	$MnC_2O_4 \cdot 2H_2O$

Textures, Epitaxies, and Adhesive Bonds: "Metal–Oxalate." Oxalation layers on rolled or drawn textured metals were always oriented, and therefore, at the intergrowth surface one always observed the monoclinic (010) surface of the oxalate. This led to the epitaxies recorded in Table IX (Fig. 4).

The intergrowths listed in Table IX can be understood if one considers the structure of the oxalate dihydrates. $FeC_2O_4 \cdot 2H_2O$ (humboldtine) is monoclinic and possesses the space group $C2/c$ and has for four formula units per elementary cell the following lattice constants: $a_0 = 12.060$ Å; $b_0 = 5.550$ Å; $c_0 = 9.085$ Å; and $\beta = 127°58'$. Every iron ion in humboldtine is surrounded octahedrally by four oxygen atoms of two planar C_2O_4 groups and two H_2O molecules (Fig. 5). Since every C_2O_4 ion is connected with two

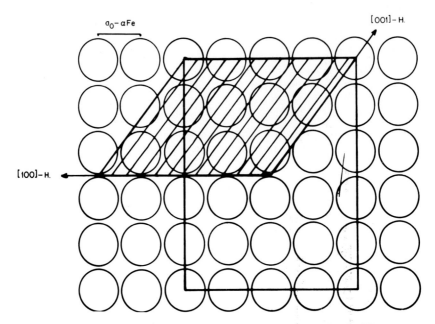

Fig. 4. Affinitive network sections: (010)-humboldtine on (001)-α-Fe.

iron ions, chains are formed, $\cdots Fe \cdots C_2O_4 \cdots Fe \cdots C_2O_4 \cdots$; these are positioned in parallel with the crystallographic b axis, thus [010], while the direction $H_2O \cdots Fe \cdots H_2O$ of the crystallographic a axis is parallel with [100].

The dihydrates of manganese(II), zinc, and magnesium oxalates have

TABLE IX
Epitaxies in Oxalation Layers

Phases		Epitaxial Laws		Periods (Å)		Misfit (%)
Host	Guest	Host	Guest	Host	Guest	
α-Fe	$FeC_2O_4 \cdot 2H_2O$	(001) [100]	(010) [100]	11.46	12.06	+5
Zn	$FeC_2O_4 \cdot 2H_2O$	(0001) [12.0]	(010) [010]	9.34	9.09	−3
Mg	$FeC_2O_4 \cdot 2H_2O$	(0001) [10.0]	(010) [001]	9.60	9.09	−6
Al	$FeC_2O_4 \cdot 2H_2O$	(001) [100]	(010) [100]	12.15	12.06	−1

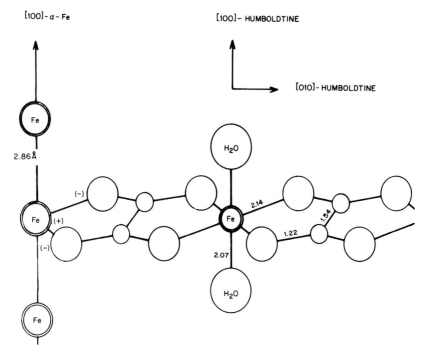

FIG. 5. Molecular intergrowth union: α-Fe-humboldtine.

naturally somewhat different lattice constants and thus chain periods. Magnesium oxalate has the larger cell constant, and zinc oxalate the smaller.

The adhesive bond according to the above law is characterized by a perpendicular orientation of the individual oxalate chains of humboldtine to the metallic surface. From this, there results for the individual oxalate chain an atomic adhesive and valence bond according to Fig. 5. Both terminal-positioned oxygen ions of the oxalate chains lie in the potential wells of iron structural units, assumed to be polarized, in the metal surface. This results in a main valence connection $Fe^+ \cdots {}^-O$ of the chain ends with the polarized iron atoms, and thus a very strong adhesive bond between the individual oxalate chain and the metal surface. The latter is supported by good correlation of the intergrowing host–guest lattice sections, which effect a solid lateral dipole and van der Waals' connection of all chains with one another to the oxalate lattice, with the formation of a densely packed oxalate layer in the solid state.

Phosphating Layers on Iron, Zinc, Copper, Brass, and Aluminum

Phosphating is defined as the production of crystalline protection coatings from baths, which consist predominantly of phosphoric acid or alkali phosphate with varying additions of heavy metal phosphates. So far we have investigated iron, zinc, manganese, calcium, (calcium + zinc), and alkali phosphate layers on the metals iron, zinc, copper, brass, and aluminum. Two of the findings made were verified (5, 6), and, instead of the data of the latter reference, additional studies have led to a large number of new phases.

Construction of Layers. The layer construction of the phosphating layers investigated will be demonstrated on the basis of two examples with other results of the phase analysis for the metals iron and zinc given in Table X.

TABLE X

COMPOSITION OF PHOSPHATE COATINGS ON METALLIC IRON AND ZINC

Heavy Metal Portion in Bath	Metallic Substrate	
	Fe	Zn
Alkali	$Fe_3(PO_4)_2 \cdot 8H_2O$ (vivianite—monoclinic)	$Zn_3(PO_4)_2 \cdot 4H_2O$
Fe	$Fe_5H_2(PO_4)_4 \cdot 4H_2O$ (Fe–hureaulite—monoclinic) $FePO_4 \cdot 2H_2O$* (strengite—monoclinic)	$Zn_3(PO_4)_2 \cdot 4H_2O$ $Zn_2Fe(PO_4)_2 \cdot 4H_2O$ $Fe_5H_2(PO_4)_4 \cdot 4H_2O$
Mn	$(Mn, Fe)_5H_2(PO_4)_4 \cdot 4H_2O$ (hureaulite—monoclinic) (with Mn–Fe–diadochite)	$Zn_3(PO_4)_2 \cdot 4H_2O$ $Mn_5H_2(PO_4)_4 \cdot 4H_2O$ (Mn–hureaulite—monoclinic)
Zn	$Zn_2Fe(PO_4)_2 \cdot 4H_2O$ (phosphophyllite—monoclinic) $Zn_3(PO_4)_2 \cdot 4H_2O$ (hopeite–rhombohedral)	$Zn_3(PO_4)_2 \cdot 4H_2O$
Zn + Ca	$Zn_2Fe(PO_4)_2 \cdot 4H_2O$ $Zn_2Ca(PO_4)_2 \cdot 2H_2O$ (scholzite—rhombohedral) $Zn_3(PO_4)_2 \cdot 4H_2O$	$Zn_3(PO_4)_2 \cdot 4H_2O$ $Zn_2Ca(PO_4)_2 \cdot 2H_2O$
Ca	$CaHPO_4 \cdot 2H_2O$ (brushite—monoclinic) $CaHPO_4$ (monetite—tetragonal)	$Zn_2Ca(PO_4)_2 \cdot 2H_2O$ $CaHPO_4 \cdot 2H_2O$ $CaHPO_4$

* Only from HNO_3–H_3PO_4 baths.

a. In phosphating from the "bath which does not build a layer"—that is, a primary heavy-metal-free bath—on iron a very dense layer of $Fe_3(PO_4)_2 \cdot 8H_2O$ (vivianite) forms, the growth of which practically stops at a layer thickness of approximately 0.1 micron. Since the phosphate layer of the required heavy metal can be obtained in such a nonlayer-forming bath only from the substrate metal with the iron in the form of Fe^{++}, two electrons diffuse from the iron lattice through the phosphate layer to the particular phase boundary "solid–fluid." The cause for the termination of growth then is obviously the result of the rapidly increasing hindrance of the diffusion of iron. This also holds for the alkali phosphating of other metals.

b. In phosphating baths which contain iron or (iron + zinc), there is formed on zinc one of the crystallite structures in the case of the higher layer portions of three-phase layer portions: $Zn_3(PO_4)_2 \cdot 4H_2O$ (hopeite), $Zn_2Fe(PO_4)_2 \cdot 4H_2O$ (phosphophyllite), and $Fe_5H_2(PO_4)_4 \cdot 4H_2O$ (iron–hureaulite). This structure is transformed, depending on the substrate (zinc), to a more and more pure zinc phosphate layer as the primary layer. On the contrary, in the case of very iron-rich baths the primary layer appears also to be of two phases, consisting of hopeite and phosphophyllite. The bath metal in this case has already, at the moment of the primary layer formation, diffused to the metal surface. With progressive growth thickness of the layer, the portion of the substrate in the phosphate layer diminishes more and more so that the external portions reflect the conditions in the phosphating solution. The diffusion secondary supply of the substrate to the phase boundary "solid–fluid" with increasing thickness of the layer is involved in phosphate layer formation, and this process slows down and finally ceases.

Textures, Epitaxies, Nucleation, and Adhesive Bonds: Metal-Layer Phases. All the phosphating layers shown in Table X, as well as the phosphating layers on copper, brass, and aluminum, are known to develop texture under appropriate experimental methods, and indeed the quality of the orientation of the phosphate crystals increases clearly with the quality of the orientation of the metal crystallites. It was necessary to investigate texture correlations between metal and phosphating crystallites. The texture correlations demonstrated so far in phosphating layers are the following: hopeite grows with (010) to (001)-α-Fe, (0001)-Zn, (001)-Cu, (011)-α-brass, (001)-Al, and (011)-Al; phosphophyllite with (100) to (001)-α-Fe and (0001)-Zn; scholzite with (100) to (001)-α-Fe, (0001)-Zn, (001)-Cu, (011)-α-brass, (001)-Al, and (011)-Al; and hureaulite with (100) to (001)-α-Fe, (0001)-Zn, (001)-Cu, (001)-α-brass, (001)-Al, and (011)-Al. (010)-Vivianite intergrows with (011)-α-Fe.

This means that the phosphating systems employed are characterized through nonequivocal texture correlations of the phosphate crystallites (guest phase) to the metal crystallites (host phase). To seek out first the genuine crystal structures—that is, the epitaxial ones out of the many textural correlations of host and guest crystallites—the corresponding host–guest surface pairs must be fixed at least with a pair of corresponding host–guest lattice directions within these intergrowth levels. Table XI and Figs. 6 and 7 show directly that the corresponding guest–host lattice surfaces of vivianite, hopeite, phosphophyllite, scholzite, brushite, and hureaulite on the one hand, and of iron and zinc (similar to copper, brass, and aluminum) on the other hand, possess to a high degree the mono- and two-dimensional

TABLE XI

Epitaxies in Phosphate Coatings on Iron and Zinc

Phases		Epitaxial Laws		Periods (Å)		Misfit (%)
Host	Guest	Host	Guest	Host	Guest	
α-Fe	Vivianite	(011)	(010)			
		[1$\bar{1}$1]	[001]	4.97	4.70	−5
		[11$\bar{1}$]	[100]	9.94	10.08	+1
α-Fe	Hopeite	(001)	(010)			
		[100]	[001]	11.46	10.64	−7
α-Fe	Phosphophyllite	(001)	(100)			
		[100]	[001]	11.46	10.49	−8
α-Fe	Scholzite	(001)	(100)			
		[210]	[001]	6.42	6.61	+3
α-Fe	Brushite	(001)	(010)			
		[100]	[100]	5.73	5.89	+3
α-Fe	Hureaulite	(001)	(100)			
		[210]	[001]	9.63	9.51	−1
Zn	Hopeite	(0001)	(010)			
		[10.0]	[100]	10.66	10.64	<1
Zn	Phosphophyllite	(0001)	(100)			
		[10.0]	[001]	10.66	10.49	−1
Zn	Scholzite	(0001)	(100)			
		[12.0]	[001]	6.95	6.61	−5
Zn	Brushite	(0001)	(010)			
		[12.0]	[001]	6.95	6.38	−8
Zn	Hureaulite	(0001)	(100)			
		[12.0]	[010]	9.26	9.11	−2

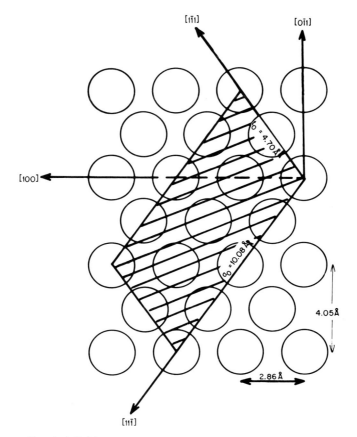

FIG. 6. Affinitive network sections: (010)-vivianite on (011)-α-Fe.

lattice correlations generally required for epitaxies (*1*). There only remained to prove that the corresponding lattice points of the metal–phosphate intergrowth planes are also suitably distributed with respect to valence. This proof is obtained through direct structural comparison of the intergrowing host–guest lattice planes (contact planes). Since up till now only the structures hopeite, phosphophyllite, and vivianite (Fig. 8) have been determined, this could be done only for these phosphates.

The intergrowth planes of the participating host lattices form extremely simple lattice samples packed with metal atoms. The lattice planes of hopeite, phosphophyllite, and vivianite intergrowing with it form with the corresponding host lattices good "polymer analog" lattice samples with exclusive

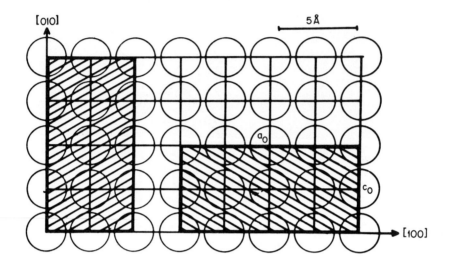

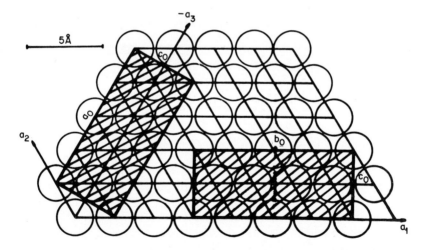

Fig. 7. Affinitive network sections. (Top) (010)-hopeite or (100)-phosphophyllite on (100)-α-Fe. (Bottom) (010)-hopeite or (100)-phosphophyllite on (0001)-Zn.

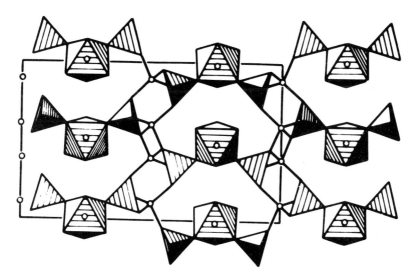

FIG. 8a. Structure of hopeite [$Zn_3(PO_4)_2 \cdot 4H_2O$], projection on (001) according to Liebau (7). Tetrahedron = [PO_4]. Octahedron = [$MeO_2(H_2O)_4$].

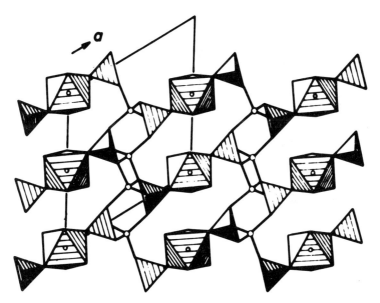

FIG. 8b. Structure of phosphophyllite [$Zn_2Fe(PO_4)_2 \cdot 4H_2O$], projection on (010) according to Liebau (7). Tetrahedron = [PO_4]. Octahedron = [$MeO_2(H_2O)_4$].

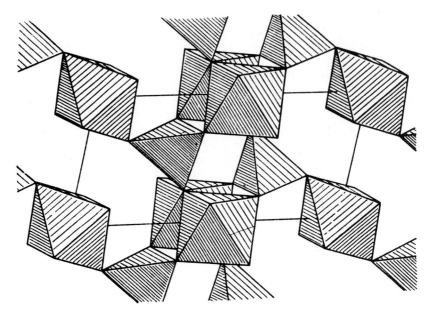

Fig. 8c. Structure of vivianite [Fe$_3$(PO$_4$)$_2$·8H$_2$O], projection on (010) according to Mori and Ito (8). Tetrahedron = [PO$_4$]. Octahedron = [MeO$_2$(H$_2$O)$_4$].

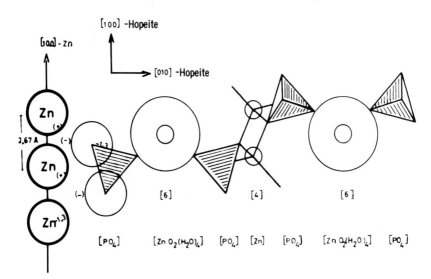

Fig. 9. Molecular intergrowth union: zinc–hopeite.

[PO$_4$] distribution, whereby in every case a corner of the [PO$_4$] tetrahedron points to an Me$^{(+)}$ lattice building stone of the host lattice (Fig. 9). The following "host–guest" structures result:

$$\text{Me}^{(+)} \cdots {}^{(-)}\text{O}{-}\overset{\displaystyle\overset{\text{O}}{|}}{\underset{\displaystyle\underset{\text{O}}{|}}{\text{P}}}{-}\text{O} \quad \text{or} \quad \text{Me}{-}\text{O}^{(-)} \cdots {}^{(+)}\overset{\displaystyle\overset{\text{O}}{|}}{\underset{\displaystyle\underset{\text{O}}{|}}{\text{P}}}{-}\text{O}$$

or a transition form of both valence states.

Building Mechanism and Growth Structure of Phosphating Layers on Metals

If one considers as the simplest case an ideal (001)-surface network plane of α-Fe, then each of its building blocks is surrounded by the four next ($\sqrt{3}/2 \cdot a_0$) and five following next (a_0) neighbors (in lattice 8 or 6). The undisturbed potential field over this surface has the same rigorous periodical construction as the surface network plane itself (9). This is expressed by curve a (Fig. 10)

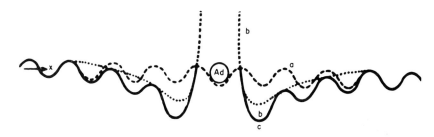

FIG. 10. Potential profile of a (001)-α-Fe surface in the direction [100] according to Mayer (9). (a) Undisturbed potential field directly above the surface lattice plane of the carrier. (b) Potential of a single adsorbed atom. (c) Resulting potential course from (a) to (b).

in the shape of the potential profile of (001)-α-Fe in the direction [100]. In case an oxygen ion of a [PO$_4$] group in the phosphating process reaches into one of these potential troughs, an "adatom" (adsorbed atom) is formed and therewith a locally limited disturbance of the potential field (curve c in Fig. 10). Within the range of this disturbance, the potential field therefore loses its strictly periodic structure. The potential troughs in the vicinity of the "adatom" become deeper than normal and at the same time change their local positions. They coincide within a certain range about the "adatom" and not any more with the lattice positions of the potential minima between, in each case, four iron atoms of the pure host surface. With this distortion of the potential field through the first "adatom," under equilibrium conditions, the position of the second "adatom" is also fixed. In the deepest trough (Fig. 10) immediately adjacent to the first "adatom" or in one voluminous "adatom," as in the case of [PO$_4$] ions, the second next, somewhat more shallow trough is occupied, and therewith the symmetry of the overall field is deformed further.

Through this, rigorously ordered ranges are formed of [PO$_4$] ions on the α-Fe surface, from which (in the case of zinc phosphate layers), through addition of zinc ions, H$_2$O molecules and more [PO$_4$] ions form hopeite nuclei, which are necessarily auto-oriented. The tangential growth of these hopeite nuclei into microcrystallites and thus the active host–guest growing together follow to an extent such that the deviation of the [PO$_4$] positions of the potential minima of the host network becomes so great that a new primary "adatom" is formed. This generates a new chain section of equal orientation which is energetically more favorable than continuation of the old chain. These chain breaks should, in general, occur after a few chain periods. Naturally they require a correspondingly dense arrangement of loose positions, according to Fig. 11.

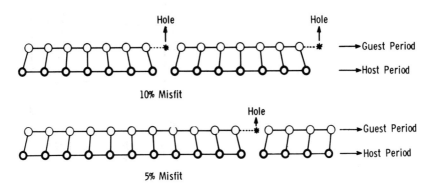

FIG. 11. Rupture of an active intergrowth chain in the case of distinct difference of corresponding host–guest periods.

Thus, an ultra-dispersed primary layer results, strongly oriented to the carrier lattice but not necessarily oriented in parallel. Further growth of this primary layer then ensues in the shape of a short period sequence of epitaxial nuclei formation which is also diffusion-controlled.

The regularly repeated, epitaxially directed nuclei formation, growth, and growth-retarding mechanisms assure the transmittal of the substrate orientation to the primary phosphate layer and to all the subsequent crystallite deposits. Thus, there results at the same time for the layer a properly adherent, fine-grained, and about equally sized crystallite structure.

Time-Law of the Layer Growth

According to the preceding findings, there results a course of the formation of the layer represented as thickness, d, of the layer as $f(t)$, according to Fig. 12. Figure 12a represents the case of layer formation from baths made of

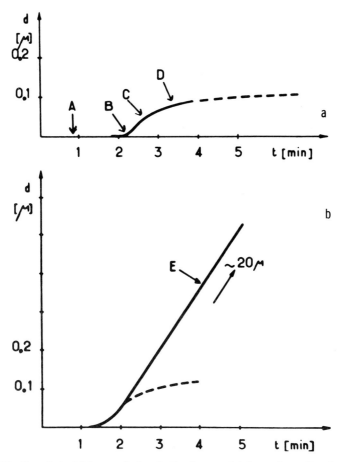

FIG. 12. Growth in thickness of phosphating layers. (*a*) Layer formation from heavy-metal-free bath. (*b*) Layer formation from heavy-metal-containing bath. A = etching period. B = nuclei formation period. C = growth thickness through solid body reaction ($\sqrt{t}$ law). D = autoretardation of diffusion. E = thickness growth through cations and anions supplied from the bath (approximate linear time law).

materials which are not layer-forming. Etching is the initial reaction: $Fe^0 + 2H^+ \rightarrow Fe^{++} + H_2$, which on the one hand reacts with a pure iron surface for the primary separation through dissolution or insufficient solution of the primary oxides, and on the other hand generates the heavy metal ions required for the formation of the layer. This etching reaction continues until the solubility product of the particular stable phosphates is obtained. For iron as substrate metal, vivianite forms in the case of a fresh bath and a short period of attack. Iron–hureaulite forms when reaction occurs in the bath over a long period or in old baths already containing iron (etching period, A). The nuclei formation of the substrate phosphate (vivianite or iron–hureaulite) continues through epitaxial growth from substrate to the phosphate coating without hindrance from nucleation. This first step of layer formation leads in the case of rolled polycrystalline metal surfaces to the formation of a multitude of nuclei sections which are oriented on the substrate. This results in the formation of a closed, oriented polycrystalline primary phosphate layer (formation of nuclei and primary layer formation period, B). The further growth in thickness of the layer can now, in the case of heavy-metal-free baths, result only through the diffusion addition of iron substrate metal in the form of $Fe^{++} + 2e^-$ through the primary phosphate layer to the phase boundary phosphate-layer bath (period C of growth process). The diffusion velocity of Fe^{++} through the uniformly thickening phosphate layer to the particular phase boundary solid solution is decisive for this growth in thickness of the layer. This results in a $\sqrt{t}$ law. Since very thin phosphate layers, in general, are very dense, the growth rate diminishes with increasing thickness of the layer. At a layer thickness of approximately 0.1 micron, it comes practically to a stop (curve portion D). Linear growth in thickness would at present become possible only if the bath itself could make available heavy metal ions when the phosphating is continued in a "layer-forming" bath (Fig. 12b). In phosphating from a layer-forming bath, it will also be valid that structural and succession observations give originally the layer-forming steps B to D shown in Fig. 12a, although they may be obscured.

The portion of the thickness determined through the $\sqrt{t}$ law in general is only a small fraction of the overall thickness of the protection layers (5 to 20 microns, layer formation period E). There occurs for the formation of phosphating layers on metals a thickness-time curve in two parts with a change of the reaction mechanism at a layer thickness of about 0.1 micron.

Conclusion

All the protective layers discussed above, such as oxide, nitride, boride, carbide, oxalate, and phosphate coatings, are built up from phases which under normal pressure-temperature-concentration conditions are either thermodynamically stable or extremely inert. The layer crystallites are linked epitaxially or topotaxially in all cases with the host crystallites via strong reciprocal reaction forces.

References

1. A. Neuhaus, *Fortschr. Mineral.*, **29–30,** 136 (1950–1951).
2. A. Neuhaus and M. Gebhardt, *Werkstoffe Korrosion*, **17,** 567 (1966).
3. M. Gebhardt, "Protection against Corrosion by Metal Finishing," Förster-Verlag, Zurich, 1967, pp. 323–330.
4. G. V. Samsonov and A. P. Epik, "Coatings of High-Temperature Materials," Part 1, Plenum, New York, 1966.
5. A. Durer and E. Schmid, *Korrosion Metallschutz*, **20,** 161 (1944).
6. J. Saison, *Compt. Rend.*, **248,** 2586 (1959).
7. F. Liebau, *Acta Cryst.*, **18,** 352 (1965).
8. H. Mori and T. Ito, *Acta Cryst.*, **3,** 1 (1950).
9. M. Mayer, "Physik dünner Schichten," Stuttgart, 1955.

Surface Finishes of Metals from the Electrochemical Point of View

A Contributed Discussion

V. Cupr and B. Cibulka

Brno, Czechoslovakia

The importance of both chemical and electrochemical surface treatments of metals merits a discussion of certain general and common factors. These include:

1. The endeavor to achieve a unified explanation of their mechanisms, including as far as possible all the accessible details that are of theoretical or practical significance.

2. The requirement to obtain good adhesion of the formed protective layers and coatings to the metal base, which serves to guarantee the adhesion of a protective coating with the metal by means of forces which are more pronounced than forces causing either mere adherence or adhesion.

3. The requirement of their sufficient continuity and entirety, which guarantee, in addition to chemical and other resistances, the capability of acting as an interface on the boundary of the metal for the paint system or another protective system.

4. The study of the formation of coatings to an expedient degree for the maintenance of the required mechanical properties, such as, for example, elasticity and mechanical strength of the protective layers.

5. The safeguarding of the protective coating surface in such a way that it is suitable for the subsequent application of coating or conservation compounds.

Surface treatments begin with the bare metal, most commonly iron. Here the following should be considered initially: Iron atoms in the substrate can participate in aqueous reactions only if they have lost or exchanged two valence electrons. This is manifested either by the formation of Fe^{++} cations in solution or by the formation of compounds in which we are concerned with bivalent iron, such as iron hydroxides, phosphates, oxalates, and tannates.

Bonding Forces on the Metal Surface

Metals are known to form space lattices because of electrostatic forces between atoms and electrons. In doing so, part of the electrons—for example, valence electrons—is separated from the atoms so that in the metal lattice, the remaining atomions are stabilized into certain positions. The separated electrons have a great mobility which causes the well-known marked conductivity of metals, and they take part in chemical reactions which are occurring on the metal surface, or in electrochemical reactions when the metal is in contact with an electrolytically conductive medium.

In investigations of the electrochemical behavior of metals in electrolytes, particularly in corrosion and passivation phenomena, and certainly also in the surface treatments of metals, it is necessary to consider the nature and formation of their surfaces and furthermore adsorption and bonding forces which act between the metal surfaces and the components of the adjacent electrolytic solution. It can be stated from electron microscope observations with a sufficient magnification that even on perfectly polished metal surfaces very numerous, submicroscopic edges and corners, as well as depth unevennesses, are present, although the crystal structure of the metals is being maintained.

It is known that the bonding forces of the atoms within the metals are mutually saturated and equilibrated, whereas from the atoms situated on the surface, or closely under the surface, unsaturated bonding forces emanate which cause on the metal surface states of tension and strength fields. These forces take part either in adsorption or in chemical bonding with particles which are in the vicinity of the metal surface.

The strength fields and bonding forces differ in accordance with the geometrical shape of the surface, especially when the surface is under the influence of an outside electrical field or local cells. Under such conditions, there occurs at the points and edges that protrude into the solution an accumulation of lines of force and an increase in the current or charge density, respectively. At these points, electrochemical reactions are being preferentially realized which are coupled according to the direction of the current with the exclusion of ions or other electrically charged particles. However, not only the geometrical shape is decisive here, but also the values of the bonding forces.

Thermodynamic investigations bear out the fact that those points exhibit a higher surface activity which, when covered with a foreign substance from the solution, liberate a larger surface energy than other points. This energy change can be brought into association with forces by which the above substances are attracted and bonded in the surface. If the bonding energy is small or if in the metal surface van der Waals' forces exist which are of small significance, it is possible to speak about the physical adsorption of foreign substances with the manifestation of minute free bonds. For the breaking of these bonds, contact with a pure solvent suffices, with the result that substances adsorbed from solution can be removed—for example, by flushing with water. Therefore, the physically adsorbed films on the metal surface are not continuous and uniform, and, furthermore, they are preferentially formed on the above-mentioned active points. These inhibition phenomena have practically no significance in the surface treatment of metals, and the same holds true for corrosion protection. Conversely, if the bonding energy of the active points is large, then on the metal surface chemisorption of substances takes place from the solution in which valence forces participate, so that it is possible to consider the formation of chemical compounds with a certain stoichiometric ratio of their components.

Because of a nonuniform distribution of bonding forces on the surface of metals, it can be expected, in the majority of cases, that physical adsorption and chemisorption simultaneously appear to varying degrees and that they overlap in their effects. The fundamental difference between both types of bonding forces lies, first of all, in the irreversibility of chemisorption processes and, second, in the larger heat of reaction and the energy of reaction of chemisorption processes.

For the generation of the energy of chemisorption, the specific affinity between the metal and the chemisorbate is decisive. Nevertheless, the activation energy required during the chemisorption reaction for the formation of a new lattice also is decisive. The electron configuration of the metal is undoubtedly of fundamental significance, not only for its activation energy, but also for its specific affinity to the substance which is being attached to the metal surface. Understandably, the decision is also influenced by the arrangement of electrons in the chemisorbed molecules, especially with those groups which, for electrostatic reasons, are diverted toward the metal surface.

If the bond between the foreign substance and the metal is very pronounced, then on the metal surface a saturation of the bonding forces will occur so that the metal becomes inert; this is otherwise expressed by the term passivity of the metal; that is, the metal loses its capability to partake in chemical reactions, as would otherwise be expected. Substances that create such a condition on the metal are called passivators. These can be substances of widely different chemical nature, above all oxygen, either atmospheric or *in statu nascendi*—for example, in solutions of alkali nitrites or chromates—also oxy-salts such as phosphates and oxalates, and certain effective corrosion inhibitors.

The above definition of the term passivation should be supplemented, and of great concern is the immobilization of the anodic reaction Me → Me^{n+} + ne^-. This, however, does not mean that on the passive metal surface no other reactions can proceed, one of the reasons being that the passive metal exhibits electrolytic solution potentials, the value of which depends on the composition of the solution adjacent to the passive surface. A small dependence of these potentials on the nature of the metal indicates that in their electrochemical reactions there is a common basis for various metals. These are, for example, oxide layers, on which in an appropriately acid solution the reaction $2H^+ + 2e^- \rightarrow H_2$ is realized, or, with a lack of hydrogen ions, the reaction $2H_2O + 2e^- \rightarrow H_2 + 2OH^-$. To these electrochemical reactions correspond potentials of more positive values than the potential of the active metal. Therefore, the passivation of metals or their activation—that is, the transition from the passive into the active state, or vice versa—is characterized by significant potential and current changes, if simultaneously resistance changes of the passivation layers take place.

Formation and Significance of Natural Oxide Layers

Surface treatments are generally carried out either on parts which during the mechanical machining operations have been exposed to the effect of atmospheric oxygen, or on pickled parts after their flushing with water or with some other solution with a deliberately selected composition. In both cases, after the drying of the parts, it is not possible to expect a perfect baring of the lattice with steel or iron surfaces, but there is at least partial covering of the metal with passivation layers. When the action of oxygen is involved, the well-known natural oxide layers are considered. Their formation is normally explained by chemical oxidation—that is, a reaction between the iron and oxygen atoms.

Under elevated temperature conditions, there are certainly no objections to this reaction mechanism. But under normal temperatures it is necessary to take into account atmospheric moisture which can initiate chemisorption of water molecules on the surface of the bare metal lattice. Therefore, it is worth considering whether the chemisorbed water molecules are more active in this case with the metal lattice than the oxygen molecules.

Discussions concerning energy relations in electron configurations of water and oxygen molecules are beyond the scope of this paper. It is sufficient here to start out with the assumption that, since the water molecule has a pronounced dipole, its negative charge concentrated in the oxygen atom is turned toward the atomions at the lattice surface. Starting out with the simplified diagram,

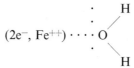

it can be accepted that the chemisorption bond formed by the two free electron pairs between the oxygen atoms of the water molecule and the iron atomion in the lattice leads necessarily to the loosening of the oxygen bond to the H^+ protons. Within the intermolecular transformation and because of the asymmetry of the water molecule, the liberated protons can react with the electrons repelled by the chemisorption bond from the iron atomions. This can lead to the formation of hydrogen by the reaction $2H^+ + 2e^- \rightarrow H_2$, or even to the formation of water by the reaction $4H^+ + 4e^- + O_2 \rightarrow 2H_2O$, if there is a sufficient supply of oxygen near the iron lattice. The oxygen anion formed from the chemisorbed water molecule by the loss of two protons is bonded onto the iron atomion in the lattice; this is accompanied by the formation of an oxide or even a hydroxide, if the intermolecular disintegration of the chemisorbed water molecule leads to the formation of hydroxyl anions.

It clearly follows that the action of oxygen on the metal lattice does not involve a direct oxidation effect which would lead to the formation of an oxide on the lattice surface, but its participation is in the reaction between the protons and the electrons liberated by the chemisorbed oxygen bonds, which thus accelerates the intermolecular disintegration of the chemisorbed water molecules. The latter can, however, proceed also with hydrogen without the presence of oxygen. This is the situation in the formation of oxides or hydroxides on a pickled iron surface during water rinses, even though it is not possible to exclude hydrolysis-like processes here—that is, the reactions

$$[Fe(OH_2)_x]^{++} \rightarrow [Fe(OH_2)_{x-1}OH]^+ + H^+ \rightarrow [Fe(OH_2)_{x-2}(OH)_2] + 2H^+$$

The hydrated bivalent iron cation enters into the reaction while it is still within the range of the lattice forces, so that it actually reacts with chemisorbed water molecules which suffer the above intermolecular disintegration and is accompanied by the formation of hydroxides. Depending on the experimental conditions, their partial dehydration can occur.

It is important to observe what happens to the natural layers in the formation of conversion coatings. It can first be admitted that they are largely dissolved in the treatment baths with the formation of bare metal surface; on this surface, the anodic process immediately operates, with the formation of the conversion substance. The remaining part of the oxide coatings thus assumes the role of the inert base for the cathodic reaction that depolarizes the anodic action. It is therefore not impossible that the natural layers are overgrown by a conversion substance, especially by fractions of the latter which are formed outside the scope of local cell action and which grow by crystallization on reaching the solubility product of the conversion substance.

Much depends on the selection of the treating solution, the composition of which should be balanced in order to fulfill the following conditions:

1. Appropriate dissolution of natural oxide layers.
2. Formation of the conversion coating because of local cell action, which results from heterogeneity of the treated metal surface.
3. Possibility of formation of the conversion coating even without electrochemical reactions—that is, through certain chemical reactions which act only to strengthen the conversion coating already anchored in the metal surface.

Formation of Coatings on Metal Surfaces

In discussions regarding both the activity of metal surfaces and the possibilities of reactions of the latter with certain components from the solution, the important question arises whether there are attraction and bonding forces acting between atomions placed in certain positions in the metal lattice and the considered foreign component, or whether this is the result of bonding forces of such a component to the free electrons which appear in the metal lattice.

Thus, basically two marginal cases can be distinguished:

1. Bonding action of atomions being formed from atoms by the loss of valence electrons and by the formation of an electron cloud assumes a negative charge of the attracted foreign component. Through the realized bond and the electrostatic action of the relevant bonding forces, electrons are thus being freed that must be consumed by certain electrochemical reactions in which the appropriate components of the solution participate. This must occur, if the electroneutrality of the considered metal in contact with the solution is to be maintained. Stated more accurately, this electroneutrality will occur only within such a range that corresponds to the creation of a potential difference between the metal and the electrolytically conducting solution.

These manifestations of bonding forces should be taken into account in the formation of conversion coatings which include phosphatization and chromate treatment, or newly introduced surface treatments, such as oxalate treatments and tannate treatments (1). It can be said that the latter are formed within the scope of electrochemical heterogeneity—that is, by anodic action: $m\text{Me} + n\text{A}^{z-} \to \text{Me}_m\text{A}_n + 2me^-$, with its simultaneous depolarization realized by a suitable cathodic reaction. A^{z-} is an anion obtained from the solution.

2. For the application of electrons from the lattice, eventually also as valence electrons, foreign components with a positive charge should be considered, or neutral components which can take up into their electron configuration additional electrons and thus form negatively charged compounds which are electrostatically bonded to the remaining atomions forming the metal lattice. To this method of bonding force application, the action of oxygen can be added, leading to the formation of passivation layers, as well as the effect of substances which supply atomic oxygen through their decomposition in the solution.

Anchoring of Conversion Coatings in the Metal Base

The anchoring of conversion coatings depends on their adhesion to the metal, as well as on the life of the coating and the metal conservation systems (2–4). This indisputable advantage of conversion coatings can be explained as follows: The above-considered affinity of atomions toward anions in the immediate vicinity of the metal surface is not limited only to atomions in the lattice surface. The same affinity, even if in a graded measure, manifests itself also in atomions present at an appropriate depth in the lattice. Only in this way is it possible to explain that the atomions can diffuse to the anions from the lattice of the Fe_mA_n substance, the former being renewed by anions which diffuse to the substance from the solution. The result of these diffusion processes is the strengthening of the layer of the Fe_mA_n substance by the undergrowing of the latter at the expense of the metal base.

The possibility of an anodic continuous growth of the Fe_mA_n substance into the metal base is directly related to the anchoring of the substance and also to the coherence of the substance with the metal base. For the sake of completeness, it should be pointed out here that crystalline coatings, such as phosphate coatings grown from components of the phosphatizing solution, are obtained for this purpose if the required values of the solubility products of the crystallizing substance are obtained. Nevertheless, the coherence of crystalline phosphate coatings to the metal base can be explained in a similar way as for amorphous coatings, since these are formed immediately after contact with the phosphatizing solution, and only through the further course of phosphatization there occurs the formation of crystallizing nuclei in the amorphous coating (5) and its overgrowing by a tangle of growing crystals of the phosphate substance (6).

References

1. V. Cupr and J. B. Pelikan, *Metalloberfläche*, **20,** 471 (1966).
2. V. Cupr and B. Cibulka, *Deut. Farben-Z.*, **18,** 442, 476 (1964).
3. V. Cupr and B. Cibulka, *Deut. Farben-Z.*, **19,** 15 (1965).
4. V. Cupr and B. Cibulka, *Deut. Farben-Z.*, in press.
5. V. Cupr and J. B. Pelikan, *Werkstoffe Korrosion*, **12,** 475 (1961).
6. V. Cupr and J. B. Pelikan, *Metalloberfläche*, **19,** 187, 230 (1965).

The Kinetics of the Formation of Phosphate Coatings

W. Machu

Technische Hochschule
Vienna, Austria

The Electrochemical Nature of the Phosphating Processes

The kinetics of the processes involved in the formation of phosphate coatings upon iron, steel, zinc, etc., are much more complicated than the kinetics of similar reactions in heterogeneous systems. It is true that many factors, such as concentration and viscosity of the solution, reactivity of the base metal, temperature, diffusion, and agitation, which are decisive for the rate of chemical reactions in heterogeneous systems, are also considerably important in phosphating processes. In addition to these factors, a number of other factors are specific for phosphating processes.

It is a basic feature of phosphating processes that coating formation does not take place over the entire metal surface but only at the local cathodes, by reason of electrochemical phenomena. As was shown by Machu (*1*), phosphating processes must be clearly distinguished between those taking place at the local cathodes and those taking place at the local anodes. The initial pickling process of the metal in dilute phosphoric acid, leading to the formation of soluble, primary heavy metal phosphates at the phase boundary layer, consists in the formation of positively charged iron or zinc ions, as with any dissolution of a metal, according to the equation

$$Fe^0 = Fe^{++} + 2e^-$$

at the local anodes, and the simultaneous discharge of hydrogen ions at the metal cathodes, according to the equation

$$2H^+ + 2e^- = H_2$$

At the beginning of the phosphating process, the metal surface is free from grease and oxides, after the necessary careful cleaning step. The cleaned metal surface is rather reactive in the acidic phosphating solution. Therefore, the major proportion of this surface will exhibit anodic behavior, while the cathodic regions are confined to the grain boundaries, oxide residues, heterogeneities of the crystal texture, and places exhibiting physical and electrochemical differences in the structure of the metal, such as, for example, cold-worked parts.

As a result of the stronger corrosion currents at the metal surface, at first relatively large amounts of primary soluble iron or zinc orthophosphates will be formed at the metal anodes. At the same time, however, the hydrogen ions corresponding to the corrosion current will be discharged at the local cathodes, also in appreciable amounts. This leads to a reduction of the concentration of hydrogen ions in the solution and a shift of the pH value to more alkaline values at the local cathodes in the phase boundary layer. The diffusion processes taking place at the phase boundary layer cannot supply fresh hydrogen ions rapidly enough to replace the discharged ones, or remove the dissolved iron salts from the phase boundary layer, not even under intensive agitation or stirring—for example, spraying a phosphating solution onto the metal surface. Since the tertiary heavy metal phosphates are soluble only at low pH values, the more difficultly soluble tertiary heavy metal orthophosphates are deposited in the phase boundary layer at higher pH layers. Hence, the formation of the phosphate coating takes place only at the cathodic areas of the metal surface, since the same factor which initiates coating formation will at the same time impoverish the solution with respect to the hydrogen ion concentration, since the hydrogen ions are electrolytically discharged at the metal cathodes. Actually, as Machu (1) has shown, a cathodic polarization favors and expedites coating formation, since hydrogen ions are discharged in the process, while an anodic treatment of the metal only intensifies the dissolution of the metal, but has practically no influence on the deposition of crystallized metal phosphates.

It is noteworthy that a treatment with alternating current will also favor coating formation, since the cathodic half-wave of the alternating current leads to the formation of the insoluble phosphates, which will then block the transmission of current of the anodic half-wave. Actually a phosphating process was practiced (2) wherein coating formation was accelerated by a treatment with alternating current. In the course of quantitative investigations of the influence of alternating currents on the phosphating process, Machu (3) has been able to show that the time of formation of phosphate coatings is reduced to about one half by treatment with alternating current.

Therefore, the phosphating process represents an electrochemical process, wherein the deposition of the phosphate coatings takes place primarily at the local cathodes. It is therefore also a topochemical process, which is limited to the cathodic areas of the metal surface and the size thereof.

Self-Passivation and Shift of the Potential toward More Noble Values during

the Phosphating Process. A further essential feature of the phosphating processes consists in the fact that the reactions necessary for coating formation are inhibited by the formation of the layer of insoluble phosphates. At those places where the insoluble and electrically nonconductive phosphate layer was deposited, the discharge of ions has ceased, and no longer is there any formation of iron ions. These cathodic areas, which are covered by the phosphate layer, are passivated. Therefore, the phosphate coating itself prevents the electrochemical processes associated with the primary pickling reaction and the deposition of the phosphate crystals.

The fact that the deposition of the phosphate coating is a self-passivating process follows from the observation, for example, that the phosphating process is associated with a shift of the potential toward more noble values. This potential shift toward more noble values corresponds to the so-called self-passivation law by Müller and Konopicky (4). According to this self-passivation law,

$$t = M + N \log \frac{E_{Me} - E}{E - E_s}$$

the plot of the function

$$\log \frac{E_{Me} - E}{E - E_s}$$

against time should give a straight line. In this formula t is the time in seconds, M and N are constants, E_{Me} is the metal potential (namely, the potential at the beginning of the coating process), E is the potential measured at any moment, and E_s is the coating potential, corresponding to the constant potential reached when the phosphating process is finished.

Thus, if, for example, the change of the potential at ground (polished) iron electrodes with time is recorded, in a nitrate-containing zinc phosphate bath of 70 points at 95°C, the curves A, B, C, and D as shown in Fig. 1 are obtained for four different phosphating processes. It is seen that the potential approaches a constant value when the phosphating process is finished after about 5 minutes.

For curve C of Fig. 1, the values of the above function of the self-passivation law have been calculated for different times and plotted as curve E. It is seen that the self-passivation law for this phosphating process is strictly fulfilled.

Furthermore, as can also be seen from Fig. 1, the potential shift toward more noble values in the deposition of the phosphate coating amounts to only about 0.15 to 0.2 volt. If a really passivating layer had been formed, the potential shift toward more noble values would necessarily have been much greater and of the order of magnitude of about 1 volt. The fact that the shift of the potential toward more noble values of only about 0.15 to 0.2 volt does not correspond to a passivating layer shows that the phosphate coating is always very porous. According to quantitative investigations of the pore areas of phosphate coatings (5, 6), phosphate coatings always have porosities

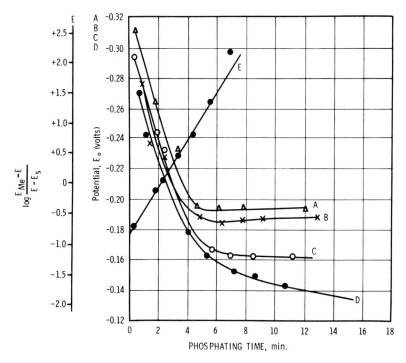

FIG. 1. Potential curves (A, B, C, D) and self-passivation law (curve E) in a zinc phosphate bath, 70 points at 95°C.

of about 0.1 to 1%, which are thus 100 to 1000 times as large as those of really passivating coatings.

Formation of Phosphate Coatings as a Crystallization Process

A further essential difference between normal reactions in heterogeneous systems and phosphating processes also consists in the fact that the initially formed, soluble heavy metal orthophosphates are converted into insoluble, tertiary heavy metal phosphates in a secondary reaction which follows shortly after. In most cases, in the formation of phosphate coatings, crystallized well-defined phosphates (7) are deposited, such as phosphophyllite, $Zn_2Fe(PO_4)_2·4H_2O$; hopeite, $Zn_3(PO_4)_2·4H_2O$; scholzite, $Zn_2Ca(PO_4)_2·2H_2O$; vivianite, $Fe_3(PO_4)·8H_2O$; and iron–hureaulite, $Fe_5H_2(PO_4)_4·4H_2O$.

Therefore, the rate of formation of phosphate coatings is also influenced by the crystallization processes. For each crystallization process there must first be a supersaturated solution, so that crystallization may begin. Moreover, there must be crystal nuclei, around which crystalline substances may be deposited and microphosphate crystals and finally the normal phosphate crystals can be formed. Actually these processes concerning the formation of crystal nuclei and crystallization processes have a great influence on the rate of formation of phosphate coatings.

Since the phosphating processes are also crystallization processes, the influence of the pretreatment of the metal surface also extends to the kinetics of the formation of the phosphate coatings. Also, the nature of the coating formation and the rate of the deposition of the phosphate coatings depend on the physical consistency and the nature of the metal surface. These processes are associated with the crystallization of the metal phosphates.

The number of nuclei and the crystallization rate are factors which control the crystallization process. It is therefore clear that the mechanical or chemical pretreatment of the metal surface prior to phosphating has an appreciable influence on the size of the crystal grains and hence on the formation of thinner and fine-grained or thicker and coarse-grained phosphate coatings.

Until recently it was not possible to explain the various accelerating or inhibiting mechanisms of the pretreatment of the metal surface prior to phosphating. It has been known for some time that, for example, an alkaline degreasing treatment or a chemical pickling treatment in acids will frequently lead to the formation of coarse-grained, heavy, slow-forming phosphate coatings. On the other hand, it was also known that brushing, polishing, grinding, sandblasting or wiping with petroleum, emulsion cleaning, or a pretreatment with solutions of dilute oxalic acid or of complex titanium phosphates and the like favors and permits the formation of fine-grained and thinner phosphate coatings in shorter treating times.

This positive or negative influence on the grain size of the crystals of the phosphate coating to be formed and on the rate of formation of the said coating can be explained as follows, according to Machu (8):

There must first be a supersaturated solution of the primary soluble metal phosphates in the boundary layer at the metal surface. This supersaturated solution is formed as a result of the primary pickling process in coating-forming as well as in noncoating-forming phosphating systems, owing to the faster rates of chemical and electrochemical processes and the slower rates of diffusion processes. At the same time, the pH value in the phase boundary layer at the local cathodes must be shifted to higher values.

However, to permit the onset of crystallization, there must be crystal nuclei or preferred places at the metal surface where the crystallization may start. Now, the average number of nuclei at the iron surface of a cold-rolled sheet iron, which is frequently used, can be increased or decreased by mechanical or chemical treatments. Preferred places, from which the reactions and the crystallization processes begin, are primarily the active centers of the metal surface, such as sharp edges, points, heterogeneities, breaks in the crystal texture and the like, which centers have a higher energy level than the mean surface energy. These active centers also have a more negative potential.

The number of nuclei is obviously related to the degree of the heterogeneities of the surface. Obviously, potential differences at the metal surface are also decisive for the number of nuclei. The larger the number of places having different surface conditions or energy levels, and hence potentials, the greater is the number of nuclei.

If a metal surface is treated—for example, with an aqueous, strongly alkaline hot cleansing solution—a large number of the previously present active centers are inactivated, owing to the formation of hydroxide or oxide coating layers, whereby the number of nuclei is reduced. A treatment with acids may have a similar effect. The action of the acid first takes place at the active centers of the surface, removes or dissolves these, and thereby also reduces the number of crystal nuclei.

Therefore, in these two cases crystallization can then start only at a diminished number of nuclei. Since the rate of crystallization can be assumed to be nearly constant, provided the composition of the bath, the temperature of the bath, the concentration of the ingredients of the bath, the agitation of the liquid, etc., remain the same, only a few crystals will grow if the number of nuclei is reduced. It will then take a longer time for the crystals formed to come into contact with each other and to form a continuous layer. The crystals will therefore be larger and coarser-grained, which leads to the formation of a thicker phosphate coating. The time of coating formation on pickled iron surfaces will be twice or three times as long as the normal value. Therefore, the rate of formation of a phosphate coating to a great extent depends on the number of nuclei and the related grain size of the phosphate crystals. Thin, fine-grained coatings form faster than thicker coarser-grained coatings.

On the other hand, it is also known that the pickling sludge which is sometimes formed in pickling operations will prevent the formation of coarser and thicker phosphate coatings after pickling. Therefore, in this case the pickling sludge takes over the function of the crystal nuclei, which were reduced in number as a result of the pickling operation. There will then again be a heterogeneous surface condition at the metal surface. Thus, in spite of the decreased number of active centers as a result of the pickling operation, the presence of pickling sludge of a certain structure again provides a sufficient number of nuclei for the fast formation of a thin phosphate coating.

If an iron or zinc surface is brushed, ground, or sandblasted, this mechanical treatment will more or less remove the inactivating oxide or hydroxide coating layers from the active centers, which will be activated again. In part, a fresh metal surface having numerous new, active centers is also formed as a result of the roughening of the surface in grinding or sandblasting; hence, again a heterogeneous structure of the surface is produced. Therefore, a mechanical treatment favors the formation of fine-grained or thin phosphate coatings by increasing the number of nuclei and/or the free energy level of the active centers.

However, as Ghali et al. (9) have shown, the degree of surface roughness alone is not decisive for a high number of nuclei. Rather, it is the number of nuclei which is decisive, since fine and intermediate sand grains produce phosphate coatings which are up to three times as stable and shock-resistant as phosphate coatings produced on metal surfaces treated with coarser-grained blasting agents.

Heterogeneous surface conditions having different metal potentials, and hence new crystallization centers, may also be produced by a chemical pretreatment. If, for example, an iron surface is treated with a dilute solution of oxalic acid, the metal surface does not contain a uniform and fully coated oxalate layer; only numerous very finely divided oxalate crystals are formed, and these numerous small crystals will act as crystallization nuclei for the formation of a thin and fine-grained phosphate coating. Pretreatment with a dilute solution of complex titanium phosphates was found to be particularly effective, leading also to the formation of a large number of very small titanium phosphate crystals on the metal surface and hence to an increased number of nuclei. A rinsing treatment with a dilute titanium phosphate solution after a cleansing treatment with aqueous, hot, alkaline cleansing solutions or after a pickling treatment with mineral acids will even eliminate the disadvantageous effect or influence of these cleansing and pickling treatments on the number of nuclei and hence on the formation of thicker and coarser-grained phosphate coatings.

Thus, a metal surface having numerous places of different surface structure and potential, which are apparently necessary for the formation of crystal nuclei, is formed when the said metal surface is treated with dilute oxalic acid or with a dilute titanium phosphate solution.

The fact that these pretreatments merely produce a great number of very small heterogeneities, and hence places having different energy levels or different potentials—namely, coated and uncoated surface areas—also follows from the observation that, for example, an emulsion cleaning treatment favors the formation of fine-grained phosphate coatings. After the emulsion cleaning treatment, a great number of fine droplets of the organic cleaning agent, such as petroleum, which are separated from each other, are left behind on the metal surface. These fine droplets cause the formation of surface areas which are uncoated and have a higher energy level, and of surface areas which are coated and have a lower energy level, the different areas having different potentials.

In the same way, the wiping of the metal surface with a cloth moistened with some petroleum, or touching of the metal with fingers, has a grain-refining effect on the formation of the phosphate coatings. In these cases very finely divided petroleum grease droplets are left behind on the metal surface. These will then result in places having different surface consistencies and different coating conditions. These areas will then have different potentials. Thereby, the number of nuclei is increased and the rate of formation of fine-grained phosphate coatings is increased.

If there is a continuous uniform oil, fat, or petroleum layer on the metal surface, the metal surface can be considered to have a homogeneous consistency without greater differences in the structure and in the potentials. On such a metal surface, the heterogeneities and potential differences, which are necessary for the formation of crystal nuclei are lacking. Such continuous and homogeneous fat or petroleum layers therefore interfere with the formation of the phosphate coatings and can even completely prevent their formation.

Hence, there is a great difference between a homogeneous and a heterogeneous fat, oil, or petroleum layer on a metal surface. Surface areas having different and heterogeneous structures and different potentials are present in greater numbers only on metal surfaces having very finely divided fat or petroleum droplets with uncovered microareas of the metal in between. These active areas then increase the number of crystal nuclei and favor the crystallization process. A uniform oxalate coating also interferes with the phosphating process, while finely divided oxalate crystals favor the phosphating process.

The influence exerted by a mechanical or chemical pretreatment on the crystallization processes connected with phosphating operations is of considerable technical importance, since thin phosphate coatings have numerous advantages over thicker coatings. Thinner and finer-grained coatings are not only denser and more corrosion-resistant, they also require less chemicals for their preparation. The gloss of paints and varnishes is also not diminished by thinner phosphate coatings. In many cases it is then sufficient to apply one single paint layer. Thinner phosphate coatings are also more easily deformable and have better resistance to shock and impact strength.

Machu's electrochemical theory concerning the formation of phosphate coatings and the influence of the surface condition on this process was recently confirmed experimentally by Cheever (*10*). Cheever showed that the formation of phosphate coatings starts primarily at the local cathodes—namely, the grain boundaries of the iron crystals. The number of phosphate crystals, which is normally between 10^5 and 10^6 per square centimeter of the iron surface, is clearly increased by a mechanical treatment or a chemical treatment of the iron surface with titanium phosphate solutions. However, the number of phosphate crystals is substantially reduced as a result of an etching or pickling treatment with acids.

Kinetics of the Formation of Phosphate Coatings

So far, three different methods have been used for investigating the kinetics of the formation of phosphate coatings, namely: (1) the gravimetric method by quantitative determination of the phosphate quantity deposited per unit time; (2) the electrochemical method of Machu (*11*), by determining the changes with time of the uncoated metal surface; and (3) the radiographic method of Saison (*12*).

In the gravimetric method, several samples of sheet metal, pretreated in the same way and phosphated in the same phosphating bath, were taken out of the phosphating solution at different times and rinsed, and then the weight of the phosphate layer formed was determined by removing the phosphate coating. In the primary pickling reaction, iron is dissolved. However, only a part of this iron appears as tertiary iron phosphate in the phosphate coating in the form of vivianite or phosphophyllite. The other part of the dissolved iron migrates by diffusion into the bath solution. Actually, more iron can be dissolved than would correspond to the phosphate coating deposited. Therefore, if only the change in weight of the phosphated sheet samples with time is recorded, it may happen that one eventually records not a weight gain but a loss in weight (*13*). Therefore, it is necessary in the gravimetric method, when investigating the kinetics of the formation of phosphate coatings, to take the dissolution of iron into account and to directly determine the coating weight.

As can be seen from the gravimetric investigations (*13*), both the largest change in weight of the sheet metal samples and the heaviest deposition of the phosphate coating take place at the beginning of the phosphating process. The weight gains then decrease asymptotically. This behavior is fully comprehensible and readily explained, since the proportion of the reactive metal surface steadily decreases as a result of being coated with the insoluble phosphate coating. Coating formation can then take place only at the available uncoated metal surface. In the subsequent course of the phosphating process, only the pores of the phosphate coating are available for the primary pickling process and the deposition of the tertiary metal phosphates, and the proportion of pores likewise decreases. Finally, the crystallization process practically ceases, since no crystallizable iron or zinc salts are supplied, the intensity of the local current steadily decreases, and the shift of the pH value takes place only to a minor extent.

The electrochemical method for the investigation of the kinetics of the formation of phosphate coatings by Machu (*11*) is based on the fact that the free, still reactive and uncoated iron surface is determined after different periods of the phosphating treatment by the electrochemical method for testing the porosity by passivation (*14*). In this method, several cylindrical electrolytic iron samples are phosphated at the same time, in the so-called "protected form," after being ground on a 00 grit disk. At various time intervals, samples are taken out, rinsed in cold water, and anodically passivated in N-sodium sulfate solution, with a platinum wire net as cathode. The area of the pores, F_A, can then be calculated by means of the initial current density i_0, in amperes per square centimeter, and the time of passivation, t_p, in seconds, according to the so-called surface-coating law (*14*):

$$\log F_A = \frac{\log t_p - \log B + n \log i_0}{n}$$

where B and n are constants.

In Fig. 2, for example, the change of the free and uncoated iron surface, F_A, with time after a phosphating treatment in a zinc phosphate solution free from accelerators, having 70 points at 95°C, is indicated as curve A. In the same Fig. 2, the change of the coating degree with time for a zinc phosphate solution of 40 points containing nitrate and nitrite as accelerators at 75°C is also indicated, in curve B (*14*). As can be seen from these curves, the phosphating process in the zinc phosphate solution which is free from accelerators takes place very slowly. The coating formation is finally finished in about 30 minutes at 95°C. In the zinc phosphate solution which is accelerated with nitrate and nitrite, a sufficient coating formation takes place in only 3 to 4 minutes, in spite of the low temperatures. The coating formation does not take place in a linear fashion, but initially very fast, and then slowly decreases with time.

If it is assumed that the iron surface is free from any phosphate coating or other coating at the beginning of coating formation—that is, at time $t = 0$—the rate of formation of the phosphate coating is fastest at the beginning of

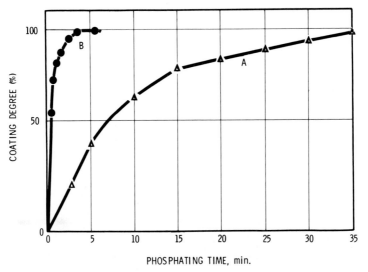

FIG. 2. Change in the coverage in a zinc phosphate bath. Curve A at 95°C and 70 points, without accelerator; curve B at 75°C and 40 points, with nitrate and nitrite as accelerators.

the coating-forming process, when the anode area is greatest. The rate of formation of the coating decreases as the anode surface becomes smaller and the cathode surface (that is, the phosphate coating) becomes larger. The rate at any time is proportional to the anode surface present in each case. On the basis of these assumptions, a law was derived by Machu (*14*) concerning the formation of phosphate coatings and similar conversion coatings on metals. This law is as follows:

$$k = \frac{2.3}{t} \log \frac{F_{A_0}}{F_A}$$

Thus, the factor which determines the rate of reaction in the formation of the phosphate coating upon metal is therefore the ratio of anode surface which was initially present, F_{A_0}, to the anode surface at any given moment, F_A. The rate of formation of the conversion coating, which takes place only at the local cathodes, is therefore proportional to the extent of the free reactive metal surface, since this area is decisive for the magnitude of the local current on the basis of the primary pickling reaction.

It follows, therefore, that the driving force in the formation of the phosphate coating is the dissolution of the metal at the local anodes—hence the primary pickling reaction. Since the discharge of hydrogen ions and the rise of the pH value associated therewith at the local cathodes also depend on the magnitude of the local anodes and the local current, the formation of the phosphate coating is governed primarily by the anodic reaction.

If the rate of formation of the coatings is determined by the Müller-Machu electrochemical method for testing the porosity in phosphating solutions containing accelerators, it is found, as is to be expected, that the free uncoated area of the pores disappears faster with the more effective accelerator.

If the constant k is calculated for the coating-forming processes in phosphating solutions containing accelerators, it is found that these constants pass through a maximum. The factor which determines the rate of coating formation is then no longer only the anode surface present at any moment. Very probably in this case other factors, such as processes taking place at the local cathodes—for example, a depolarization of the hydrogen discharge, a greater number of nuclei, or a faster formation of nuclei at the local cathodes—also influence the rate of formation.

The radiographic x-ray method of investigating the kinetics of the phosphating processes is theoretically based on the fact that the changes in the intensity of a characteristic x-ray of the resulting compound are determined. Saison (12) found, by using the radiographic method, and in agreement with Machu's investigations, that coating formation takes place initially very fast. However, the rate of coating formation, as well as the coating weight, is retarded by the phosphate coating formed. Machu's determination of the free iron surface and the experiments conducted by Saison showed that the coverage of the surface shows deviations about a mean value. The curves for coating formation are practically identical, irrespective of whether they are arrived at by the radiographic or the electrochemical method.

As can be seen from Saison's (12) data in Fig. 3, the variations in the coating thicknesses are due to a redissolution and reprecipitation of the phosphates. Figure 3 also shows that the thickness of the phosphophyllite coating always reaches its maximum sooner than that of the hopeite coating. Therefore, the iron-containing phosphophyllite begins to precipitate first. However, both types of crystals always participate in the construction of the coating.

It is seen, therefore, that all three methods of investigating the kinetics of the formation of phosphate coatings give a similar picture of these processes.

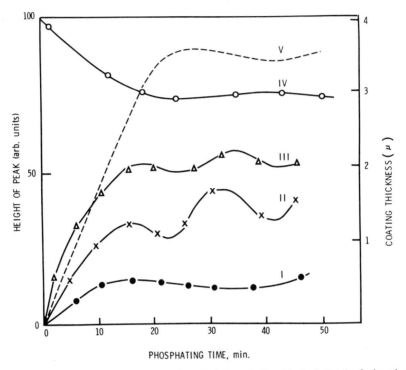

FIG. 3. Typical kinetic curves for a zinc phosphate bath. (I) Peak (004) of phosphophyllite. (II) Peak (241) of hopeite. (III) Peak (040) of hopeite. (IV) Peak (110) of the ferrous substrate. (V) Thickness, deducted from peak (110) of iron (*12*).

The Part Played by Accelerators

It is remarkable that substances that are chemically very different, such as oxidizing agents, reducing agents, metal salts, the cation of which is more noble than the base metal, organic compounds, and inhibitors, in like manner may have an accelerating effect on the phosphating processes. In the case of oxidizing agents, the explanation of the accelerating effect is relatively simple, since it consists primarily in a depolarization of the discharge of the hydrogen ions. Actually, the visible evolution of hydrogen during the phosphating process disappears almost completely in phosphating solutions containing oxidizing agents. Therefore, the cathodic process of a discharge of hydrogen ions and, hence, the coating formation are favored by oxidizing agents.

In the case of reducing agents, metal salts such as copper compounds, organic substances such as nitroguanidine, and some inhibitors, the accelerating effect also consists in a promotion of the cathodic process. Possibly the accelerators have a promoting influence on coating formation also, in so far as they favor the formation of heterogeneous surface conditions and thus increase the number of nuclei. This also follows from the fact that the phosphate coatings produced are the finer-grained, the higher is the concentration of the oxidizing agent.

Also, the addition of metal salts, such as copper compounds, leads to the formation of heterogeneous surfaces. The copper is electrochemically deposited in very finely divided form on the iron surface and results in a surface having areas of different potentials. Only a few milligrams of copper per liter are sufficient for accelerating the phosphating process, since it is necessary only to increase the number of areas having different potentials. The very finely divided copper nuclei on the iron surface then act as crystallization nuclei. If the copper concentration is too high, a homogeneous copper layer is formed, which interferes with the phosphating process.

One of the most favorable factors for coating formation is the facilitation of diffusion processes of the soluble reactants associated with spraying, brushing, and rolling methods of applying the phosphating solution. Similarly, coating formation takes place faster, the more concentrated the baths are, since then the supersaturation, which is of decisive importance for the crystallization, is reached more rapidly. Of course, a rise in the temperature will also accelerate the rate of the coating formation.

The pH value of the baths is important in so far as supersaturation is reached more rapidly at higher pH values of the solution, and then crystallization can begin earlier. Also, the number of nuclei increases with increasing pH values. This leads to the effect that the coatings will be the finer-grained, the higher are the pH values of the baths. While hot phosphating solutions operated at temperatures higher than 50°C generally have a pH value of about 2, cold phosphating baths at temperatures below 50°C are operated at pH values of about 3, and noncoating-forming alkali metal phosphate baths are operated at higher pH values of about 4 to 6. With increasing pH values, the grain size of the phosphate crystals is strongly decreased. Coatings produced with cold phosphating solutions are always very fine-grained and thin. Iron phosphate coatings are practically amorphous to microcrystalline.

Processes in the Formation of Phosphate Coatings. On the basis of the electrochemical interpretation of the phosphating processes, coating formation in a zinc or manganese phosphate bath can be explained as follows. The primary pickling reaction starts at the existing active centers of the metal surface having a more negative potential. This primary pickling reaction is more extensive and intensive, the more nuclei are present and the greater are the potential differences and energy levels, respectively, of the various areas of the metal surface. Initially, crystal nuclei are formed by deposition of primarily vivianite and phosphophyllite, and iron–hureaulite, since the phase boundary layer contains more primary iron phosphate than zinc or manganese phosphate in dissolved form, as a result of the primary pickling reaction of the bath with iron. This deposition takes place in accordance with electrochemical processes—only at the local cathodes, at which the pH value has been shifted upward by discharge of hydrogen ions. This layer nuclei is very thin and does not cover the iron or zinc in a uniform manner, but only incompletely.

Since the diffusion processes take place slowly, while the electrochemical processes and the crystallization from the supersaturated solution in the presence of crystal nuclei occur very rapidly, the pH value rapidly increases at the local cathodes. If there is insufficient primary iron, zinc, or manganese phosphate, there may even be an excess of hydroxyl ions. Iron hydroxides will then be precipitated in the phase boundary layer. This iron hydroxide precipitation is particularly pronounced in alkali phosphate solutions.

The higher the concentration of iron in the solution, the more iron the first crystallization layer can contain in the form of vivianite and phosphophyllite. Thinner layers are also more concentrated than thicker layers with respect to iron.

The continued dissolution of iron at the local anodes and the steady rise of the pH value, or the already reached high concentration of hydroxyl ions, will then lead to the further precipitation of insoluble, tertiary heavy metal phosphates at the local cathodes, which heavy metal phosphates thus participate in the coating formation.

Since only minor amounts of primary iron phosphate are supplied by diffusion from the metal surface, the amount of zinc or manganese phosphate crystals which crystallize out will steadily increase, since these substances are already present in the phosphating solution. However, the crystals still contain some iron.

Thus, the cause of the crystallization is a supersaturation of the solution in the vicinity of the metal surface, due to the rise of the pH value at the cathodic areas. The direct driving force for this rise in the pH value is the primary pickling reaction, which still goes on.

Thus, crystallization of zinc or manganese phosphate must be attributed only to the rise of the pH value in the phase boundary layer as a result of electrochemical processes. However, as time passes, the iron surface which is still available for this pickling attack steadily decreases, as a result of its becoming covered with insoluble heavy metal phosphates. Thereby, the electrochemical processes and the intensities of the local currents also decrease, and less and less hydrogen ions will be discharged. The further rise of the pH value at the local cathodes will then be slower and less significant. Finally, the deposition of crystallized, insoluble trizinc phosphates and, hence, the coating formation are gradually stopped.

If the zinc phosphate solution also contains primary calcium phosphate, the coating formed will contain scholzite, in addition to phosphophyllite and hopeite. Scholzite also crystallizes in insoluble form as the pH value rises.

In the noncoating-forming alkali metal phosphate solutions, the rise of the pH value in the phase boundary layer as a result of the said electrochemical processes associated with the primary pickling reaction is also to be regarded as the cause of the deposition of the protective coating. However, the attack of the weakly acidic alkali metal phosphate solution having a pH value of about 4.0 to 5.5—that is, a pH value which is already rather high per se—on the iron will take place more slowly and not so intensively as is the case in the more strongly acidic zinc phosphate baths. However, the alkali metal phosphate solution does not contain any heavy metal cations which could be used for the deposition of trimetal phosphates. Only the ferrous ions formed in the pickling reaction can build up the coating. Similarly as with the coating-forming phosphating baths, at first a very thin crystal layer of vivianite is formed around the nuclei. Since the pH value of the solutions is already rather high, a slight shift of the pH value toward higher values will then suffice for further coating formation from the ferrous phosphate formed in the pickling reaction to take place. Therefore, crystallization takes place very rapidly and leads only to very small microcrystals.

Since the corrosive attack at the higher pH values of the bath is rather slight, only a small quantity of ferrous phosphate is formed. The excess supply of hydroxyl ions will then even lead to a precipitation of iron hydroxide, which is converted into ferric oxide on drying in air. Iron phosphate coatings always contain larger amounts, about 30 to 50%, of ferric oxide.

Thus, coating formation is to be regarded as an electrochemical process in the coating-forming heavy metal phosphate solutions as well as in the noncoating-forming alkali metal phosphate baths. In the coating-forming solutions, crystallization of the tertiary heavy metal orthophosphate takes place primarily as a result of the rise in the pH value at the local cathodes from the primary heavy metal orthophosphates which are present in the solution. In the noncoating-forming alkali metal phosphate solutions, which are practically free from iron and other heavy metal phosphates, the rise of the pH value will even lead to the deposition of iron hydroxide. Since experience has shown that the crystals become smaller, the higher is the pH value of the solution from which crystallization or deposition takes place, the coatings become finer-grained, the higher is the pH value of the baths initially at the beginning of the coating-forming process. Therefore, the size of the crystal grains decreases steadily in the order of hot to cold heavy metal phosphating baths, and finally to the alkali metal phosphate solutions.

In the later course of the coating-forming processes, there is also a redissolution and reprecipitation of metal phosphates, as was shown by Saison (*12*) and Jimeno and Arevalo (*15*), respectively. Redissolution and reprecipitation of the phosphate crystals are also of practical importance, since these processes also cause a recrystallization of the phosphate crystals. The crystals thereby become more even and uniform.

Summary

The kinetics of the phosphating processes is unique in certain respects, as a result of the electrochemical character of these processes. Coating formation as such proceeds only at the local cathodes of the metal surface, while the rate of formation of phosphate coatings is determined primarily by the processes taking place at the local anodes. Since phosphating is also associated with crystallization processes, the factors which influence the rate of coating formation and the kinetics of the processes include the number of nuclei and the rate of formation of the nuclei, and thus also the mechanical and chemical pretreatment of the metal surfaces. Any steps taken to increase the number of nuclei, by producing surface conditions as heterogeneous as possible, having surface areas of different potentials, will favor the formation of fine-grained and thin phosphate coatings, which are particularly valuable properties from a technical standpoint.

The kinetics of the formation of phosphate coatings is also influenced by the self-passivation of the coating formation, in accordance with the self-passivation law of Müller-Konopicky. The formation of phosphate coatings from aged baths which are free from accelerators was found to follow certain laws, and it could be demonstrated that the driving force for coating formation is the primary pickling reaction and the dissolution of the metal at the metal anodes.

The manner in which the coatings are formed with time was found to be identical when investigated by the gravimetric, the electrochemical, and the radiographic methods for testing the kinetics of the coating-forming process. Coating formation proceeds most rapidly at the beginning and then decreases gradually asymptotically with increasing coverage of the free metal surface as a result of the insoluble phosphate coating.

The rise of the pH value at the local cathodes is of decisive importance for the deposition of the phosphate coatings, since this rise in the pH value leads to a supersaturation of the solution with regard to heavy metal phosphates. This leads to a crystallization of the tertiary metal phosphates at the crystal nuclei of the metal surface. The rise of the pH value at the local cathodes also brings about the growth in the thickness of the coatings.

In the later course of the phosphating processes, there is also redissolution and reprecipitation of the heavy metal phosphates, which leads to recrystallized, more homogeneous crystal grains.

References

1. W. Machu, *Korrosion Metallschutz*, **17**, No. 5, 157 (1941).
2. U. S. Patents 2,132,438 and 2,132,439 (1939).
3. W. Machu, *Arch. Metallk.*, **3**, No. 8, 278 (1949).
4. W. J. Müller and K. Konopicky, *Monatsh. Chem.*, **52**, 463 (1929).
5. W. Machu, *Chem. Fabrik*, **13**, 461 (1940).
6. W. Machu, *Korrosion Metallschutz*, **18**, 89 (1942).
7. A. Neuhaus and M. Gebhardt, *Werkstoffe Korrosion*, **17**, No. 7, 567 (1966).
8. W. Machu, *Werkstoffe Korrosion*, **14**, 566 (1963); *Rev. Ind. Chim. Belge*, **T28**, No. 3, 1 (1963).
9. E. Ghali, J. Voeltzel, and A. Hache, *Compt. Rend.*, **Ser. C263(17)**, 1001 (1966).
10. G. D. Cheever, *J. Paint Technol.*, **39**, No. 504, 1 (1967).
11. W. Machu, *Arch. Metallk.*, **3**, No. 6, 203 (1949); No. 1, 1 (1949); *Österr. Chemiker-Ztg.*, **50**, No. 7, 148 (1949).
12. J. Saison, Thèse à la Faculté des Sciences de l'Université de Paris (1962).
13. W. Machu, *Chem. Fabrik*, **13**, 461 (1940); *Korrosion Metallschutz*, **18**, No. 3, 89 (1942).
14. W. Machu, *Korrosion Metallschutz*, **20**, 1 (1944).
15. E. Jimeno and A. Arevalo, *Corrosion Anti-Corrosion* **6**, 51 (1958).

Discussion

FOWKES: In the precipitation of zinc phosphate and the formation of a tightly adherent insulating film, obviously it becomes more noble. The surface is no longer an anodic area when the phosphate is precipitated in the region of higher pH. This, perhaps, is some form of electrodeposition of this material.

MACHU: No, it is not an electrodeposition, but it is a conversion with the iron surface converted into a phosphate compound *in situ*. In the formation of a phosphate coating, no electrodeposition takes place.

FOWKES: I see. You mean it is not iron in the solution that is reacting with the acid zinc phosphates?

MACHU: Yes, there is a reaction, but it occurs in a certain way. All ferrous ions, in a very short time, are precipitated. Most of the coating consists of zinc phosphate and not iron phosphate; only a very little part of the coating is iron phosphate. Only the immediate contacting layers are rich in iron, but the upper ones are rich in zinc.

MICHAELS: I presume that the propagation of the crystalline or amorphous phosphates must be nearly two-dimensional; that is, nucleation takes place on the surface, and the process tends to propagate laterally along the surface, rather than out into solution, because the pH is highest near the surface. Is this correct?

MACHU: Yes. The growth is more lateral than vertical. However, when there are only a few nuclei, you have pronounced vertical growth too.

H. MARK: In the surface layer, are there actually mixed crystals of iron and zinc phosphate, or are there only iron phosphate crystals and zinc phosphate crystals? Do mixed crystals exist?

MACHU: Yes, but the amount of iron phosphate in the coating is very, very small because in modern phosphating baths a very high concentration of oxidizing agents is present which results in the precipitation of ferric phosphate. The insoluble ferric phosphate then collects in the phosphating bath. Also, some of the iron is immediately consumed in the crystallization of phosphophyllite which is $Zn_2Fe(PO_4)_2 \cdot 4H_2O$—that is, monoferrous zinc phosphate. Phosphophyllite and hopeite ($Zn_3(PO_4)_2 \cdot 4H_2O$) are monoclinic and rhombic, respectively, and they are so intimately mixed with each other that it is impossible by means of a normal microscope to distinguish between them on the metal. It can be done by x-ray diffraction.

H. MARK: Maybe that is just the reason for the firmness of the zinc phosphate layer in the coating.

MACHU: The lattice parameters are very, very close to each other, so they can grow into each other very easily. Only in the color of the coating do you have a sign of whether there is more or less iron. If it is dark gray, there is more phosphophyllite and more iron, and if it is light gray, there is more hopeite.

P. MARK: You say that this process begins at certain active sites. Do you know what these are?

MACHU: The grain boundaries are the most active. Dr. Cheever has investigated this and found that there are 10^5 to 10^6 such sites per square centimeter. Practically every iron crystal is a base for a phosphate crystal.

P. MARK: Does a particular surface plane in the iron crystals work better than others? Supposing one did this with a single crystal of iron?

MACHU: Neuhaus has reported that there is no difference in the epitaxy when there is a difference in the surface plane.

FOWKES: How can you distinguish between sites for nucleation of crystal growth and local cathodic regions?

MACHU: You cannot see, of course, the substrate on which a certain crystal grows, but you can count the number of crystals, say, after a certain treatment. If you acid pickle, you have only 10^2 or 10^3 sites per square centimeter. If you treat initially with titanium phosphate, you have 10^6 sites per square centimeter. One treatment decreases and the other treatment increases the number. If a sandblasted surface is kept in air for 15 minutes or dipped in water, many of these active centers react with water or oxygen and become deactivated. These active centers must be present on entering the phosphating bath.

CHEEVER: We took steel, phosphated it, and mounted and polished it by conventional metallographic techniques, and we examined across the zinc phosphate coating with the electron probe for iron and zinc. And, indeed, we found that very close to the steel interface we had a high concentration of iron in the phosphate coating, but away from the iron interface we saw mostly zinc. This is exactly what Professor Machu was saying. In the electron probe, there is another picture which comes from the back-scattered electrons, and this gives the picture of the entire system; therefore, we know where the iron interface is and where the zinc phosphate coating begins. To comment about the effect of grain boundaries on the phosphating process, through conventional polishing techniques, we found approximately, on the average, that 75% of the zinc phosphate crystals began to nucleate at the grain boundaries of the steel.

MAURER: In the what might be called conventional phosphatizing baths, I certainly agree that strong acids or strong alkalies cause larger crystal structures. How do you explain the fine-grained crystal structure obtained with pickling or with strong alkali ahead of the calcium-modified zinc phosphate process?

MACHU: With calcium, there is a disturbance of the zinc phosphate lattice during crystallization which gives a much smaller crystal size, about one twentieth of that from the normal zinc phosphate baths.

MAURER: Why does the intermediate pH range (for example, 7 to 10) pretreatment result in larger crystal structure than either extreme?

MACHU: In a slightly alkaline solution, there is less reaction with the active centers. The effect is small at pH 7 to 9. Only strong alkaline solutions give the blocking of the active centers.

EIRICH: Pertaining to the question of adsorption on phase boundaries and the adherence of the deposited material, would it not be also possible that one gets, apart from deposition and growth on the surface, a penetration by growth into the underlying crystallites? Most surfaces are rich in defects, and one must count on ion exchange by diffusion of zinc and other ions into the original lattice. One might even get a formation of phosphate in depth below the original surface, rather than growth only from the original surface outward. Such processes may be the main source of the observed strong adherence of the phosphate coating.

MACHU: Looking at the coating-steel cross section, there is never a uniform boundary between metal and phosphate phases. There is an anchoring of the phosphate into the metal phase, and this is why one usually does not calculate the coating thickness.

ZIMMT: If the deposition occurs on the cathodic areas, why does the anodic area control the kinetics of coating formation?

MACHU: At the beginning of the process, about 99% of the area is anodic. In fact, on the surface, only 1% of the area is cathodic, but the current is the same, both at the anodes and at the cathodes. The current density is very high at the cathodic areas because the area is small; therefore, the intensity of the pH rise is very rapid at the cathodes. Pickling starts immediately, and within a few seconds the coating is formed. Although the current at the cathodes and anodes is the same, the current density at the cathodes is much higher.

LAUKONIS: I would like to comment on several points which have been brought up in the discussion of this paper. First, I want to comment on the suggestion of phosphating single crystals to study orientation effects. I have phosphated several different kinds of iron single crystals—single crystal iron whiskers and thin sheets of strain-anneal grown single crystals of iron. To date my phosphating studies are inconclusive and probably for reasons that Dr. Machu has pointed out in his presentation, namely, that single crystals have an entirely different history than do polycrystalline samples. For example, both iron whiskers and strain-anneal single crystals are grown in a hot hydrogen environment. My paper will indicate the existence of a hydrogen effect on phosphating. Also, in the case of iron whiskers the ratio of edges to plane surface is unusually large, so that an "edge effect," which is probably one example of an "active site," must be unraveled in the interpretation of their phosphating behavior. Further, strain-anneal single crystals frequently must be electropolished before phosphating, and the effects and variables of electropolishing must be incorporated into any explanation of phosphating behavior. Other complications also exist, but without further elaboration I am making the point that great care should be exercised in comparing the phosphating behavior of single crystals to that of polycrystalline materials.

Second, the same considerations suggest that the role of the grain boundary in the nucleation of phosphate crystals must be examined closely. For the practical surfaces which Dr. Cheever studied, 75% of the phosphate crystals were nucleated at grain boundaries. In the phosphating of large-grained strain-annealed iron sheets I have found a few grain boundaries to be extremely active in the nucleation of phosphates, while most are completely inactive. In addition to other differences in the histories of these materials the strain-annealed sheets are much purer than the cold-rolled polycrystalline sheets, and these differences undoubtedly account for the differences observed in phosphating. However, I would expect that conventional materials could be given heat or chemical pretreatments to either depress or enhance the activity of grain boundaries, so that any measure of activity should be closely coupled with the history of the surfaces that were phosphated.

EIRICH: It has been mentioned that phosphates have a tendency to form amorphous compounds. Thus, one might form a phosphate glass over the top of your pickled surface. I could imagine that the interaction of the next layer, let us say the polymer that is laid down over this would be very much influenced by the nature of the phosphate, whether it is glassy or not. Do you know of any experiments which describe how much amorphous material is there, and what its effect is on subsequent coatings?

MACHU: There are two different things to take into consideration. First, iron phosphate coatings produced from alkaline phosphate baths are more or less amorphous. Second, there are processes which employ hexametaphosphate as a constituent in the bath which give very thin coatings of, say, ferrous phosphate, and either have dissolved complex chelating calcium phosphates or some strontium phosphates. If you now add a chelating agent like hexametaphosphate to the same bath, you decrease the coating crystal size, and you decrease the coating weight very much. There is a process known in the art which adds a small amount of such a chelating agent to produce thin, fine crystalline coatings, but you must be very careful. If you add more than one gram per liter, for instance, you no longer get a phosphate coating, because it is completely dissolved.

Wetting of Phosphate Interfaces by Polymer Liquids

G. D. CHEEVER

General Motors Research Laboratories
Warren, Michigan

Metals that find application in the automobile, appliance, and building trade industries often are treated with a series of solutions which both clean and react with the metal surface to form a new inorganic interface. The new interface, generally known as a conversion coating, may be composed of metallic phosphates, oxides, chromates, or oxalates, depending on the chemical composition of the reacting solution. In practice, protective polymer coatings are applied over the conversion coating.

Although the chemical compositions of conversion coatings may be very different, their functions are the same—namely, to protect and improve the performance of the metal substrate. One major property which must be developed in order for the conversion coating–polymer system to perform satisfactorily is effective adhesion between polymer and conversion coatings. It is necessary that the polymer film wet and spread over the inorganic interface and displace adsorbed gases and organic contaminants which are always present; therefore, it would be valuable to consider some of the basic principles through which liquids wet solid substrates. In these discussions, "wetting" will be defined as that process which occurs when a solid substrate and a liquid come in contact and form a solid–liquid interface (*1, 2*).

Fundamentals of Wetting

Smooth Surfaces. Since Zisman (*3*) and other workers (*1, 2, 4–10*) have extensively reviewed wetting phenomena on smooth surfaces, it suffices to summarize these earlier publications. On solid substrates, three types or degrees of wetting have been observed. They are simple liquid attachment, partial or equilibrium wetting, and complete wetting or spreading. These regions or types of wetting are defined by mathematical relationships involving γ_{SV}, γ_{SL}, and γ_{LV}, where γ_{SV} is the solid–vapor interfacial tension, γ_{SL} the solid–liquid interfacial tension, and γ_{LV} the liquid–vapor interfacial tension. Gray (*6*) gives the requirements for the three regions of wetting as follows.

Type of Wetting	Requirement	
Attachment	(a) $K = \gamma_{LV} + \gamma_{SV} - \gamma_{SL} > 0$	(1)
	(b) $B = \dfrac{\gamma_{SV} - \gamma_{SL}}{\gamma_{LV}} > -1$	(2)
Equilibrium	(a) $K > 0$	(1)
	(b) $B > -1$	(2)
	(c) $B < 1$	(3)
Spreading	(a) $K > 0$	(1)
	(b) $B > -1$	(2)
	(c) $B > 1$	(4)

where K = the attachment coefficient (6).
B = the ratio of the interfacial tensions, the value of which determines the region of wetting.

The expression for K is the same as that for the work of adhesion, W_{SL}, which is the energy required to separate the liquid from 1 cm² of solid. In each region of wetting, K is positive, and B is greater than -1. When B is less than 1, the liquid will exhibit equilibrium wetting, and when B is greater than 1, the liquid will spread over the solid; therefore, the ratio of the surface tensions as expressed by the quantity B determines whether the attached liquid will give equilibrium or spreading wetting.

In attachment wetting, when the respective interfacial tensions generate a positive value for K, liquid wets the solid and becomes attached to the surface. For most liquids in air, K is positive—that is, most solids are wetted by liquids; however, with liquids of extremely high surface tensions, for example liquid metals or molten salts (11), K is negative. In these particular cases there is no attachment, and complete nonwetting of the solid by the liquid occurs.

In partial or equilibrium wetting, the drop rests on the solid substrate, and equilibrium is established between the solid, the liquid, and the vapor according to Young's equation:

$$\cos \theta = \dfrac{\gamma_{SV} - \gamma_{SL}}{\gamma_{LV}} \qquad (5)$$

where θ = the contact angle. Comparison between equation 5 and equation 2 shows that, in the equilibrium region of wetting, the value of the cosine of the contact angle is the same as B. The equilibrium region of wetting investigated by Zisman (3) and co-workers led to the concept of the critical surface tension of wetting, γ_c, of the solid. Studies of contact angle behavior of pure liquids on both low-energy and high-energy solid surfaces to which organic substances had been adsorbed established a rectilinear relation between the cosine of the contact angle, θ, and the surface tension of both homologous and nonhomologous series of organic liquids. The physical meaning of γ_c is that liquids spread on low-energy solids when γ_{LV} is less than γ_c. With pure liquids, critical surface tensions of wetting have been used to compare and characterize the wettabilities of a wide variety of surfaces (3). Also, in more complex systems, Levine et al. showed the relation between the critical surface tension of polymers to adhesion (12) and between wettability of surface-treated metals and the effect on lap shear adhesion (13).

In the third region of wetting (spreading), liquid spreads over the surface to form a continuous film. A liquid will spread over a solid if the work of adhesion, W_{SL}, of the solid for the liquid is greater than the work of cohesion, W_{LL}, of the liquid for itself. The difference between these two quantities is Harkins' spreading coefficient, S (14).

$$W_{SL} = \gamma_{LV} + \gamma_{SV} - \gamma_{SL} \tag{6}$$

$$W_{LL} = 2\gamma_{LV} \tag{7}$$

$$S = W_{SL} - W_{LL} = \gamma_{SV} - \gamma_{SL} - \gamma_{LV} \tag{8}$$

Thus, for spreading, S must be greater than 0. A number of workers have investigated dynamic spreading of liquids over smooth solids (15–19).

Rough Surfaces. Wenzel (20) showed that the effect of surface roughness is to increase the departure of the contact angle, θ, found on smooth surfaces from 90° (7). Mathematically,

$$\cos \theta' = f \cos \theta \tag{9}$$

where $\theta =$ the contact angle on a smooth surface of the solid.
  $\theta' =$ the apparent contact angle on the rough solid.
  $f =$ the roughness coefficient or the ratio of the actual area to the projection area of the solid.

On a smooth surface, if θ is less than 90°, roughening the surface will result in θ' being even smaller; that is, the extent of wetting is increased. Therefore, the degree of wetting could even go from that of equilibrium to complete spreading. However, if θ is greater than 90° on a smooth surface, roughening the surface will increase the contact angle θ' still further and decrease the degree of wetting. In the case of rough surfaces, the conditions for three degrees of wetting become (6)

$$K = \gamma_{LV} + f(\gamma_{SV} - \gamma_{SL}) \qquad (10)$$

$$S = f(\gamma_{SV} - \gamma_{SL}) - \gamma_{LV} \qquad (11)$$

$$B = \frac{f(\gamma_{SV} - \gamma_{SL})}{\gamma_{LV}} \qquad (12)$$

From equations 10 to 12, it is seen that, when the difference $(\gamma_{SV} - \gamma_{SL})$ is positive, K, S, and B will increase when the solid surface is roughened.

A second factor on rough surfaces results from the action of capillary forces. Because of surface tension, there is a pressure difference generated whenever a concave curvature is developed in a liquid–vapor interface such as that achieved by a liquid flowing into a crack or a pore in the rough surface. The pressure is greater on the concave side of the interface, and this pressure difference, ΔP, is given by the Young-Laplace equation,

$$\Delta P = \gamma_{LV} \left(\frac{1}{r_1} + \frac{1}{r_2} \right) \qquad (13)$$

where r_1 and r_2 are the principal radii of curvature.

When the contact angle of the liquid on the walls of the capillaries is less than 90°, the action of capillary forces will result in liquids spreading more quickly on rough surfaces than on smooth ones. For example, Bascom et al. (*18*) reported that random surface scratches increased as much as 50% the spreading rate of some liquids and that open capillaries filled well ahead of the diffusional advance of the primary film. Similarly, Cottington et al. (*21*) reported that some liquids spread along capillaries in the form of scratches produced during polishing and that liquids flowed in these capillaries while they were nonspreading on a plane surface. When the liquids reached dead ends of the scratch structures, flow slowed or stopped. These discussions show that capillary structures can markedly affect spreading of liquids and can either help or hinder the degree of wetting desired. A number of workers have investigated penetration of polymeric materials into solids having capillary structures (*2, 6, 22–25*).

Although conversion coatings are used industrially as substrates for polymers, the fundamental wetting processes occurring during the application of the polymer liquids have not been reported. Because the physical and chemical processes occurring during polymer film formation are so complex, it is appropriate to begin with simple polymer liquids of known chemical structure and study dynamic wetting on controlled zinc phosphated steel conversion coatings.

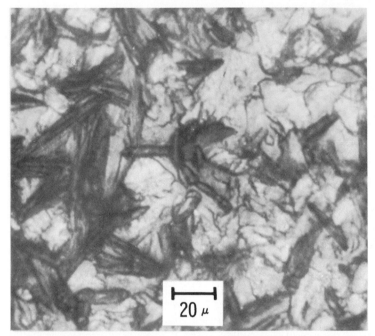

FIG. 1. Nitrocellulose replica of zinc phosphate coating taken in a light microscope.

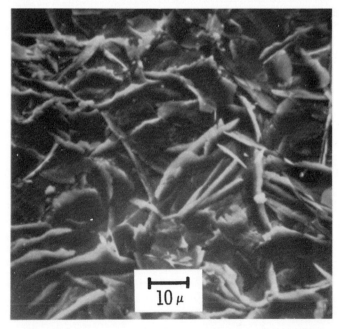

FIG. 2. Zinc phosphate coating taken in a scanning-beam electron microscope.

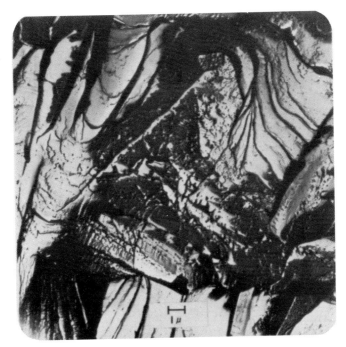

FIG. 3. Carbon replica of zinc phosphate coating taken in a transmission electron microscope.

This experimental study has two objectives:
1. To construct a mathematical and physical model which describes the observed dynamic wetting and spreading of polymer liquids on zinc phosphated steel interfaces.
2. To use the model to establish the relative wettability of zinc phosphate coatings by polymer liquids.

Derivation of Model

Microscopic examination of zinc phosphate coatings on steel showed that these crystalline interfaces are extremely rough and porous. Micrographs of typical zinc phosphate coatings are shown in Figs. 1, 2, and 3; Fig. 1 is a nitrocellulose replica taken in a light microscope, Fig. 2 is the actual coating taken in a scanning-beam electron microscope, and Fig. 3 is a carbon replica taken in a transmission electron microscope by means of a nitrocellulose replica technique developed by Scott and Turkalo (26). Figures 1 and 2 show that the zinc phosphate coatings are composed of crystals at surface populations of approximately 10^5 to 10^6 per square centimeter. In addition, further properties are shown in Fig. 3, where the high resolution of the electron microscope revealed that each dark line is a slit-like crevice into which the nitrocellulose replicating solution flowed during the preparation of the replica.

Since zinc phosphate coatings are porous, in the derivation of the mathematical model the zinc phosphate coating is treated as another example of a capillary matrix or capillary bed in which the capillaries assume the particular form of slits. Washburn (27) and Rideal (28) in two classical papers laid the foundation for study of liquid flow in capillary systems. For cylindrical open-ended capillaries in the absence of gravitational or kinetic forces, the Washburn-Rideal equation is

$$l^2 = \frac{\gamma_{LV} r \cos \theta t}{2\eta} \tag{14}$$

where l = the distance traveled.
r = the effective radius of the capillaries.
t = the time.
γ_{LV} = the liquid surface tension.
η = the liquid coefficient of viscosity.
θ = the contact angle of the liquid on the walls of the capillaries.

Since that time, many papers (29–40) have appeared in which the Washburn-Rideal equation has been used to study liquid flow under capillary pressure both in tubes and in porous media.

The starting point for derivation of models describing liquid flow in porous media is Poiseuille's equation which describes the laminar flow of Newtonian liquids in tubes of particular dimensions. For capillaries in the form of slits and without slip factors or correction for the viscosity of air, Manegold et al. (31) have expressed Poiseuille's equation as

$$\frac{dv}{dt} = \frac{2}{3} \frac{d^3 b P}{\eta x} \tag{15}$$

where dv/dt = the volume rate of flow of the liquid.
d = ½ width of the slit.
b = the depth of the slit.
x = the length of the slit filled with liquid.
η = the liquid coefficient of viscosity.
P = the pressure needed to force the liquid through the slit under the given conditions of flow.

As a first approximation, the capillary slits in zinc phosphate coatings are assumed to be open, with the result that atmospheric pressure will not contribute to the forces affecting flow. Also, it can be shown that forces due to changes of the momentum of the liquid can be neglected because of the small dimensions of the slits; therefore, in a single capillary slit, the accelerating forces reduce to those only of capillary forces, which in turn are equal to the retarding viscous forces due to motion of the liquid. In Fig. 4, a spreading

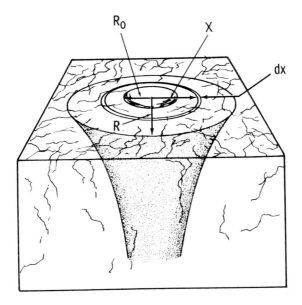

FIG. 4. Spreading drop of polymer liquid on zinc phosphate substrate.

drop of polymer liquid in the zinc phosphate coating is shown schematically. The initial radius of the drop is R_0, and the radius of the drop at time t is R. The technique of Greinacher (39) is used in the following derivation to equate the capillary forces to the viscous flow forces in the band dx at a distance x from the center of the spreading drop; the distance will be integrated from R_0 to R; and then time will be integrated from 0 to t.

Bikerman (23) derived the capillary pressure, P_c, generated by a liquid of surface tension γ_{LV} in a slit of width w as

$$P_c = \frac{2\gamma_{LV} \cos \theta}{w} \qquad (16)$$

For the particular slits in question, equation 16 reduces to

$$P_c = \frac{\gamma_{LV} \cos \theta}{d} \quad (17)$$

The capillary force, F_c, is equal to the capillary pressure times the cross-sectional area of the slit, or

$$F_c = P_c \cdot 2db = 2b\gamma_{LV} \cos \theta \quad (18)$$

Similarly, the viscous force, F_η, is equal to the product of the viscous pressure, P, obtained from Poiseuille's equation (equation 15), and the cross-sectional slit area. First, equation 15 is expressed in a different form to obtain the viscous pressure, P:

$$P = \frac{3\eta x}{2d^3 b} \frac{dv}{dt} \quad (15)$$

or

$$P = \frac{3\eta x}{2d^3 b} \frac{d}{dt}(2dbx) \quad (19)$$

and

$$P = \frac{3\eta x \dot{x}}{d^2} \quad (20)$$

where $\dot{x} = dx/dt$. Finally,

$$F_\eta = P \cdot 2db = \frac{6\eta x \dot{x} b}{d} \quad (21)$$

Since expressions have been derived for the capillary and viscous forces in a single capillary, they now will be totaled throughout the entire capillary matrix.

In the band dx, v_x is the velocity of liquid flow, $\dot{x}$, and n_x is the total number of capillaries in which flow is occurring. Then, the total viscous forces in dx, $dF_{\eta x}$, is

$$dF_{\eta x} = n_x \frac{6\eta b v_x \, dx}{d} \quad (22)$$

At the flow front R, v_R is the velocity of flow, and n_R is the total number of capillaries. It is possible to show that at x

$$n_x = n_R \left(\frac{x}{R}\right) \quad (23)$$

and

$$v_x = v_R \left(\frac{R}{x}\right) \quad (24)$$

It is seen from equations 23 and 24 that it is possible to relate the number of capillaries and flow rate at x with the same quantities at R. Replacing the values of n_x and v_x from equations 23 and 24 in equation 22, we obtain

$$dF_{\eta x} = n_R \frac{6\eta v_R b\, dx}{d} \qquad (25)$$

Then, at R the total viscous forces, $F_{\eta R}$, are found by integrating equation 25 from R_0 to R, or

$$F_{\eta R} = \int_{R_0}^{R} dF_{\eta x} = \frac{6\eta n_R v_R b}{d} \int_{R_0}^{R} dx \qquad (26)$$

and

$$F_{\eta R} = \frac{6\eta n_R v_R b}{d}(R - R_0) \qquad (27)$$

Since the total viscous forces are known, the total capillary forces next will be found.

Let F_1 equal the capillary forces per square centimeter of cross section of capillary matrix, and n_1 equal the number of capillaries per square centimeter of cross section of the matrix. The total capillary forces at R, F_{cR}, are given by

$$F_{cR} = 2\pi R \delta F_1 \qquad (28)$$

where δ = the thickness of the capillary matrix.
n_R = the total number of capillaries at R, which is

$$n_R = 2\pi R \delta n_1 \qquad (29)$$

Now,

$$F_{\eta R} = F_{cR} \qquad (30)$$

That is, the total viscous forces are equal to the total capillary forces. Replacing the value of n_R in equation 27 by its equivalent, equation 29, and then setting equation 28 equal to equation 27, we find that

$$2\pi R \delta F_1 = \frac{6\eta 2\pi R \delta n_1 v_R b(R - R_0)}{d} \qquad (31)$$

Dividing both sides of equation 31 by $2\pi R \delta n_1$ gives

$$\frac{F_1}{n_1} = \frac{6\eta b v_R (R - R_0)}{d} \qquad (32)$$

but F_1/n_1 is the capillary force per capillary (equation 18). Then, equation 32 becomes

$$\frac{6\eta b v_R (R - R_0)}{d} = 2b\gamma_{LV} \cos \theta \qquad (32')$$

Since v_R equals dR/dt, equation 32' can be integrated to give

$$\frac{6\eta b}{d} \int_{R_0}^{R} (R - R_0) \, dR = 2b\gamma_{LV} \cos\theta \int_0^t dt \qquad (33)$$

or

$$\frac{3\eta b}{d} (R - R_0)^2 = 2b\gamma_{LV} \cos\theta t \qquad (34)$$

and finally

$$(R - R_0)^2 = \frac{2}{3} \frac{\gamma_{LV} \, d \cos\theta \, t}{\eta} \qquad (35)$$

Washburn (27) suggested that $\gamma_{LV}/\eta \cos\theta$ measures the penetrating power of a liquid into a porous solid and called it the "coefficient of penetrance" or the "penetrativity" of the liquid. Also, Studebaker and Snow (40) noted that $\cos\theta$ is a measure of wettability of a porous body by a liquid; that is, the smaller the value of $\cos\theta$, the more difficult it is to wet a given substrate with the liquid.

Equation 35 describes the size of the spreading drop in terms of the following entities:

γ_{LV}/η = a property of the spreading liquid which can be measured independently.

d = a measure of the capillary structure or porosity of the zinc phosphate substrate.

$\cos\theta$ = a measure of the wettability of the capillary matrix by the polymer liquid, or, to state it differently, the interaction between the liquid and the porous substrate.

Since both the capillary structure and the wetting processes are extremely complex in zinc phosphate coatings, it is recognized that both d and $\cos\theta$ probably are composed of other terms and that they are only measures of the capillary structure and of wetting.

Concerning the units in equation 35,

$$\eta = \text{coefficient of viscosity in g/cm-sec or dynes-sec/cm}^2$$

$$\frac{\eta}{\rho} = \text{kinematic viscosity} = \nu$$

where ρ = liquid density in g/cm³.

ν = kinematic viscosity in cm²/sec.

$$\frac{\gamma_{LV} \cos\theta}{\eta} = \text{velocity in cm/sec.}$$

Testing Model

Taking logarithms of both sides of equation 35 gives

$$\log(R - R_0) = \frac{1}{2}\log t + \frac{1}{2}\log\frac{2}{3}\frac{\gamma_{LV}\, d\cos\theta}{\eta} \tag{36}$$

or

$$y = mx + c \tag{37}$$

where $y = \log(R - R_0)$.
$m = \frac{1}{2}$.
$x = \log t$.

$$c = \frac{1}{2}\log\frac{2}{3}\frac{\gamma_{LV}\, d\cos\theta}{\eta} = \frac{1}{2}\log A.$$

In this investigation, the quantity A is called the penetration coefficient. From equations 36 and 37, plotting $\log(R - R_0)$ versus $\log t$ should be linear with slope $\frac{1}{2}$ and y intercept $\frac{1}{2}\log\frac{2}{3}(\gamma_{LV}\, d\cos\theta/\eta)$. Since γ_{LV}/η is known, the y intercept then yields both A and $d\cos\theta$.

Experimental Procedure

The steel substrates on which zinc phosphate conversion coatings were formed were obtained from cold-rolled SAE 1010 steel. The formation of zinc phosphate coatings proceeded by a number of steps in which aqueous solutions were sprayed onto the steel surface. First, steel panels were cleaned with a hot alkaline solution followed by a hot-water rinsing step. Then, the acidic zinc phosphating solution was applied. During this step, zinc phosphate crystals formed and grew on the steel substrate. After the zinc phosphating step, the phosphated steel panels were rinsed with cold water and then with a hot dilute chromic acid solution which imparted added corrosion protection to the steel substrate. Finally, the phosphated panels were dried in an oven.

Since zinc phosphate coatings are porous, some organic contamination always will be present. A procedure which obtained zinc phosphate coatings at a constant low level of contamination was developed by placing 1-inch-square pieces of the zinc phosphated steel in a Soxhlet extractor column and extracting with boiling trichloroethylene for 3 hours.

The trichloroethylene extract was evaporated to dryness, and the residue, which was contamination extracted from the zinc phosphated steel pieces, was examined by infrared. It was found that the contamination was either an amine or amide which was similar to corrosion inhibitors placed on steel.

To determine the effect of the trichloroethylene extraction step on the crystal properties of zinc phosphate coatings, the solvent-extracted pieces were examined both by x-ray diffraction and by electron probe techniques. The results are given in Table I. The heights of the most intense x-ray diffraction peaks were measured for hopeite, $Zn_3(PO_4)_2 \cdot 4H_2O$, phosphophyllite, $Zn_2Fe(PO_4)_2 \cdot 4H_2O$, and zinc phosphate dihydrate, $Zn_3(PO_4)_2 \cdot 2H_2O$, and expressed in counts per second (cps). The electron probe measurements for zinc and chromium x-rays also are given in counts per second. The data

TABLE I
Effect of Cleaning with Trichloroethylene

Substrate	x-Ray-Diffraction			Electron Probe	
	$Zn_3(PO_4)_2 \cdot 4H_2O$ (cps)	$Zn_2Fe(PO_4)_2 \cdot 4H_2O$ (cps)	$Zn_3(PO_4)_2 \cdot 2H_2O$ (cps)	Zn (cps)	Cr (cps)
Original	4675	138	325	1705	135
Solvent-cleaned	4713	138	338	1900	87

in Table I show that from solvent extraction there was no significant alteration, either of the crystalline composition of the coatings or of the amount of residual chromium from the chromic acid rinse. The only known effect of solvent cleaning was to remove organic contamination.

Data were obtained from polymer liquids spreading both on the solvent-cleaned substrates and on the zinc phosphate coatings in their original condition, which is the state in which they primarily are used in practice.

The model polymer liquids studied were obtained from the respective manufacturers and were used without additional purification. Drops 1.0 μl in size delivered with an automatic dispenser were placed on zinc phosphated steel samples. Time was started when the drop came in contact with the surface. The diameter of the drop was measured as a function of time with a 7× microscope. All measurements were made in air at 24° ± 1°C, and samples were covered with a glass petri dish during spreading of long duration to prevent airborne contamination.

The viscosities of the liquids were measured with a Cannon-modified Oswald kinematic viscometer, and the surface tensions were obtained with a duNouy tensiometer to which appropriate corrections were made (*41*). The densities were measured in a pycnometer. These measurements similarly were performed at 24° ± 1°C.

TABLE II
Identification and Physical Properties of Liquids

Liquid Number	Identification	γ (dynes/cm)	ρ (g/cm^3)	ν (cs)	η (cp)	γ/η (cm/sec, $\times 10^{-2}$)	η/γ (sec/cm, $\times 10^3$)
	Polyalkylene glycol*						
1	LB65	32.8	0.955	17.1	16.3	2.01	4.98
2	70% LB65 + 30% LB135†	33.2	0.959	23.8	22.8	1.45	6.90
3	30% LB65 + 70% LB135	33.6	0.966	32.4	31.3	1.07	9.35
4	LB135	34.1	0.975	47.8	46.6	0.73	13.70
5	LB165	34.4	0.979	59.0	57.8	0.60	16.80
6	LB385	35.0	0.987	148.8	146.9	0.24	42.10
	Polyoxyethylene‡						
7	15-S-7	31.9	0.991	50.3	49.9	0.64	15.60
8	15-S-9	32.2	1.003	60.4	60.6	0.53	18.80
	Polydimethylsiloxane§						
9	1.5 cs	18.0	0.853	1.5	1.3	14.10	0.71
10	2.0 cs	18.7	0.873	2.1	1.8	10.30	0.97
11	3.0 cs	19.2	0.900	3.1	2.8	6.98	1.43
12	5.0 cs	19.7	0.920	5.3	4.9	4.05	2.47
13	10.0 cs	20.1	0.940	10.1	9.5	2.12	4.71
	Chlorinated biphenyl¶						
14	1221	40.8	1.190	6.8	8.1	5.06	1.98
15	1232	42.1	1.270	12.3	15.6	2.70	3.71
16	70% 1232 + 30% 1242	42.3	1.310	16.7	21.9	1.93	5.17
17	50% 1232 + 50% 1242	42.4	1.330	21.6	28.7	1.48	6.78
18	30% 1232 + 70% 1242	42.8	1.350	28.7	38.8	1.10	9.06
19	1242	43.1	1.390	48.0	66.7	0.65	15.50

* Union Carbide Ucon Series. † All percentages are by volume. ‡ Union Carbide Tergitol Series. § Dow Corning 200 Series.
¶ Monsanto Aroclor Series.

All liquids studied were nonvolatile, one-component systems which contained no additives. Several mixtures of a single type of liquid were prepared to obtain a larger range of viscosities. The chemical identification of the liquids and their physical properties are given in Table II. All liquids studied were polymers. Chlorinated biphenyl compounds were included for comparison. Vapor-phase osmometry number average molecular weights are given for some of the polymer liquids in Table III. In addition, the general appearances of the gel permeation chromatography spectra are given in Table IV.

TABLE III
Liquid Molecular Weights

Liquid Number	Identification	$\bar{M}n$
	Polyalkylene glycol	
1	LB65	410
4	LB135	675
5	LB165	725
6	LB385	1350
	Polydimethylsiloxane	
9	1.5 cs	375
11	3.0 cs	810
13	10.0 cs	1090

The gel permeation chromatography spectra were obtained in tetrahydrofuran at 25°C, with a flow rate of 1 ml/min, and with four columns in series consisting of one 250-Å, one 100-Å, and two 60-Å columns.

Results and Discussion

A cross-sectional profile of a spreading 5.0-cs polydimethylsiloxane drop as a function of time is shown schematically in Fig. 5. The drop is of the type described by Schonhorn et al. (22) as a "capped drop"—that is, the drop has a "cap" in the center with an advancing "foot." When the drop is first placed on the zinc phosphate substrate, the drop has the shape seen in Fig. 5a, which Schonhorn et al. describes as a "spherical segment drop." Within 10 seconds, the spreading foot appears around the center cap. The center cap remains essentially constant in diameter but decreases in height until 400 seconds, when the cap begins to decrease also in diameter and finally disappears after 1500 seconds. It appears that the center cap of the drop supplies liquid by diffusion and transport through the capillary matrix to the advancing foot. Although the times at which the above events occurred for the spreading drop varied according to liquid type and viscosity, the sequence of events with regard to the drop cap and foot were the same for

TABLE IV

GEL PERMEATION CHROMATOGRAPHY RESULTS

			Peak Descriptions	
Liquid Number	Identification	Number of Peaks	Major Peak Position*	Shape of Major Peak
1	Polyalkylene glycol LB65	1	21.70	Broad + four shoulders
4	LB135	1	20.24	Narrow
5	LB165	1	20.10	Narrow
6	LB385	1	19.00	Narrow
7	Polyoxyethylene 15-S-7	1	21.00	Narrow
8	15-S-9	1	20.70	Narrow + shoulder
9	Polydimethylsiloxane 1.5 cs	1	23.68	Very narrow
12	5.0 cs	1	20.82	Narrow
13	10.0 cs	2	18.90 and 20.80	Broad
14	Chlorinated biphenyl 1221	1	28.95	Narrow + two shoulders
15	1232	2	27.90 and 29.00	Narrow
19	1242	1	28.60	Narrow

* Milliliters of effluent ÷ 5.

all liquids. There is a marked difference in appearance of the spreading foot and the stationary cap. While the cap is shiny in appearance, the foot is dull. This behavior indicates that the spreading foot is penetrating into the phosphate coating during flow and not just spreading on top of the coating.

To obtain R_0, the initial drop radius, the diameters of the spreading drops

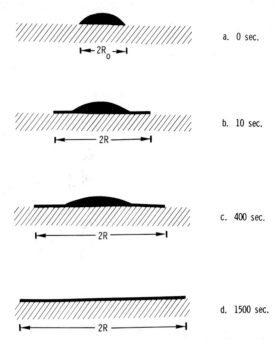

FIG. 5. Cross-sectional profile of spreading drop on zinc phosphate substrate.

were plotted against time and extrapolated to time zero, a procedure which required extrapolation only over 2 to 3 seconds. Plots of $\log(R - R_0)$ versus $\log t$ were made for each liquid, both on original and on solvent-cleaned zinc phosphate substrates. Each spreading experiment was performed a minimum of three times; typical results are shown in Figs. 6 through 11. Figure 6 is the spreading behavior of the polyalkylene glycol liquids on original zinc phosphate substrates; Fig. 7 shows similar data on solvent-cleaned zinc phosphate coatings, Fig. 8, polydimethylsiloxane liquids on original substrates; Fig. 9, the same liquids on solvent-cleaned zinc phosphate substrates; Fig. 10, chlorinated biphenyl liquids on solvent-cleaned zinc phosphate coatings; and Fig. 11, the spreading data for the polyoxyethylene liquids on both original and solvent-cleaned zinc phosphate substrates. The average values of $1/n$, n, A, and $d \cos \theta$ of the zinc phosphate substrates taken from the type of data shown in Figs. 6 through 11 are summarized in Table V. Generally, three regions are seen in the spreading data shown in Figs. 6 through 11 and Table V. First, for the initial 10 to 20 seconds, the slopes are greater than 0.5. Second, for the next several hundred seconds of spreading time,

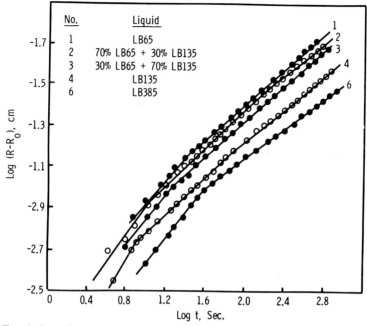

FIG. 6. Spreading of polyalkylene glycol liquids on original zinc phosphate substrates.

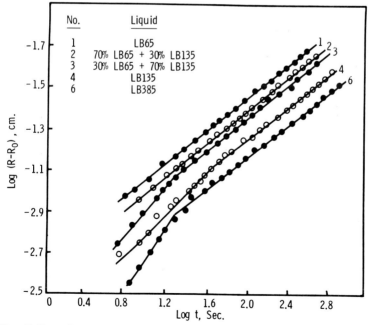

FIG. 7. Spreading of polyalkylene glycol liquids on solvent-cleaned zinc phosphate substrates.

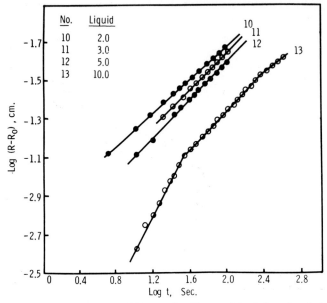

FIG. 8. Spreading of polydimethylsiloxane liquids on original zinc phosphate substrates.

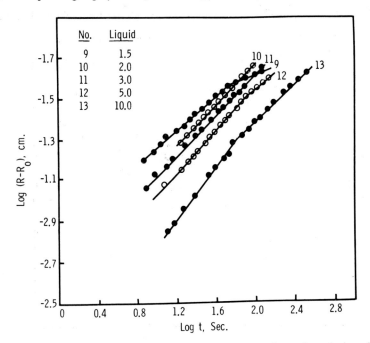

FIG. 9. Spreading of polydimethylsiloxane liquids on solvent-cleaned zinc phosphate substrates.

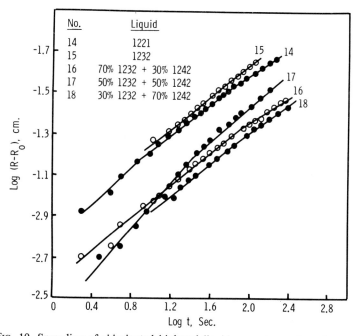

FIG. 10. Spreading of chlorinated biphenyl liquids on solvent-cleaned zinc phosphate substrates.

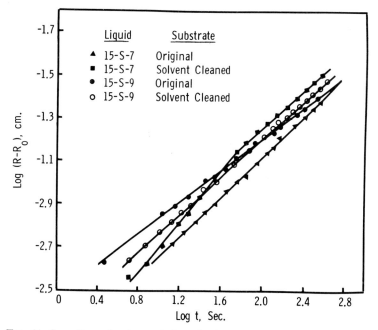

FIG. 11. Spreading of polyoxyethylene liquids on both original and solvent-cleaned zinc phosphate substrates.

TABLE V

PENETRATION, POROSITY, AND WETTABILITY OF ZINC PHOSPHATE SUBSTRATES

Liquid Number	Identification	Original Substrates*				Solvent-Cleaned Substrates*			
		Slopes, $1/n$	n	A (cm^2/sec, $\times 10^4$)	$d \cos \theta$ (cm, $\times 10^6$)	Slopes, $1/n$	n	A (cm^2/sec, $\times 10^4$)	$d \cos \theta$ (cm, $\times 10^6$)
	Polyalkylene glycol								
1	LB65	0.45	2.2	5.85	4.33	0.40	2.5	4.22	3.14
2	70% LB65 + 30% LB135†	0.40	2.5	3.98	4.12	0.40	2.5	3.17	3.28
3	30% LB65 + 70% LB135	0.45	2.2	3.09	4.33	0.40	2.5	2.52	3.53
4	LB135	0.45	2.2	1.94	3.98	0.40	2.5	1.59	3.27
5	LB165	0.43	2.3	1.19	3.00	0.40	2.5	1.19	3.00
6	LB385	0.38	2.6	0.56	3.49	0.38	2.6	0.53	3.34
	Polyoxyethylene								
7	15-S-7	0.48	2.1	1.54	3.60	0.42	2.4	1.02	3.19
8	15-S-9	0.43	2.4	1.00	2.78	0.48	2.1	1.70	4.84
	Polydimethylsiloxane								
9	1.5 cs	0.47	2.2	18.60	2.00	0.44	2.3	35.90	3.83
10	2.0 cs	0.43	2.3	26.00	3.53	0.50	2.0	21.90	3.29
11	3.0 cs	0.48	2.1	20.00	4.29	0.50	2.0	17.70	3.80
12	5.0 cs	0.44	2.3	10.00	4.00	0.50	2.0	8.71	3.23
13	10.0 cs	0.48	2.1	6.65	5.70	0.46	2.2	4.79	3.39
	Chlorinated biphenyl								
14	1221					0.45	2.3	9.82	2.91
15	1232			Does		0.48	2.1	10.80	6.01
16	70% 1232 + 30% 1242			not		0.42	2.4	5.18	4.02
17	50% 1232 + 50% 1242			spread		0.44	2.3	5.23	5.32
18	30% 1232 + 70% 1242					0.40	2.5	1.66	2.26
19	1242					0.41	2.5	2.37	5.51

* All spreading data are taken from the second region. † All percentages are by volume.

$1/n$ is equal to or slightly less than 0.5. Finally, at long spreading times, $1/n$ is still smaller.

In the present investigation, n is the exponent to which the difference $(R - R_0)$ is raised to maintain the linear relationship with time, and for the Washburn-Rideal model for liquid flow in capillary systems, n is exactly 2, or $1/n$ is 0.5; therefore, the departures of $1/n$ from 0.5 (or n from 2.0) in Table V represent the lack of agreement of the spreading data with the derived model. Departures of spreading data from the Washburn-Rideal model have been reported earlier (25, 36, 39). When $1/n$ is greater than 0.5, spreading is occurring at rates faster than those predicted by the model for a balance of capillary and viscous forces. There has not been enough time for the spreading drop to penetrate into the capillaries and bring the viscous forces into play. Apparently, the liquid is simply skimming over the surface and not penetrating into the substrate.

In the second region, when $1/n$ is less than 0.5, the converse is true; that is, the drop is spreading at rates slower than those predicted by the model. It is generally agreed (25) that the decrease of $1/n$ from 0.5 in the second region results from a buildup of air pressure in blocked capillaries, which slows the rate of liquid penetration. From Table V, we see that $1/n$ for polydimethylsiloxane liquids on solvent-cleaned substrates was approximately 0.50, while $1/n$ for the remaining liquids varied from 0.38 to 0.48; therefore, flow slowdown on the same substrate varied according to the particular polymer liquid. It is seen that, within a given family of polymer liquids, the value of $1/n$ was reasonably constant. This individual behavior of polymer liquids with regard to $1/n$ also was reported by Mack (25) from a study of the penetration of paint media into bricks. With some paints, $1/n$ was 0.50, while with other paints $1/n$ was 0.21, all in the same type of brick substrate. The individual behavior of each polymer liquid suggests that polymer liquids penetrate the capillaries and displace air to different extents. It is not known what properties of the liquid determine this effect; therefore, at the present time, the effect cannot be incorporated into the spreading model.

Further slowdown of flow in the third region also was observed by Greinacher (39) in paper substrates. At large distances from the center of the spreading drop, the material supply process may break down. Not enough liquid has been transported to the spreading liquid front both to fill existing capillaries and to flow into new regions. In other words, material is being depleted from the flowing front faster than it can be replenished. Of course, this effect is just the opposite of the first region, where liquid may be supplied faster to the spreading front than it can be removed by the substrate capillaries. To determine if flow slowdown in the third region might be caused by a shortage of liquid, a drop of 5.0-cs polydimethylsiloxane was placed on an original zinc phosphate substrate. The spreading behavior is shown in Fig. 12. The first spreading region had a slope of 0.64 and existed from 4 to 20 seconds, the second was 0.48 from 20 to 100 seconds, and the third was 0.38 beyond 100 seconds. The experiment was repeated with another drop and substrate except that a second drop was added to the center after 76 seconds. It is seen that the second region continued beyond 100 seconds at

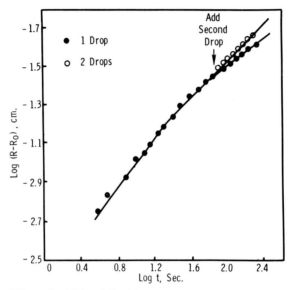

FIG. 12. Effect of additional liquid on spreading behavior of polydimethylsiloxane liquid on zinc phosphate substrates.

the same slope of 0.48; therefore, additional material was able to maintain the liquid flow at a constant rate.

Comparison of Tables II and V shows that some chlorinated biphenyl, polydimethylsiloxane, and polyoxyethylene liquids spread faster; that is, A was larger, with higher viscosities than with some liquids of lower viscosities of the same liquid families. This behavior is contrary to that of the derived model and is contrary to experience. This viscosity effect and the individual behavior of the liquids with regard to the regions of spreading behavior may be explained by results reported by Bascom *et al.* (*18*). It was found that spontaneous spreading of nonpolar liquids on metal surfaces was characterized by the spreading liquid existing in several structures. A primary film of less than 1000 Å advanced from the bulk liquid and was usually followed by a thicker secondary film. The movement of the secondary film was caused by a surface tension gradient formed in the transition zone between the primary and secondary films. The surface tension gradient was produced by the unequal evaporation from these two regions of a volatile contaminant having a lower surface tension than the bulk liquid. The primary film in this case had a higher surface tension than the bulk, and spreading was enhanced. When the volatile contaminant had a higher surface tension than the bulk, its evaporation left the surface tension of the primary film lower than that of the bulk. In this case, spreading was slowed and in some cases actually was stopped. These results show that dynamic liquid flow could be completely controlled by trace amounts of volatile impurities.

Comparison of Tables II and V shows that, except for the chlorinated biphenyl liquids, solvent cleaning did not have an appreciable effect on the penetration coefficient, A, or on the product of the substrate porosity and wettability, $d \cos \theta$. The chlorinated biphenyl series did not spread on the original substrate but rather remained fixed on the surface as lens-shaped drops. The reason for this nonspreading behavior may be explained by the spreading coefficient, S, for rough surfaces given in equation 11. The presence of adsorbed organic contamination probably lowered the magnitude of $f(\gamma_{SV} - \gamma_{SL})$ to the point where it was smaller than γ_{LV}, which is 41 to 43 dynes/cm for the biphenyl series. The difference $f(\gamma_{SV} - \gamma_{SL})$ is greater than 35 dynes/cm for the polyalkylene series because spreading readily occurs with the LB385 liquid. Since spreading is determined by the difference $(\gamma_{SV} - \gamma_{SL})$, which on polar substrates varies markedly according to the degree of liquid adsorption on the surface, it is not possible to assign an exact upper limit to the size of the surface tension of polymer liquids which spread over original zinc phosphate coatings. Also, the degree of surface contamination conceivably could vary from lot to lot, which would affect $(\gamma_{SV} - \gamma_{SL})$; however, of the nineteen polymer liquids studied, for spreading on original zinc phosphate substrates, the liquids must have a bulk surface tension of less than 41 dynes/cm. When the substrates were solvent-cleaned, chlorinated biphenyl liquids spread easily over the substrates such that the highest $d \cos \theta$ found was 6.0×10^{-6} cm for the 1232 liquid.

In Tables II and V, it is seen that, with the polyalkylene glycol liquids on both original and solvent-cleaned substrates and with the polydimethylsiloxane liquids on solvent-cleaned substrates, there was no marked change in substrate wettability, although the kinematic viscosity of the liquids varied from 17 to 149 cs for the polyalkylene glycols and from 1.5 to 10 cs for the polydimethylsiloxanes. The rate of spreading as indicated by the penetration coefficient was inversely proportional to the viscosity, but the degree to which the substrates were wetted as indicated by $d \cos \theta$ did not vary as a function of viscosity. These liquids may have spread at varying rates over the substrates, but the interaction between the spreading liquid and the substrate appears to have remained essentially constant.

To determine the effect of the chromic acid rinse on the penetration coefficient and on wettability, two series of zinc phosphated panels were prepared simultaneously, except that the chromic acid step was omitted from one series. The spreading behavior of 5.0-cs polydimethylsiloxane was determined on both sets; the spreading data from three runs are presented in Table VI. It is seen that the chromic acid rinse step gave no detectable difference in either A or $d \cos \theta$.

TABLE VI
EFFECT OF CHROMIC ACID RINSE

	$1/n$	n	Spreading Data	
			A (cm²/sec, × 10⁴)	$d \cos \theta$ (cm, × 10⁶)
With chromic acid	0.46	2.2	12.20	4.52
Without chromic acid	0.45	2.3	11.30	4.20

Applications of Spreading Model

From the spreading data and the application of the model which was derived from a capillary bed structure, it is possible to obtain an estimate of the porosity, d, of the zinc phosphate substrates and the capillary pressure generated in them. If it is assumed that $\cos \theta$ for the 1232 chlorinated biphenyl liquid on solvent-cleaned substrates is essentially equal to 1, it follows that d is 6.01×10^{-6} cm or that the width of the capillaries in the zinc phosphate crystals is $2 \times 6.01 \times 10^{-6}$ cm, or 1200 Å (0.12 micron). It is difficult to obtain the width of the capillaries in nitrocellulose replicas viewed in the electron microscope, but many of the capillaries which can be measured are of this order of magnitude; therefore, the spreading model gave a reasonable estimate of the size of the capillaries in the zinc phosphate substrate. Equation 16 gives the capillary pressure, P_c, generated by slits of width $2d$. On the basis of this equation,

$$P_c = \frac{\gamma_{LV}}{d} = \frac{42.1 \text{ dynes/cm}}{6.0 \times 10^{-6} \text{ cm}} = 7.0 \times 10^6 \text{ dynes/cm}^2$$

for the 1232 chlorinated biphenyl liquid. This capillary pressure is equal to 7 atmospheres in zinc phosphate substrates, which is an appreciable effect.

Examination of Table V reveals that some polymer liquids give essentially the same values of A or $d \cos \theta$. Since the substrates are the same for all liquids studied, the value of d is essentially constant, and only $\cos \theta$ varies in the product $d \cos \theta$. The application of the spreading model permits one to measure both A and $d \cos \theta$ in polymer liquids, tabulate these properties, and then choose those liquids that best meet the performance criteria for a given application. Therefore, polymer liquids can be grouped into corresponding states of either substrate penetration or substrate wettability.

Conclusions

Zinc phosphate conversion coatings have been treated theoretically as a capillary matrix in which the capillaries assume the particular shape of slits. A mathematical and physical model was derived from Poiseuille's equation, which adequately describes the flow of polymer liquids in zinc phosphated steel substrates. The flow equation is

$$(R - R_0)^n = \frac{2}{3} \frac{\gamma_{LV} \, d \cos \theta t}{\eta}$$

where R_0 = the radius of the drop at time 0.
R = the radius of the drop at time t.
n = a number varying from 2.6 to 2.0.
γ_{LV} = the bulk liquid surface tension.
η = the coefficient of viscosity.
d = a measure of the porosity of the substrate.
$\cos \theta$ = the substrate wettability.

It was possible to express flow in terms of liquid parameters only, γ_{LV} and η, a substrate parameter, d, and the interaction of the substrate and the liquid, $\cos \theta$.

Polymer liquids were classified according to the rate at which they penetrated zinc phosphate coatings and the extent of wetting. These two physical properties in zinc phosphate–polymer liquid systems may be used to select the particular system which best meets the performance requirements for a given application. Utilization of the model gave the capillary pressure which was generated in the zinc phosphate coating and was estimated to be 7 atmospheres.

Acknowledgments

It is a pleasure to thank G. L. Leithauser for many helpful suggestions; E. R. Cprek for the scanning beam electron microscope photograph of the zinc phosphate coating; W. A. Florance for the vapor-phase osmometry molecular weights and infrared spectra; J. M. Montgomery for the kinematic viscosity measurements; T. O. Morgan and E. L. White for the gel permeation chromatography results; and T. P. Schreiber for the electron probe data.

References

1. H. J. Osterhof and F. E. Bartell, *J. Phys. Chem.*, **34**, 1399 (1930).
2. J. R. Huntsberger, *J. Paint Technol.*, **39**, 199 (1967).
3. W. A. Zisman, "Adhesion and Cohesion," ed. by P. Weiss, Elsevier, Amsterdam, 1962.
4. J. L. Moilliet and B. Collie, "Surface Activity," Van Nostrand, New York, 1951.
5. R. E. Johnson, Jr., *J. Phys. Chem.*, **63**, 1655 (1959).
6. V. R. Gray, *J. Oil Colour Chemists' Assoc.*, **44**, 756 (1961).
7. S. J. Gregg, "The Surface Chemistry of Solids," Reinhold, New York, 1961.
8. G. R. Lester, *J. Colloid Sci.*, **16**, 315 (1961).
9. E. A. Boucher, *J. Paint Technol.*, **38**, 329 (1966).
10. H. E. Garrett, "Aspects of Adhesion," Vol. 2, ed. by D. J. Alner, University of London Press, London, 1966.
11. A. Bondi, *Chem. Rev.*, **52**, 417 (1953).
12. M. Levine, G. Ilkka, and P. Weiss, *J. Polymer Sci.*, Part B, **2**, 915 (1964).
13. M. Levine, G. Ilkka, and P. Weiss, *Adhesives Age*, **7**, No. 6, 24 (1964).
14. W. D. Harkins, "The Physical Chemistry of Surface Films," Reinhold, New York, 1952.
15. E. B. Bielak and E. W. J. Mardles, *J. Colloid Sci.*, **9**, 233 (1954).
16. W. B. Hardy, "Collected Works," Cambridge University Press, Cambridge, 1936.
17. D. H. Bangham and Z. Saweris, *Trans. Faraday Soc.*, **34**, 554 (1938).
18. W. D. Bascom, R. L. Cottington, and C. R. Singleterry, *Advan. Chem. Ser.*, **43**, 355 (1964).
19. E. D. Shchukin, Yu. V. Goryunov, G. I. Den'shchikova, N. V. Pertsov, and B. D. Summ, *Kolloid. Zh.*, **25**, 108 (1963).
20. R. N. Wenzel, *Ind. Eng. Chem.*, **28**, 988 (1936).
21. R. L. Cottington, C. M. Murphy, and C. R. Singleterry, *Advan. Chem. Ser.*, **43**, 341 (1964).
22. H. Schonhorn, H. L. Frisch, and T. K. Kwei, *J. Appl. Phys.*, **37**, 4967 (1966).
23. J. J. Bikerman, "The Science of Adhesive Joints," Academic Press, New York, 1961.
24. K. Kanamura, *Kolloid-Z.*, **192**, 51 (1963).
25. G. W. Mack, *J. Oil Colour Chemists' Assoc.*, **44**, 737 (1961).
26. R. L. Scott and A. M. Turkalo, *Am. Soc. Testing Mater., Proc.*, **57**, 536 (1957).
27. E. W. Washburn, *Phys. Rev.*, **17**, 273 (1921).
28. E. K. Rideal, *Phil. Mag.*, **44**, 1152 (1922).
29. C. H. Bosanquet, *Phil. Mag.*, **45**, 525 (1923).
30. R. L. Peek, Jr., and D. A. McLean, *Ind. Eng. Chem. Anal.*, **6**, 85 (1934).
31. E. Manegold, S. Komagata, and E. Albrecht, *Kolloid-Z.*, **93**, 166 (1940).

32. D. D. Eley and D. C. Pepper, *Trans. Faraday Soc.*, **42**, 697 (1946).
33. D. Tollenaar and G. Blokhuis, *Appl. Sci. Res.*, **A2**, 125 (1949–1951).
34. D. Tollenaar, *Appl. Sci. Res.*, **A3**, 451 (1951–1953).
35. D. Tollenaar, *Appl. Sci. Res.*, **A4**, 453 (1953–1954).
36. G. F. N. Calderwood and E. W. J. Mardles, *J. Textile Inst. Trans.* **46**, T161 (1955).
37. T. Gillespie, *J. Colloid Sci.*, **13**, 32 (1958).
38. T. Gillespie, *J. Colloid Sci.*, **14**, 123 (1959).
39. H. Greinacher, *Z. Phys. Chem. (Frankfurt)*, **19**, 101 (1959).
40. M. L. Studebaker and C. W. Snow, *J. Phys. Chem.*, **59**, 973 (1955).
41. H. H. Zuidema and G. W. Waters, *Ind. Eng. Chem., Anal. Ed.*, **13**, 312 (1941).

Discussion

GRISKEY: On the chlorinated biphenyls, you did not show a slide of the plotted data. Was it a straight line, or did it have the bends and hooks in it?

CHEEVER: It had the bends and hooks in it, the same as the others—that is, on the solvent-cleaned surface where it spread.

GRISKEY: You listed viscosities of these materials, but three or four of them, at least, will be non-Newtonian fluids. You mentioned one viscosity as being characteristic for each of them. How did you measure this viscosity, and what did you pick as the shear rate?

CHEEVER: The viscosities were measured with the Cannon-modified Oswald viscometer; therefore, the viscosities are obtained with a very slow shear rate. We measured each of the viscosities of the liquids, and, as we pointed out, they varied from 2 up to 150 centistoke.

GRISKEY: Did you reverse the process and use your form of the Hagen-Poiseuille equation to see if you were in agreement with the shear rate?

CHEEVER: No, we did not. This is a very good point, though, and this is something we could certainly look into.

GRISKEY: Well, the reason I suggest this is because of the shape of the curves, especially these bends. However, the biphenyl which is Newtonian refutes my original argument. But maybe some of the bends do occur because of variations in the viscosity behavior.

CHEEVER: All these liquids have very similar molecular weights. We hope that they will not be too different, one from another, in molecular structure and therefore will not give us non-Newtonian flow in the measurements.

PETERLIN: Chlorated biphenyls certainly do not show non-Newtonian flow properties. One has to add very high-molecular-weight solute in order to obtain easily detectable deviations from Newtonian viscosity. Therefore, I do not think that the reported experiments can be explained by an increase of viscosity with decreasing gradient.

CHEEVER: In our derivations, as Professor Mark said, we are assuming that these liquids are Newtonian, and our flow rates through the phosphate coating are very slow; they are, perhaps, only 1 centimeter per second at most.

GRISKEY: I tend to completely disagree with your turbulence explanation. In fact, if you are working with your data, even as the liquid became more viscous, you would seem to have more turbulence. Actually, as the liquid gets more viscous, it tends to displace turbulence.

CHEEVER: I am using the word turbulence to describe the liquid as merely flowing over the surface and not penetrating into the capillaries in the phosphate coatings.

HERMANS: Wouldn't it be much simpler to say that it is not so much a matter of turbulence, but the height of the drop is well above the capillary bed, so that the drop flows partly over this bed, and this is the reason for the rapid flow in the beginning. Once the top of the drop has reached the level, then the flow proceeds more slowly.

CHEEVER: I think this is a better way of expressing this first region.

ASBECK: Your curve, of course, has to level out. As soon as the drop spreads to approach a monomolecular layer, it can no longer spread further.

MACHU: In regard to the chromic acid treatment, I have made quantitative measurements of porosities after treatment with chromic acid and found that it actually alters the porosity by only a few percent. The main function of chromic acid is to act as an inorganic corrosion inhibitor. Did you use only one type of phosphate coating?

CHEEVER: Yes, we studied only zinc phosphate coatings.

MACHU: Did you try to find any influence of, say, coarser- and finer-grained crystals?

CHEEVER: Yes, but this was somewhat accidental. We noticed that the crystal structure of zinc phosphate coatings can vary tremendously, depending on their respective corrosion resistances. This is something that Mr. Leithauser pointed out many months ago. We have found that highly crystalline zinc phosphate coatings perform very well. On the other hand, we have found that other coatings prepared from the same bath, but with different steels, do not perform well and are rather amorphous in appearance. We have measured the wetting rates on these two types of zinc phosphated crystal surfaces, and, of course, all the data presented here apply to the very nice plate-like crystalline structures, but on the ones which are more grainy, and less well-defined, the spreading rates are much lower for the same liquid. This would indicate a poorer capillary structure in the amorphous-like zinc phosphate coatings.

HELLER: With regard to your equation, do you assume that the capillaries are cylinders of constant diameter along the length?

CHEEVER: We assume that they are slits of constant width along the length.

HELLER: If that is so, it could be that the fast flow at the beginning is due to capillaries that are conical. In that case, you would not have to bring in turbulence.

CHEEVER: This is an alternate explanation for the first spreading region. The paper chemists have examined the effect of flow rate as a function of type of capillary. We will certainly take a look at that. All the reported measurements are two-dimensional measurements. The three-dimensional measurements should give us a better idea of the shapes of these capillaries.

HELLER: When you speak of wettability, are you referring to what is generally called the contact angle?

CHEEVER: That is correct. We are referring to the contact angle of the liquid within the capillaries.

HELLER: Now then, you should, in principle, be in a position to measure the contact angle and see if the d values are reasonable. This will give you a check for the equation. Have you done that?

CHEEVER: We have not measured any contact angles yet, but we will extend the work that has been done on measuring the effect of roughness on contact angles, and we can obtain another check for d. As we mentioned, we obtained an estimate of d from the electron microscope examination of the coatings which was consistent with the d found by wetting measurements.

LEE: The obvious effect, here, is not due to molecular weight, not due to turbulence, but is due to molecular weight distribution of the high-viscosity material. Therefore, my comment is twofold. Number one, you should take your current liquids and, by gel permeation chromatography, see how many peaks you do get for each liquid. Number two, carry out your experiments with some narrow-molecular-weight distribution liquids to test the equation.

CHEEVER: I would like to comment on three additional experiments that we did. First, let us talk about the effect of liquid droplet size. (Fig. 13). Here, we are looking at three drop sizes, 0.5, 1.0, and 1.5 μl, of 5.0-cs polydimethylsiloxane. The first result with the 0.5-μl drop is that we obtain only one spreading region. With the 1.0-μl drop, we obtain the first break, and then it follows along in the second region as with the 0.5-μl drop. With the 1.5-μl drop, the first break is even more pronounced. Again, the latter stages of growth were similar to those of the 0.5-μl and 1.0-μl drops. This behavior supports Dr. Hermans' explanation for the initial rapid spreading. Second, we have the gel permeation chromatograph results on all the liquids; therefore, we know how many peaks are present. Finally, we have prepared narrow-molecular-weight fractions. In particular, Mr. John Laverty, of the Research Laboratories, vacuum-distilled 10.0-cs Dow Corning 200 fluid.

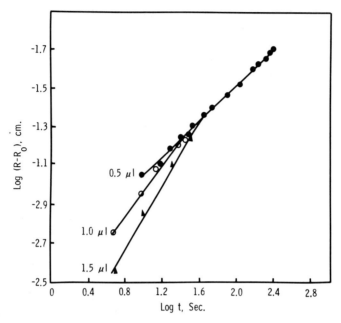

FIG. 13. Effect of droplet size of 5.0-cs polydimethylsiloxane on spreading.

The spreading behavior was determined for each fraction (Table VII). The spreading rate was much higher with the lower-molecular-weight fractions, but again there was not a marked difference in wettability between the various fractions. The spreading droplet, as we mentioned, is shaped like a hat. Perhaps we obtained some separation of the lower-molecular-weight species

TABLE VII
Effect of Molecular Weight on Spreading

Sample	η (cp)	γ (dynes/ cm)	$\overline{Mn}$	$\overline{\dfrac{1}{n}}$	n	A (cm^2/sec, $\times 10^4$)	$d \cos \theta$ (cm, $\times 10^6$)
First cut*	4.1	20.2	530	0.50	2.0	22.00	6.62
Second cut†	6.2	20.7	800	0.47	2.1	11.90	4.92
Original	9.5	20.1	1090	0.48	2.1	6.65	5.70
Flask residue	13.2	20.1	1240	0.45	2.2	4.68	4.37

* Boiling point 155°C at 0.35 mm Hg.
† Boiling point 180°C at 0.30 mm Hg.

which might give interface molecular organization. If, indeed, the small molecules move much faster in the spreading droplet front, in the center of the drop that remains behind, we could have an enrichment of the more viscous, longer-chain molecules.

MICHAELS: I am puzzled about one thing. You put a drop on an impermeable surface, and you will get spreading and advance and an increase in diameter of the drop, provided that the contact angle is finite, less than 90°. It seems to me that when the drop is small, and at the very start, the contribution of the capillary pressure within the drop upon the surface may govern the rate of change of the radius more than the rate of penetration within the surface. This could account for the higher slope. I was wondering how you factored out spreading out versus spreading within.

CHEEVER: Bascom, Cottington, and Singleterry microscopically examined the spreading liquids on surfaces which contained microscratches. They found that the liquids flowed preferentially in these scratches, rather than on the flat portion of the surface. Also, Huntsberger suggested that, with adhesives, interstices of the order of 50 to 1000 Å are going to dominate the performance on these surfaces. Then, too, our studies indicate that on smooth surfaces the rate of spreading is much slower. Where the scratches exist, they should dominate the flow because of the high capillary pressure. As was pointed out, we believe that the first rapid-spreading region represents spreading upon the surface of the capillary bed and not in it.

The Role of Oxide Films in the Zinc Phosphating of Steel Surfaces

J. V. LAUKONIS
General Motors Research Laboratories
Warren, Michigan

Introduction

Most steel surfaces to be phosphated are already covered with a thin, more or less continuous, film of iron oxides. It is generally thought that such oxides must be removed before phosphate coatings can be deposited (*1*). This paper shows that this is not the case in zinc phosphating and will report on how the oxides influence the phosphating.

The zinc phosphating of three different types of iron oxide surfaces was studied. The studies included optical microscopic examination of both the form and the density of crystals in phosphate coatings, gravimetric determination of their growth kinetics, *x*-ray diffractometer analyses of their *in situ* composition, and the standard salt-spray evaluation of the corrosion resistance which they impart. The conclusions reached in these studies are all incorporated into a comprehensive growth mechanism. Within the framework of the growth mechanism, the following important questions can be answered.

1. Independent of their relative adhesion to the substrate, why do thin complete phosphate coatings confer much greater corrosion resistance to a steel substrate than do thick complete coatings?

2. Why do thin complete coatings consist of smaller crystals than those of thick complete coatings?

3. Why do later stages in the growth of many immersion-formed phosphate coatings show a tendency to become richer in iron? Presumably the later stages of formation take place when a minimum of iron is being dissolved from the substrate.

Background Information

The essential constituents of a zinc phosphating bath are primary zinc phosphate ($Zn(H_2PO_4)_2$) and an excess of phosphoric acid (H_3PO_4). The principal chemical reaction is (*1, 2*)

$$3Zn(H_2PO_4)_2 \rightleftarrows Zn_3(PO_4)_2 + 4H_3PO_4 \tag{1}$$

wherein primary zinc phosphate is hydrolyzed to insoluble tertiary zinc phosphate and the excess of "free" phosphoric acid must be maintained for the equilibrium to exist.

When a steel surface interacts with the equilibrium solution (equation 1), the following reaction takes place.

$$Fe + 2H_3PO_4 \rightarrow Fe(H_2PO_4)_2 + H_2 \qquad (2)$$

That is, iron is attacked by the free phosphoric acid. This attack is equivalent to raising the pH of the solution, especially in the immediate vicinity of the surface. Reaction 1 is driven to the right, and the insoluble tertiary phosphate precipitates on the steel surface. In the process of precipitation the tertiary phosphate picks up four water molecules of hydration and becomes attached to the substrate. This hydrated complex is one of the principal constituents of a zinc phosphate coating. Its chemical formula is $Zn_3(PO_4)_2 \cdot 4H_2O$, and it is commonly known as the mineral hopeite. The water for the hydration is readily available, since the phosphating solution is about 95% water (3).

While the essentials of zinc phosphating are illustrated by equations 1 and 2, no commercial solution has the simple composition shown in equation 1. Commercial solutions contain additives, and the complex mixtures used are frequently proprietary. The chemical complexities of zinc phosphating were reviewed by Bloor (2) and by Gilbert (4). Zinc phosphating is further complicated by the fact that the initial condition of the surface has a strong influence both on how the phosphating proceeds and on the properties of the coating formed. Machu (1) studied the effects of sandblasting, mechanical polishing, acid pickling, alkaline degreasing, and the deliberate deposition of trace impurities of different metal ions on the surface to be phosphated. He showed that all these pretreatments leave their own characteristic imprint on a metal surface in sufficient degree to affect phosphating. When, as is usually the case, several of these operations are performed prior to phosphating, their net effect depends on the order in which they are performed.

Experimental Techniques

The Metal and the Preparation of Surfaces

Only one type of steel was used in this study, cold-rolled SAE 1010. The phosphated panels were 1 inch × 4 inches × 35 mils, all cut from one "heat" of the 1010 steel. The test panels for salt-spray corrosion resistance determinations were the same steel, but they varied in size from a minimum of 4 inches × 5 inches to a maximum of 4 inches × 8 inches. All panels had a

$3/16$-inch hole stamped into one end to accommodate a glass hook from which they were suspended while being phosphated.

Before being phosphated, the panels were rinsed and scrubbed with oleum spirits (a mixture of aliphatic hydrocarbons) until the rinse solution showed no discoloration. The panels were then scrubbed with a fiber brush and/or a soft rag in a warm water solution of Dreft. The detergent-scrubbed surface was rinsed and then checked with cold water for water breaks. If the cold water film did not break, the surface was considered clean. The occasional panel that did not respond to repeated scrubbings was discarded.

After being cleaned and cut to size, the panels were stored in a desiccator until use.

The Chemicals and the Phosphating Method

All panels were phosphated with Bonderite-37 (hereafter referred to as B-37) stock chemicals provided by the Parker Division of the Hooker Chemical Corporation (hereafter called Parker). Just before phosphating the panels were alkaline-cleaned (and activated) with Parker's 341 cleaner, one of the mildest alkaline solutions recommended for use in the B-37 system. The usual titrations were made for activator (in the 341 cleaner, a dilute addition of titanium phosphate), accelerator (in B-37, sodium nitrite), free acid content, and total acid content. The phosphating bath was kept at a temperature of 67°C (usually ±1°C). The total acid content of the bath was kept between 12 and 15 "points," and the free acid content was kept between 0.4 and 1.0 "point."

Most production zinc phosphating of sheet metal is spray phosphating. Experimental difficulties make it impractical to duplicate a spray facility in the laboratory. Therefore, all our panels were phosphated by immersion. The usual similarities and differences between immersion-formed and spray-formed phosphates were recognized, and when any question has arisen regarding the applicability of immersion results to the spray process they have been checked. This was done by comparing the salt-spray corrosion resistance of a gilsonite-coated immersion coat with that of a similarly coated spray coat (5). All the B-37 spray phosphating was done by Parker.

Coating Weight Determinations

When coating weights are given in this report, they represent strip and weigh determinations. The coatings were stripped in a hot (90°C) chromic acid solution containing 200 grams of CrO_3 per liter of water. The phosphated panels were weighed on a microbalance, immersed for 1 minute to dissolve the phosphate coating, rinsed, dried, and reweighed.

Preparations of Iron Oxide Surfaces

The phosphating of three different states of surface oxidation was studied. The three types of surfaces were prepared as follows.

Cold-Rolled Steel Surfaces (CRS). This first state of oxidation was the one already existing on the precleaned cold-rolled steel surfaces. The oxides present are primarily those formed during the steel mill processing and storage. Such oxides are more or less uniform and of the order of 100 Å in thickness. It is likely that both of the low-temperature iron oxides $\gamma\text{-}Fe_2O_3$ and Fe_3O_4 are present, along with some of the high-temperature ferric oxide $\alpha\text{-}Fe_2O_3$. It is also likely that some of the oxides are hydrated. Hereafter such surfaces will be referred to as CRS (cold-rolled steel) surfaces.

Hydrogen Annealed-Reduced Surfaces (HYD A-R). The second state of oxidation was that developed after CRS surfaces were simultaneously annealed and reduced in a hydrogen atmosphere at 800°C. After a 4-hour anneal-reduction at temperature, the panels were cooled in hydrogen to room temperature and stored in a desiccator until phosphated. Such surfaces are known to be covered with a thin, uniform, microcrystalline oxide film. 20 to 40 Å thick, consisting solely of the oxides Fe_3O_4 and $\gamma\text{-}Fe_2O_3$ (6). These oxides are much thinner than those on a CRS surface, they are not hydrated, and the high-temperature ferric oxide is not present. Hereafter these surfaces will be referred to as HYD A-R (hydrogen annealed-and-reduced) surfaces.

Oxidized Surfaces (OXID). The third state of oxidation was obtained by deliberately oxidizing CRS surfaces until they showed a first-order blue temper color. Twenty minutes in air at 350°C will accomplish this, or shorter times at higher temperatures. On the basis of simple interference theory, and with the refractive index assumed to be 3.0, the blue color indicates a thickness of about 500 Å. Although there is some uncertainty about the thickness (7), this film is certainly thicker than either of the other two surfaces studied. Caule *et al.* (8) have shown that the outside surface of this film consists solely of the high-temperature ferric oxide $\alpha\text{-}Fe_2O_3$, and it is in this characteristic that this film most notably differs from the other two oxide films. Hereafter these surfaces will be referred to as OXID (oxidized) surfaces.

Experimental Results

In the first part the data are essentially qualitative and based on laboratory observations. The interpretation of these data requires some use of the quantitative material which is given in the second part of this section.

Part I. Qualitative Data

Do Oxide Films Survive the Phosphating Process? The most direct evidence that iron oxide films do survive the phosphating process was obtained in studying the phosphating of OXID surfaces. As stated, these surfaces start out being blue in color. After they have been phosphated, the coating is noticeably and consistently darker on OXID surfaces than are coatings of the same thickness on either CRS or HYD A-R surfaces. Both of the latter surfaces start out with the characteristic metallic color. Therefore, if it is assumed that the phosphate coating is somewhat transparent, it can be concluded that the initial oxide film with its darker color still exists under the coating on OXID surfaces.

When we use chromic acid to strip phosphate coatings from OXID surfaces to determine their weight per unit area, the resulting surfaces show the same blue colors they showed before the phosphating. This further and more directly indicates that the oxides survive the phosphating process.

Finally, if we assume that the oxide films must be dissolved before phosphating can proceed, we would anticipate incubation intervals which increase with the thickness of initial oxides. We are not certain about the existence of incubation times for all our steel surfaces, but we have seen no evidence of them when we phosphate OXID surfaces. Therefore, we conclude that the "thick" oxide films on OXID surfaces survive the phosphating process and, without evidence to the contrary, the thinner oxide films on the other two types of surfaces also survive.

The Nature of the Surface of a Zinc Phosphate Coating. Optical microscopic examination of phosphated surfaces at $500\times$ shows a typical good zinc phosphate coating to be a dense agglomeration of plate-like crystals lying both in the plane of the substrate and at all angles to it. If the coating confers appreciable corrosion resistance to the substrate, the crystals are small, the coating weight is not excessive, and the density of crystals is large. For example, a typical good spray-deposited zinc phosphate coating on a CRS surface has a coating weight of about 200 mg/ft^2 (2.16 g/m^2). Such a coating spread over the surface as a uniform film would have a thickness of 0.7 micron (0.03 mil). Instead, optical microscopic examination reveals a large number of small platelet crystals growing from 3 to 5 microns above the surface. The density of crystals is counted to be of the order of 10^6 cm^{-2}.

There is a surprising uniformity in the size of the crystals in the coating. This can be illustrated by comparing the observed parameters with those that would be observed in an ideal coating of the same weight and crystal density. Thus, 200 mg ft^{-2} = 0.2153 mg cm^{-2}, and dividing by the surface density of crystals gives the average weight of a crystal as 2.153×10^{-7} mg. The

Fig. 1. Schematic view of the arrangement of platelet crystals in a phosphate coating. The uniformity in crystal size is as per the calculation in the text.

material density is known to be 3.0 g cm^{-3}, and therefore the average crystal volume is 7.18×10^{-11} cm^3. Using the experimental observation that the crystals are about five times as long as they are wide, we can calculate the dimensions of half of a flat circular disk of the above volume. This calculation gives the disk a diameter of 10 microns and a thickness of 2 microns. Such an ideal coating, consisting of half-disk crystals of one size, would show the crystals to be growing from 3 to 5 microns above the surface. The spread would depend on whether we were viewing an upright crystal growing adjacent to one in the plane of the substrate, or an isolated upright crystal. This is exactly the spread observed in a real coating. A schematic view of the calculated coating is shown in Fig. 1, and a photomicrograph of a spray-deposited coating is shown in Fig. 2. Measurements of the thickness and

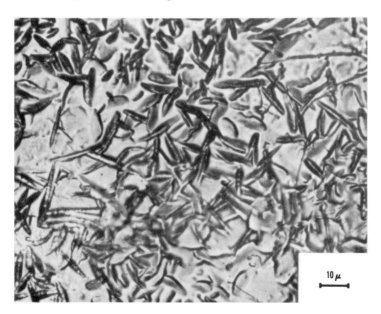

Fig. 2. Photomicrograph of a typical B-37 spray coating on steel.

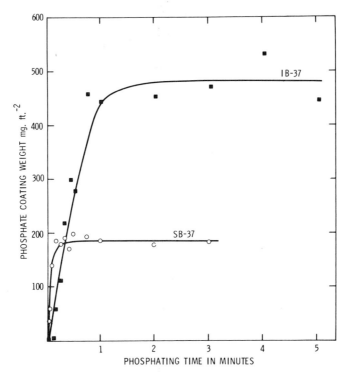

FIG. 3. Comparative kinetics of spray and immersion phosphating of SAE-1010 cold-rolled steel. IB-37 = immersion Bonderite-37. SB-37 = Spray Bonderite-37.

length of individual crystals in the photomicrograph are found to be near those calculated.

Quantitative Data

Kinetics of Phosphating Iron Oxides and Corrosion Resistance. Each of the kinetics curves given here is a visual best fit of about twelve data points. Each point is a gravimetric determination of coating weight as a function of deposition time. Since the gravimetric measurement is destructive, twelve panels and twelve growth experiments were needed for twelve data points.

Immersion versus Spray Phosphating. Figure 3 shows typical kinetics curves of coating weight versus time of phosphating for IB-37 (immersion Bonderite-37) coatings which we applied and SB-37 (spray Bonderite-37) coatings applied by Parker. Both curves show the usual "saturation" where no more coating is deposited with time. Saturation occurs at about 20 seconds for the spray coat and at about 1 minute for the immersion coat. The coating at saturation is defined as a "complete" coating. The complete immersion coat is more than twice as heavy as the complete spray coat, and this is typical. The average immersion weight is two to three times that of a comparable spray coat (9). "Comparable," in this case, means comparable corrosion resistance as determined by the salt-spray test. Salt-spray tests on both IB-37 and SB-37 coatings yield similar results in "creepback." For example, from 1 to 4 units* of creepback for 2 weeks of exposure to salt spray is typical for both coatings.

Kinetics of Immersion Phosphating CRS, HYD A-R, and OXID Surfaces. Figure 4 shows typical kinetics curves for the three iron oxide surfaces. The saturation coating weight on the OXID surfaces is about half that on the CRS and the HYD A-R surfaces. Comparing Figs. 3 and 4, one sees that the

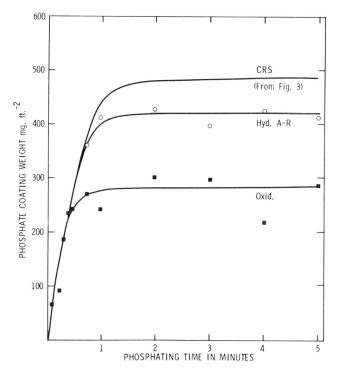

FIG. 4. Comparative kinetics of IB-37 phosphating of CRS, HYD A-R, and OXID steel surfaces.

* One unit of creepback is defined as $1/32$ inch.

immersion coating weight on OXID surfaces is nearer to the spray-coating weight than to the immersion weight on CRS surfaces. The corrosion resistance imparted by the coatings was compared by salt-spray testing of three panel sets of each of the surfaces after spray phosphating. The results shown in Table I indicate that all three surfaces show very excellent and comparable

TABLE I

Salt-Spray Corrosion Resistance of Oxidized Surfaces Whose Immersion Kinetics are Shown in Fig. 4

(All surfaces were SB-37 phosphated by Parker, and salt-spray-tested according to ASTM Standard B-117-64.)

Oxide Type	Panel No.	Units of Creepback
CRS	1	1
	2	1
	3	1
HYD A-R	4	1
	5	2
	6	2
OXID	7	2
	8	1
	9	1

corrosion resistance. Therefore, we conclude that with no sacrifice of corrosion resistance the ferric oxide of the OXID surfaces is easier to phosphate than are the mixed ferrous–ferric oxides of CRS and HYD A-R surfaces. This observation suggests that it is easier to phosphate ferric oxides than it is to phosphate ferrous oxides.

An Important Difference in the Phosphating of Ferric and Ferrous Oxides. In our earlier discussion of the chemicals used in zinc phosphating we pointed out that there are four ingredients in the system whose concentrations must be regulated to assure successful phosphating. These are the free acid, total acid, and accelerator contents of the phosphating bath and the activator content of the alkaline cleaner bath. The concentrations of all four were varied over wide limits in phosphating both the ferric and the mixed oxides surfaces. Both oxide compositions reacted similarly to variations in the concentration of the three parameters of the phosphating bath but differently to variations in the activator content of the alkaline cleaner. In the B-37 system the activator is a 0.05% by weight colloidal dispersion of a titanium salt, usually titanium phosphate, which is added to the cleaner. Thus, surfaces are activated while being alkaline-cleaned. While the nature of the activation is not well understood, the usual explanation is that small titanium particles of atomic dimensions are adsorbed on the steel surface, and these act as nucleation sites for the growth of phosphate crystals. We have generally observed in both immersion and spray phosphating that activation increases the nucleation density of crystals from five to ten times that observed without activation.

Figure 5 shows the phosphating kinetics for alkaline-cleaned (and activated) and nonalkaline-cleaned (and therefore not activated) CRS surfaces. Figure 6 shows the same comparison for HYD A-R surfaces, and Fig. 7 contains the two curves for OXID surfaces. Of obvious interest is the failure of the non-activated mixed oxides surfaces to saturate at coating weights far greater than their saturation coating weights when they are activated, and the fact that

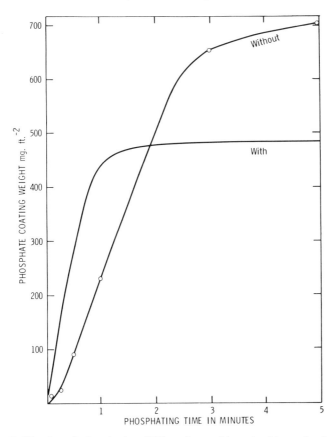

FIG. 5. Kinetics of phosphating CRS surfaces with and without titanium activation (IB-37 phosphating).

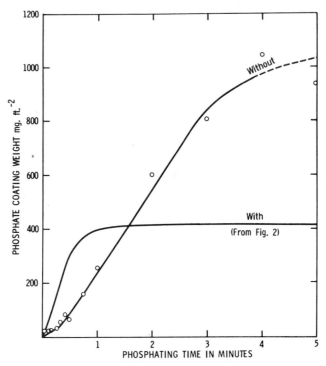

FIG 6. Kinetics of phosphating HYD A-R surfaces with and without titanium activation (IB-37 phosphating).

ferric oxide surfaces behave the same whether or not they are activated. Salt-spray testing of the very thick phosphate coatings on both types of nonactivated surfaces showed that they confer little or no corrosion resistance. After the conventional 2 weeks of exposure to salt spray, the gilsonite coating was found to be completely undermined. This corresponds to a creepback of more than 64 units, compared to the 1 to 2 units noted in Table I for the same surfaces when they are activated before phosphating. These observations indicate that one important difference between ferric and ferrous oxides is that the ferric oxides are self-activated, whereas ferrous oxides must be activated before they can be phosphated successfully.

Form and Density of Crystals in Phosphate Coatings. The observations of form and density of crystals were made with an optical microscope on complete coatings only. The following generalizations can be made.

1. The best phosphate coatings, those that bestow most corrosion resistance, consist of small compact phosphate crystals.

2. The poorest phosphate coatings consist of very large, branched dendritic crystals.

Figure 2 is an example of the appearance of our best phosphate coatings. It is a photomicrograph of a spray coating, but it equally represents the appearance of our immersion coatings on either activated or nonactivated OXID surfaces. The crystal density is of the order of 10^6 cm^{-2}. The individual crystals are small, and there is a high degree of uniformity in their size.

The crystallites in immersion coats on alkaline-cleaned (and activated) CRS and HYD A-R surfaces are also quite uniform but are larger than those on OXID surfaces. The crystal densities vary from 2×10^5 to 5×10^5 cm^{-2}. On nonactivated CRS surfaces the crystals are appreciably larger than on activated CRS surfaces. The crystal density is about 1×10^5 cm^{-2}. Finally, on nonactivated HYD A-R surfaces the crystals are very large, and they differ from other phosphate crystals in that they show very coarse surfaces. The coarseness is the result of growth proceeding dendritically with many side branches. The surface densities of crystals vary from 1×10^4 to 7×10^4 cm^{-2}.

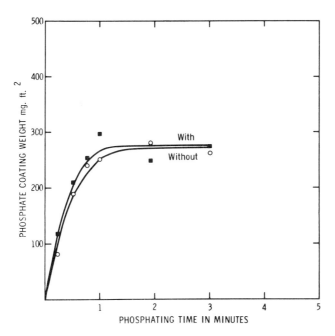

FIG. 7. Kinetics of phosphating OXID surfaces with and without titanium activation (■ = with activation, ○ = without activation) (IB-37 phosphating).

The increase in crystal size takes place not only in the plane of the substrate but also perpendicular to it. Thus, the crystals in our best coatings are quite uniform in height and vary from 3 to 5 microns. The crystals in our poorest coatings are 5 to 10 microns in height on nonactivated CRS surfaces, and individual dendritic crystals tower as much as 20 microns above nonactivated HYD A-R surfaces. Figure 8 shows the surface of a typical 3-minute coating on a nonactivated HYD A-R surface. According to Fig. 6, the 3 minutes of phosphating does not saturate this surface, since the coating weight is still increasing with time, but the dendritic nature of the coating is already well

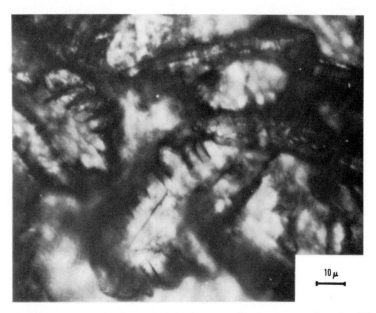

FIG. 8. Photomicrograph of a typical 3-minute coating on a nonactivated HYD A-R surface. Magnification is same as that of Fig. 2.

established. The magnification in Fig. 8 is the same as that in Figure 2. Note that, whereas hundreds of crystals are seen in Fig. 2, only the upper regions of four dendritic crystals can be resolved in Fig. 8.

X-Ray Studies. *Composition of Zinc Phosphate Coatings.* Saison (**3**) first showed that the principal constituents of a zinc phosphate coating are the two minerals hopeite ($Zn_3(PO_4)_2 \cdot 4H_2O$) and phosphophyllite ($Zn_2Fe(PO_4)_2 \cdot 4H_2O$). This was later but independently confirmed by Neuhaus *et al.* (*10*). For their analyses both Saison and Neuhaus *et al.* relied heavily on x-ray diffraction studies of zinc phosphate coatings *in situ* on iron surfaces.

Neuhaus et al. further concluded that the principal constituents are the only constituents. Saison, mostly on the basis of the color of the zinc phosphate coating along with some independent studies of iron phosphates, reasoned that trace amounts of the low-temperature cubic iron oxides (γ-Fe_2O_3 and/or Fe_3O_4) and the iron phosphate mineral vivianite are probably present in zinc phosphate coatings. Neither group found any evidence of the oxides or vivianite in their diffraction patterns, but this does not preclude the presence of small amounts of either or both. For example, extensive x-ray studies of iron phosphates by Neuhaus et al. showed that a thick coating of vivianite must be present before it can be identified by x-ray diffraction.

TABLE II
CONSTITUENTS OF ZINC PHOSPHATE COATINGS ON IRON

A. Principal constituents	
1. Hopeite	$Zn_3(PO_4)_2 \cdot 4H_2O$
2. Phosphophyllite	$Zn_2Fe(PO_4)_2 \cdot 4H_2O$
B. Minor constituents	
1. Iron oxides	γ-Fe_2O_3 or Fe_3O_4
2. Vivianite	$Fe_3(PO_4)_2 \cdot 8H_2O$
3. Lower hydrates	$Zn_3(PO_4)_2 \cdot 2H_2O$ and $Zn_2Fe(PO_4)_2 \cdot xH_2O (x < 4)$

No evidence of either the iron oxides or vivianite was found in our x-ray patterns. However, as reported, iron oxides already present on a steel surface before it is phosphated do survive the phosphating process. Therefore, if the whole coating is considered, this author tends to agree with the composition analysis by Saison. In addition, both this work and independent x-ray analysis studies by Cheever (11) frequently showed small broad peaks of lower hydrates of hopeite and phosphophyllite, and these should be included as constituents. To summarize, a more complete analysis of zinc phosphate coatings is shown in Table II.

Finally it should be noted that the final drying operation prior to painting ($2\frac{1}{2}$ minutes at 350°F is typical) will dehydrate the principal constituents (and the vivianite) so that the dihydrates listed in Table II become the major constituents (3, 11) of the coating.

Amounts of Hopeite and Phosphophyllite in Zinc Phosphate Coatings. As mentioned, the principal constituents of a zinc phosphate coating are the two minerals hopeite and phosphophyllite. The crystal structures of the two minerals are very similar (10). The principal difference between them is that in phosphophyllite one divalent ferrous ion is substituted for a divalent zinc ion in hopeite. Therefore, it would seem that if the phosphating solution could be enriched in ferrous ion the coating concentration of phosphophyllite should increase. However, such an enrichment is not easily accomplished. Saison (3) has shown that in a well "broken-in" bath, such as we used, the ratio of iron in solution remains constant, and any tendency to increase the concentration of the ferrous ion causes the precipitation of iron in the form of solids which make up much of the sludge which settles in phosphating tanks. Nevertheless, the expectation has been realized. Several authors (3, 10, 12) have used indirect methods of enriching the substrate-solution boundary layer in ferrous ion and have thereby observed an increase in the ratio of phosphophyllite to hopeite in the coating.

Early in this paper, in our discussion of background information and specifically in equation 2, we showed that iron must go into solution for phosphate crystals to grow. This dissolution is the result of "free" phosphoric acid attack of the substrate. Since there is a maximum area of substrate exposed to the phosphoric acid at the beginning of the phosphating, the dissolution rate should be largest at this stage. Therefore, if we assume that ferrous ions going into solution are the source of iron in the phosphophyllite component, we expect to find the initial growth of the coating to be phosphophyllite-rich with respect to the final stages of growth. On the contrary, we find for all surfaces and for both spray and immersion methods, that the early stages of growth are always hopeite-rich. For the spray method the coating starts and remains hopeite-rich, while for the immersion method the coatings show a pronounced tendency to become phosphophyllite-rich in late stages of growth only—at and beyond the saturation point of the kinetics curves. The partial resolution of this unexpected observation played a leading role in the development of our ideas on the growth mechanism of zinc phosphates. We say partial resolution because we cannot always reproduce such an enrichment in the late stages of all immersion coats, but it has been observed in enough cases to convince us that the effect is real. The complete resolution of this problem probably requires the identification of at least another variable in this complex process, but such an identification cannot be made at this stage of our studies.

Discussion of Results

Our results clearly show that the corrosion resistance of phosphate coatings is not solely a function of thickness. The thin coatings (those spray-deposited on CRS surfaces or immersion-deposited on OXID surfaces) bestow much greater corrosion resistance than the thick coatings (those on either nonactivated CRS or nonactivated HYD A-R surfaces). Thin coatings are made up of small platelet crystals, whereas thick coatings are made up of large dendritic crystals. These results suggest that there are two types of crystals that can make up a single coating. This conclusion is also suggested by Cheever (*12*), who states that "occasionally, additional needle-type crystals were precipitated on top of crystals already present." If we extend the range of nucleation densities that Cheever observed, (10^5 to 10^6 cm^{-2}) to the range 10^4 to 10^6 observed in this work, and generalize his observation of needle-type crystals to include more growth forms, (for example, coarse dendritic

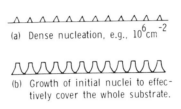

(a) Dense nucleation, e.g., 10^6 cm^{-2}

(b) Growth of initial nuclei to effectively cover the whole substrate.

FIG. 9. Growth of a complete phosphate coating with one nucleation process.

crystals), then our results are in excellent agreement with his, and we are now in a position to propose the following complete growth mechanism.

Assume that the growth of a phosphate coating can proceed to completion in one stage or, in some cases, two. Both possibilities require a nucleation process of phosphate crystals on the substrate; these are shown schematically in Figs. 9*a* and 10*a*. Only these crystals, in their subsequent lateral growth (Figs. 9*b* and 10*b*), confer good corrosion resistance on the substrate. In the second stage of nucleation, additional phosphate crystals are nucleated upon the surfaces of stage-one crystals. For this to occur the first-stage crystals must grow to a size that will provide sufficient surface area to make nucleation probable. This nucleation probability will also depend on the composition of the bath and on the method of application—spray or immersion. If the initial nucleation density is large (Fig. 9*a*)—for example, our work suggests 10^6 cm^{-2}—then before the second process becomes probable the first-stage crystals have effectively covered the whole substrate (Fig. 9*b*); no more iron goes into solution, no more growth occurs, and few if any stage-two crystals are found in the complete coating. If the initial nucleation density is small (Fig. 10*a*)—our results suggest 10^4 cm^{-2}—then long before lateral growth of the stage-one crystals covers the substrate, stage-two crystals are growing and the coating takes on the appearance of dendritic growth (Fig. 10*c*). The dendritic branches grow in random directions, material normally available for lateral growth of stage-one crystals is incorporated into the branches, and the coating grows to a much greater apparent thickness. Ultimately, the dendritic branches intertwine to such an extent that they effectively enclose cells of the phosphating bath between the branches and the substrate (Fig. 10d). The outside surface of such a cellular coating will no longer grow outward, and iron going into solution will enrich the small volume of bath entrapped in the cells. This produces two results. First, additional

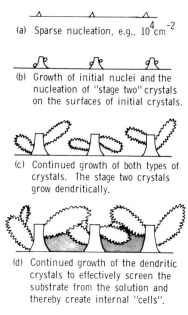

(a) Sparse nucleation, e.g., $10^4 cm^{-2}$

(b) Growth of initial nuclei and the nucleation of "stage two" crystals on the surfaces of initial crystals.

(c) Continued growth of both types of crystals. The stage two crystals grow dendritically.

(d) Continued growth of the dendritic crystals to effectively screen the substrate from the solution and thereby create internal "cells".

FIG. 10. Growth of a complete phosphate coating with two nucleation processes.

phosphophyllite-rich dendritic branches grow, but only on the inside "walls" of the cells. Such growth could ultimately fill the cell except for steric hinderances between crystals. However, such spaces could be effectively filled by the second result of the enrichment, the precipitation of sludge as was observed by Saison (3). This stage in the growth will show an apparent thickening of the film (as measured by strip and weigh methods), but it will be a relatively porous, nonadherent addition which includes entrapped corrosive (acidic) sludge. Therefore, such relatively thick films will bestow little or no corrosion protection. The thin coatings, on the other hand, that consist only of stage-one crystals, effectively cover the whole substrate with an adherent coating, with little or no entrapped sludge, and therefore give good corrosion protection.

The answers to the questions posed in the Introduction to this paper are now apparent. Thin complete phosphate coatings confer much greater corrosion resistance than do thick complete coatings because they do not incorporate corrosive sludge in the coating. The crystals in a thin complete coating are smaller because they are stage-one crystals only—that is, because they are the result of a different nucleation process from that producing the crystals that can be seen on a thick complete coating. And finally, later stages in the growth of many immersion-formed coatings tend to become iron-rich because of the two results of coating intracells, the preferred growth of the phosphophyllite component and the precipitation of iron-containing sludge.

Conclusions

The oxide film on a steel surface survives the zinc phosphating process. A typical oxide film on a steel surface contains both ferrous and ferric oxides. It is easier to deposit corrosion-resistant zinc phosphate coatings on the ferric oxides than on ferrous oxides. Ferric oxide surfaces are self-activated, and zinc phosphating of these surfaces proceeds identically whether or not they are exposed to the usual titanium phosphate activator.

Generally, zinc phosphating of steel surfaces proceeds in two stages. Both stages involve a nucleation process. The nucleation process of stage one is the initial nucleation of phosphate crystals upon the substrate surface. Stage two involves the nucleation of a second growth of crystals upon the surface of stage-one crystals. Unlike stage-one crystals, stage-two crystals grow dendritically. A phosphate coating which shows a maximum corrosion resistance is deposited under conditions which completely eliminate stage-two growth. Such elimination is accomplished by increasing the nucleation density of stage-one crystals, and for the surfaces studied 10^6 crystals/cm² are required.

Spray-deposited zinc phosphate coatings start and remain hopeite-rich. Immersion-phosphated coatings start out hopeite-rich but frequently become phosphophyllite-rich in the late stages of coating formation. This is probably the result of the formation of coating intracells, and at least one other system variable which has not been identified in this study.

Acknowledgments

I am indebted to my colleague Mr. D. C. Lechman, who helped in the performance of all phases of the work, and Dr. G. D. Cheever of the Polymers Department for many valuable discussions. I also gratefully acknowledge the valuable advice and discussions of Dr. C. E. Bleil of the Physics Department.

References

1. W. Machu, *Werkstoffe Korrosion*, **14,** 566 (1963).
2. D. W. Bloor, *Electroplating Metal Finishing*, **17,** 235 (1964).
3. J. Saison, *J. Rech. Centre Natl. Rech. Sci. Lab. Bellevue (Paris)*, **13,** No. 58, 79 (1962).
4. L. O. Gilbert, *Tech. Proc. Am. Electroplaters' Soc.*, **44,** 73 (1957).
5. Test B117-64, *Am. Soc. Testing Mater. ASTM Std.*, **21,** 1 (1966).
6. P. B. Sewell, C. D. Stockbridge, and M. Cohen, *J. Electrochem. Soc.*, **108,** 933 (1961).
7. E. J. Caule, *Corrosion Sci.*, **2,** 147 (1962).
8. E. J. Caule, K. H. Buob, and M. Cohen, *J. Electrochem. Soc.*, **108,** 829 (1961).
9. J. I. Maurer, private communication, 1967.
10. A. Neuhaus, E. Jumpertz, and M. Gebhardt, *Korrosion*, **16,** 155 (1963).
11. G. D. Cheever, private communication, 1967.
12. G. D. Cheever, *J. Paint Technol.*, **39,** 1 (1967).

Discussion

McCullough: If the initiation of the phosphating process requires dissolution of the iron from the matrix, I am curious as to how this chemical dissolution occurs with the complete survival of the oxide film, unless there must be considerable imperfections in the iron oxide.

Laukonis: First let me emphasize that the initiation mechanism proposed by others and reiterated in this paper—viz., iron dissolution, hydrogen discharge, and the pH change that this discharge produces—is so reasonable that I accept it without reservation. Therefore, I feel that the dissolution of iron is necessary. My observations suggest that the oxide film is not the source of any significant portion of this iron. Therefore, your suggestion of considerable imperfections in the iron oxide is very probably correct. Once we come to this conclusion we can speculate on the nature of the imperfections. If they were highly localized, the dissolution would be turbulent, the substrate should be pitted, and I would expect a nonuniform phosphate coating to be deposited. Although I have not studied the uniformity of the phosphate coatings in any detail, both superficial examination and their excellent behavior in the salt spray suggest that they are uniform. Therefore, I am inclined to think that the imperfections are uniformly distributed throughout oxide film so as to effectively make it porous to the phosphating solutions. If this is so, then substrate iron dissolution can proceed uninhibited by the film of iron oxides.

McCullough: It seems that a magnification examination for pits would be in order.

LAUKONIS: I agree.

DOUTY: We have been aware for a long time that phosphate coatings do grow on iron oxides. In fact, there have been commercial applications of this. However, I also am perplexed by the fact that, if panels are very carefully weighed and then phosphated under the most efficient possible conditions, we find that there is ultimately a loss of panel weight of never less than perhaps a third of the weight of the coating produced. It varies, depending on the exact circumstances, from numbers like two to perhaps five, but never more. Iron is always lost from the panels in our experiments. I do not think the oxide films on the panels that we were using weighed very much in milligrams per square foot.

LAUKONIS: I am happy to hear of your corroboration regarding the phosphating of iron oxides. The information is, of course, completely missing in the literature.

I assume that the careful weighing experiments which you mention were performed with what I have called cold-rolled steel surfaces. I did not perform such experiments, but I anticipate that they would show similar dissolution losses in my experiments also. I further anticipate, on the basis of the growth mechanism I have proposed, that panel weight loss should be proportional to coating weight. Therefore, the panel weight loss should be less for a coating on an oxidized surface and much greater for coatings on such surfaces as my unactivated hydrogen annealed-reduced surfaces. In any case the weight of iron incorporated in the oxide films on all the surfaces I phosphated is less than one-third of the coating weights. For the surfaces covered with blue oxide films the iron content is about one-fifteenth of the coating weights and much less for all the other surfaces.

DOUTY: At any rate, that observation is an old one. I have held pieces of steel in a Bunsen burner and obtained various degrees of blue and then cut up sections, and the amount of coating is actually sometimes greater on the oxidized than on the unoxidized portion.

LAUKONIS: Your observation is very interesting and probably adds to my paper without contradicting my results which show that coating weights are smaller on oxidized surfaces.

First, the surfaces you held in the flame probably were oxidized at temperatures above 570°C. Therefore, the principal oxide that formed was FeO, which disproportionated, on cooling, to Fe_3O_4 and Fe. My surfaces were generally oxidized below 400°C so that no FeO was present. Your observations suggest that the very thin α-Fe_2O_3 layer on oxides formed above 570°C either does not confer the benefits I report, or the benefit, and layer, are effectively destroyed when the thick FeO layer disproportionates on cooling—thus, the larger coating weights which you observed.

Second, unless the pieces of steel that you held in a Bunsen burner were large, it is unlikely that any of the surface remained free of some high-temperature oxide growth. The portions that looked to be unoxidized were covered with thinner oxide films which formed at lower temperatures because they were removed from the flame. In this case the composition of the thinner films was about the composition I report for my blue films, and therefore less coating was deposited.

Finally, even if the surfaces were relatively free of any high-temperature oxides, your results suggest that more coating is deposited on the FeO, etc., oxides than on room-temperature oxide films.

DOUTY: If we take into account that ferrous iron has an atomic weight of 55.9 and the zinc phosphate coating weight is about 300 mg per square foot, a substantial amount of iron must dissolve in order to cause the coating to form.

LAUKONIS: Using the figures you gave in your first question we can say that about 100 mg per square foot of iron must be dissolved. This corresponds to a layer of iron about 1400 Å thick. Therefore, the 500-Å (or thinner) blue oxide must permit the passage of 1400 Å of iron in about the 1 minute of time which it takes for the phosphate coating to form. In addition, this must be done without destroying the bonds between the oxide and the iron. The numbers look formidable but not unrealistic.

SESSION III

Chairman
P. H. MARK

A Collective Viewpoint of Surfaces
A Contributed Discussion

P. H. MARK

Department of Electrical Engineering
Princeton University
Princeton, New Jersey

During the course of this conference, I have been impressed by how many useful properties of surfaces can be understood with an atomistic model of the surface. There are, however, whole classes of physical (and perhaps also chemical) phenomena, principally electrical and optical, that are better understood by viewing the surface collectively, as the solid-state physicist views the interior of a solid. For this purpose, I thought it advantageous to give a very short review of the collective picture of a structurally perfect surface.

In the collective sense, a periodic surface is characterized by its so-called surface states. These are the energy eigenstates of the Schrödinger equation for a finite lattice whose wavefunctions are largest at the edges of the lattice and tend to vanish rather rapidly with distance toward the interior, so that, when such a state is occupied, the probability of localizing the electron at the surface is maximum. Because of this localization, occupied surface states are frequently called free valences by catalysis chemists. Although surface-state energies may coincide with band or gap energies of the bulk band structure, the term is almost always reserved for states whose energies are in the gaps. These surface states are a collective property of the lattice; they are crystal states in the sense that their wavefunctions are solutions of the Schrödinger equation for the entire lattice just as are the unattenuated wavefunctions whose eigenenergies form the bulk band structure of the lattice. The surface states originate from a different boundary condition on the wave equation. The conventional Born-von Karman cyclic boundary condition, from which the bulk band structure follows, is replaced with a boundary condition that is a more accurate description of how a finite lattice terminates.

Surface states can be intrinsic for lattices terminating at a free surface with no structural defects, impurities, or other imperfections, other than the surface itself. These states can be developed according to the various elementary approximations, such as that of nearly free electrons or that of tight binding, that have been applied to bulk solids. In the former, the lattice is terminated by a potential discontinuity and the boundary condition consists in smoothly joining the oscillatory wavefunction inside the lattice to decaying function outside the lattice. Depending on the symmetry properties of the wavefunctions, this boundary condition may cause two closely spaced (the spacing of the surface states is negligible for macroscopic solids) surface states to appear in the gap between the valence and the conduction band of a semiconductor or insulator (or in a metal in a much less meaningful sense), the upper state dropping out of the upper band and the lower state rising from the lower band as shown in Fig. 1a. The surface states must derive from the bulk bands by the conservation of states. In the tight binding approximation, no mathematical boundary condition is imposed directly on the crystal wavefunctions, since these are taken as linear

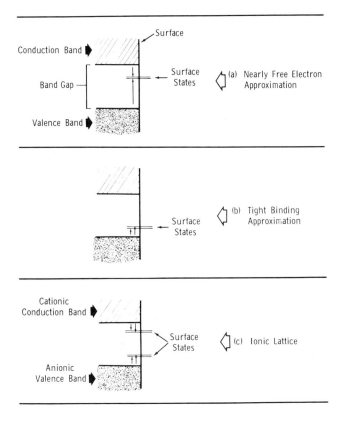

FIG. 1. Collective picture of a structurally perfect surface.

combinations of atomic orbitals. The surface is specifically included in the potential of the wave equation and ultimately reaches the wavefunction through the expansion coefficients. In this approximation, there are again two closely spaced surface states in the gap, as seen in part *b* of the figure, but now they both derive from the lower band and exist only when the surface distortion is large enough. A final class of surface states can be derived from ionic binding. As shown in part *c*, in this case there are four surface states in the gap arranged as two pairs of closely spaced states: one pair in the upper half of the gap deriving from the upper band, and the other pair in the lower gap deriving from the lower band.

The significant differences among these surface states hinge on their occupation. At $0°K$, their occupation coincides with that of the bands from which they derive. If the gap in question is that between the valence and conduction bands of a semiconductor or of an insulator, then at $0°K$ the nearly free electron surface states are half occupied (metal-like), the tight binding surface states are fully occupied (insulator-like), and the ionic surface states are half occupied with the occupied and empty states separated by an energy gap (semiconductor-like). At a finite temperature, the occupation of the surface states is always determined by the location of the Fermi level *at the surface* through the simultaneous satisfaction of the Fermi statistics, charge neutrality, and the Poisson equation. The experimental and theoretical determination of the occupation of surface states is usually a thorny problem.

In addition to the intrinsic surface state of perfect surfaces, there are surface states associated with structural imperfections, such as one sort or another of active site, and with impurities, such as adsorbed species. Such extrinsic surface states are frequently handled with a mixed model that still considers the surface collectively but views the imperfection atomistically.

In closing, I should like to give a short bibliography for those who are interested in pursuing the matter: E. T. Goodwin, *Proc. Cambridge Phil. Soc.*, **35,** 205, 221, 232 (1939); W. Shockley, *Phys. Rev.*, **56,** 317 (1939); K. Artmann, *Z. Physik*, **131,** 244 (1952); J. D. Levine and P. Mark, *Phys. Rev.*, **144,** 751 (1966) for the surface states of structurally perfect surfaces; P. Mark, *J. Phys. Chem. Solids* (in press) for a treatment of the chemisorption states of ionic lattices; and finally review papers in *Catalysis Reviews* **1,** 165 (1968) and in "Proceedings of the Conference on Clean Surfaces: Their Preparation and Characterization for Interfacial Studies" (Marcel Dekker, Inc.) (in press).

The Physical Chemistry of Surfaces and Surface Heterogeneities

A. C. ZETTLEMOYER

Center for Surface and Coatings Research
Lehigh University
Bethlehem, Pennsylvania

Introduction

Great strides have been made in adhesion science in recent years. We now know that ordinary dispersion or van der Waals' forces can be responsible for adhesive strengths far greater than experienced in practice. In many cases the dispersion forces are the only ones which interact across an interface; dipole and induced dipole forces contribute little or no interactions. Of course, sufficiently intimate contact with crevices and recesses may not be achieved. Then, the wetting process—that is, the rate of flow—may govern ultimate performance of an adhesive joint. The coating may congeal before thermodynamic equilibrium is reached.

This wetting or adsorption theory of adhesion is accepted by most workers in the field of adhesion at the expense of the diffusion and electrostatic theories which are mostly unrequired to explain results. Chemical bonding sometimes occurs in adhesion. Although the thermodynamic aspects of wetting will be emphasized here, it should be noted that the kinetic aspects may dominate practical situations. In special cases, also, chemisorption may provide the links across the interface.

The physical nature of interactions in terms of surface free energies and dispersion forces will first be examined. Then, the use of heats of immersion as a prime method of measurement will be described. A natural product of the development will be a description of work of adhesion and the maximum force of adhesion that can be expected.

The nature of chemical heterogeneities will be examined. Two techniques for examining heterogeneities—vapor adsorption and immersional calorimetry—will be described. Information concerning heterogeneities on oxide and metal surfaces will be presented.

Interactions at Interfaces

Interactions across interfaces will first be treated by the geometric means technique due to Fowkes (1). Then the results of a new arithmetic approach will be examined. One of the best techniques for estimating the part of the surface energy that interacts with a contacting phase is the heat of immersion technique; its use will be described.

Geometric Mean Averaging

The interfacial free energy or the interfacial tension can be expressed in terms of the surface tensions of the two individual phases which are placed in contact:

$$\gamma_{12} = \gamma_1 + \gamma_2 - 2\sqrt{\gamma_1^d \gamma_2^d} \tag{1}$$

This relation is parallel to the Berthelot relation for imperfect gases, and its use has been recently developed by Fowkes (1). The geometric mean contained in the last term is not unusual in physical interactions; it is based on the dispersion part of the surface free energies only, the γ^d values, and arises as indicated in Fig. 1. Take liquid 1, first without a second condensed phase in contact. Surface concentration of the molecules is lower than the bulk as a manifestation of the pull inward due to the surface tension, γ_1. At equilibrium, and except near the critical temperature, the vapor phase is obviously much less dense than the liquid. When a second condensed phase, phase 2, is brought in contact with the first, the γ_1 tension is reduced by the pull across the interface in the opposite direction. This tension is found to depend almost entirely on the dispersion force interaction. It is expressed as the geometric mean $\sqrt{\gamma_1^d \gamma_2^d}$, and so the γ_1 is reduced to $\gamma_1 - \sqrt{\gamma_1^d \gamma_2^d}$. A similar argument for the second phase leads to the new tension, $\gamma_2 - \sqrt{\gamma_1^d \gamma_2^d}$. Upon addition, equation 1 is obtained.

A saturated hydrocarbon in contact with water is a simple case to consider. The interfacial tensions of many such systems are well known, as, of course, are the γ_1 and γ_2. Furthermore, the hydrocarbon will display only dispersion force interactions; hence γ_1 and γ_1^d will be the same. Not so for water. The γ_2 for water (72 dynes/cm at 25°C) will be much larger than γ_2^d because the large dipole of water contributes heavily to γ_2.

What might be called the Law of Fowkes can now be written:

$$\gamma_{H_2O} = \gamma_{H_2O}^d + \gamma_{H_2O}^P \tag{2}$$

where $\gamma_{H_2O}^P$, the dipole or hydrogen bond contribution, is $72 - 22 = 50$ dynes/cm. Very often only the γ^d contributes appreciably across an interface with another condensed phase.

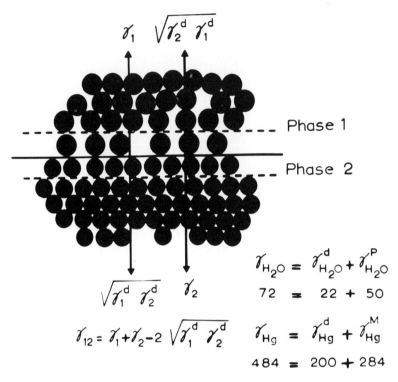

FIG. 1. Model for estimating the London dispersion force contribution to interfacial energies.

Consider the mercury–water interface. The surface tension is 484 dynes/cm at 25°C. It is the metallic bond which is responsible for the high surface tensions or surface free energies of metals; they are commonly in the thousands of dynes per centimeter or ergs per square centimeter. Here we can write

$$\gamma_{Hg} = \gamma_{Hg}^d + \gamma_{Hg}^M \qquad (3)$$

where γ_{Hg}^M represents the contribution of the metallic bond. From the measured values for the mercury–hydrocarbon interfaces, γ_{Hg}^d is found to be 200. Thus, the appropriate value for γ_{12} for the Hg–H$_2$O interface is readily calculated; it checks the value measured.

This analysis of interfacial tensions implies, as stated heretofore, that often only dispersion forces operate significantly across interfaces. Adhesion as a thermodynamic quantity derived from the wetting hypothesis need not include dipole and induced dipole interactions. We shall see that the theoretical adhesional forces which dispersion forces can produce are far larger than are ever realized in adhesional joints.

A curious comparison arises when the values of γ_s^d of bare metals and their oxides are examined. For both iron and tin, the values are compared in Table I. Both have γ_s^d values of the same magnitude. But oxides can be more complex. First, they usually possess hydroxylated surfaces in varying surface concentrations unless produced anhydrously. Second, the hydroxyls

TABLE I
VALUES OF γ_s^d FOR SOLID SURFACES

Solid	γ_s^d (ergs/cm²)
Fe	90–108
Fe$_2$O$_3$	107
Sn	101
SnO$_2$	111

possess strong dipoles which interact strongly with hydrogen-bonding molecules such as water. We shall return to this problem of omnipresent water on oxide surfaces.

Arithmetic Mean Averaging

A surprising result of using the geometric mean relationship to predict intermolecular forces acting at a metal–organic liquid interface is that the interaction of dipoles with the metal often appears to be negligibly small. Furthermore, there is no indication of an interaction of the metal (mercury) with the π-electrons of aromatic compounds.

Lavelle of our laboratories, using an arithmetic mean of the dispersion force attractions to estimate the magnitude of the interaction between dissimilar materials, has found that the following equation yields some interesting results:

$$\gamma_{12} = \gamma_1 + \gamma_2 - (\gamma_1^d + \gamma_2^d) \tag{4}$$

The two terms by which the values of γ are diminished are taken to be $(\gamma_1^d + \gamma_2^d)/2$. Of course, the arithmetic mean has no particular scientific basis for its use.

From the interfacial tension data with hydrocarbons, $\gamma_{H_2O}^d$ is still found to be 22 ergs/cm² by use of the arithmetic mean. However, γ_{Hg}^d is found to be only 108 ergs/cm² instead of 200 by this averaging technique. When these $\gamma_{H_2O}^d$ and γ_{Hg}^d values are used in equation 4 to calculate γ_{Hg/H_2O}, a value of

TABLE II
Comparison of Geometric and Arithmetic Means in Calculating the Energy of Mercury–Organic Liquid Interaction in Excess of Dispersion Forces (ergs/cm²)

Liquid No. 2	Geometric Mean	Arithmetic Mean
Hexane	0	0
Benzene	0	13
Toluene	0	17
p-Xylene	0	15
Bromobenzene	0	25
Aniline	0	34
1,2-Dibromomethane	0	29
Cyclopentanol	−9	+10
Methanol	−11	−8
n-Propanol	−8	−3

426 dynes/cm is obtained. This interfacial free energy checks the best experimental values and indicates that the arithmetic mean has the same internal consistency as does the geometric mean approach.

The arithmetic mean predicts an interaction of mercury with the π-electron system of aromatic compounds at a magnitude of about 15 ergs/cm². Table II shows a comparison of the geometric and arithmetic means in calculating the polar interfacial attractions at the mercury–organic liquid interface. It is interesting to note that both averaging techniques predict no interaction of short-chain alcohols with mercury.

TABLE III
Comparison of Geometric and Arithmetic Means in Calculating γ_s^d Values from π_e Measurements with Heptane Vapor (25°C)

Solid	π_e	γ_s^d (geometric mean)	γ_s^d (arithmetic mean)
Graphite	63, 56, 58	132, 115, 120	83, 76, 78
Copper	29	60	49.4
Silver	37	74	57.4
Lead	49	99	69.4
Iron	53	108	73.4
Ferric iron oxide	54	107	74.4
Anatase (TiO$_2$)	46	92	66.4
Silica	39	78	59.4
Stannic oxide	54	111	74.4
Tin	50	101	70.4

Table III shows a comparison of the geometric and arithmetic means in calculating γ_s^d values from the free energy of adsorption of heptane vapors on the solids. In all cases, the arithmetic mean predicts a lower contribution of the London dispersion forces to the surface free energy.

Table IV shows the work of adhesion of heptane on iron as calculated by both averaging procedures. The calculated values are identical and agree

TABLE IV
COMPARISON OF GEOMETRIC AND ARITHMETIC MEANS IN CALCULATING THE WORK OF ADHESION FOR HEPTANE ON IRON

$$W_A = \gamma_S + \gamma_L - \gamma_{SL}$$

$W_A = 2\sqrt{\gamma_S^d \gamma_L^d}$ (geometric mean)	$W_A = \gamma_S^d + \gamma_L^d$ (arithmetic mean)
$W_A = 2\sqrt{(108)(20.4)}$	$W_A = 73.4 + 20.4$
$W_A = 93.8$ ergs/cm²	$W_A = 93.8$ ergs/cm²
$W_A = \pi_e + 2\gamma_L$	[Harkins and Loesser J. Chem. Phys., **18**, 556 (1950)]
$W_A = 53 + 2(20.4)$	
$W_A = 93.8$	

with the value of 93.8 ergs/cm² measured by Harkins and Loesser. This agreement again shows the internal consistency of both averaging techniques.

Heat of Immersion Technique for Determining γ^d Values

The heat of immersion technique is one of the most direct methods for determining the γ^d values of solids. For example, an evacuated solid is broken into a liquid in a calorimeter, as shown in Fig. 2. The heat liberated is measured by a thermistor. The free energy change per unit area is

$$g_I = \gamma_{SL} - \gamma_S \tag{5}$$

Then from equation 1:

$$g_I = \gamma_L - 2\sqrt{\gamma_S^d \gamma_L^d} \tag{6}$$

Since

$$h_I = g_I - T\left(\frac{dg_I}{dT}\right) \tag{7}$$

then

$$h_I = \gamma_L - 2\sqrt{\gamma_S^d \gamma_L^d} - T\left[\left(\frac{d\gamma_L}{dT}\right) - 2\sqrt{\gamma_L^d}\frac{d\sqrt{\gamma_S^d}}{dT} - 2\sqrt{\gamma_S^d}\frac{d\sqrt{\gamma_L^d}}{dT}\right] \tag{8}$$

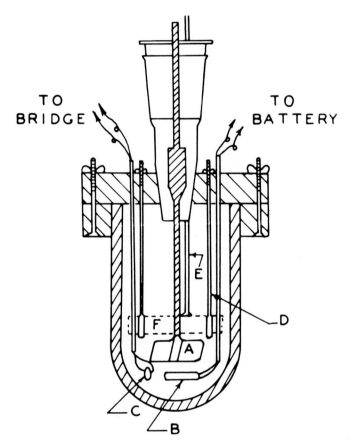

FIG. 2. Thermistor calorimeter for measurement of heat of immersion. *A*, stirrer; *B*, heater; *C*, thermistor; *D*, sample holder; *E*, breaking rod; *F*, sample tube. A sensitive resistance bridge such as the Mueller bridge and a galvanometer of sensitivity of the order of 10^{-10} ampere allows the determination of heats of the order of 0.01 calorie. The calorimeter is placed inside an air thermostat. Evolution of heat up to about 10 minutes can be followed with a simple arrangement of this type. Submarine-type calorimeters enclosed in large water thermostats are needed to follow heat effects over longer periods.

The h_I, γ_L, $\gamma_L{}^a$, and $d\sqrt{\gamma_L{}^a}/dT$ are determinable as already discussed. The $d\sqrt{\gamma_S{}^a}/dT$ values are very small, and the terms containing $d\sqrt{\gamma_S{}^a}/dT$ can be neglected. (The values of $d\sqrt{\gamma_S{}^a}/dT$ may be estimated from the fourth power of the density.)

A simple case is a graphitic solid like Graphon (a graphitized carbon

TABLE V
COMPARISON OF CALCULATED AND MEASURED HEATS OF IMMERSIONAL WETTING OF GRAPHON AT 25°C (ergs/cm²)

Liquid	h_I (measured)	h_I^d (calculated)
Heptane	-103 ± 3	-112
Butyl amine	-106 ± 6	-112
Butyl alcohol	-114 ± 5	-107
Butyl chloride	-106 ± 2	-111
Butyric acid	-115 ± 1	-110
Water*	-32.2 ± 0.1	-34.0

* $\gamma_L^d = 21.8$ ergs/cm².

black), where the interaction with liquids might be expected to be due only to dispersion forces. From a variety of sources $\gamma_s^d = 110$ ergs/cm². Together with the liquid data, the results in Table V have been calculated from equation 7. The agreement between the measured and the calculated values is good. Graphite or Graphon does not show any appreciable interaction with the liquid dipoles.

The work of adhesion can be calculated from the appropriate γ values. We can consider two extremes as paths for separation of a solid–liquid interface as shown in Fig. 3. In the first path, a film of liquid is left behind on the solid in equilibrium with the saturated vapor of the liquid, p_0, at T. In the second path, a clean separation of the solid–liquid interface is depicted.

For path I:

$$W_A = \gamma_{SV^\circ} + \gamma_{LV^\circ} - \gamma_{SL} \tag{9}$$

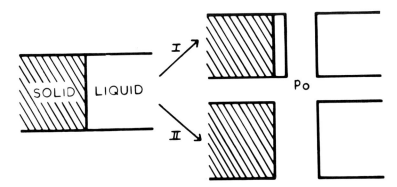

FIG. 3. The adhesion process. The solid and liquid surfaces may be separated to give a film-free solid (path II) or a solid surface with a film of adsorbed liquid (path I).

Young's equation, $\gamma_{SV^\circ} - \gamma_{SL} = \gamma_{LV} \cos \theta$, enables us to write

$$W_A = \gamma_{LV^\circ}(1 + \cos \theta) \qquad (10)$$

For complete wetting of heptane on iron, $\cos \theta = 1$, and $W_a = 2\gamma_{LV^\circ} \cong 40$ ergs/cm².

For path II:

$$W_A' = \gamma_{S^\circ} + \gamma_{LV^\circ} - \gamma_{SL} \qquad (11)$$

The difference between γ_{S°, the bare solid value, and γ_{SV°, the film-coated solid to p_0, is called the equilibrium spreading pressure, π_e:

$$\gamma_{S^\circ} - \gamma_{SV^\circ} = \pi_e \qquad (12)$$

and so we substitute $\gamma_{SV^\circ} = \gamma_{S^\circ} - \pi_e$ into Young's equation:

$$\gamma_{S^\circ} - \gamma_{SL} = \gamma_{LV^\circ} \cos \theta + \pi_e \qquad (13)$$

Then

$$W_A' = \gamma_{LV^\circ} \cos \theta + \pi_e + \gamma_{LV^\circ} \qquad (14)$$

In this case, complete wetting gives

$$W_A' = 2\gamma_{LV^\circ} + \pi_e \qquad (15)$$

The value of π_e can be even larger than the first term on the right. For the iron–heptane system, it is 53 ergs/cm², so $W_A' = 93$ ergs/cm². Obviously, it takes much more energy to separate heptane cleanly from iron than to leave a film behind. In the latter case, only the cohesive energy of the liquid is involved. Incidentally, the value of π_e will be insignificant when the solid is poorly wetted, but this situation is not important in adhesion.

These adhesional energies seem to be small until we translate them into forces. A reasonable distance over which the force of adhesion might operate is 2 Å. Then

$$f_A = \frac{40 \text{ ergs/cm}^2}{2 \times 10^{-8} \text{ cm}} = 20 \times 10^8 \text{ dynes/cm}^2 \cong 29 \times 10^3 \text{ psi} \cong 14.5 \text{ tons/in.}^2$$

and $\qquad (16)$

$$f_A' = \frac{93 \text{ ergs/cm}^2}{2 \times 10^{-8} \text{ cm}} = 46 \times 10^8 \text{ dynes/cm}^2 \cong 68 \times 10^3 \text{ psi} \cong 34 \text{ tons/in.}^2$$

$$(17)$$

It is not surprising, therefore, that there is evidence that many joints when separated leave some adhesive on the solid. Of course, practical adhesive joints are invariably much weaker than these values would suggest. Dispersion forces and wetting easily can account for the strength of adhesive joints.

To complete the thermodynamic picture of adhesion, we next consider the enthalpy of adhesion:

$$h_{A'} = h_{S^\circ} + h_{LV^\circ} - h_{SL} \tag{18}$$

or

$$h_A' = h_{LV^\circ} - h_{I(SL)}$$

where $h_{I(SL)}$ is the heat of immersion of the solid in the liquid. Measurement of this quantity will be discussed later, and h_{LV° is easily obtained from the temperature variation of the surface tension of the liquid; it is 118.5 ergs/cm² for water, and much lower for common organic liquids (20 to 50 ergs/cm²). The entropy of adhesion is then given by

$$S_A' = \frac{h_A' - W_A'}{T} \tag{19}$$

All these thermodynamic quantities are macroscopic and deal with averages. The microscopic properties will be examined shortly.

Only if the separation occurs at the solid boundary does the nature of the solid affect the result in the idealized cases we have considered. When the solid affects the adhesion, then different crystal faces can give different adhesional strengths. Schonhorn (2) has reported that surface treatment of polymeric substrates can affect adhesion to them. Also, the adhesion may be aided by a roughened surface if the adhesive flows into the crevices. Finally, and very likely most important in many practical situations, we have the presence of strains in the dried or solid adhesive. These strains, due to contraction or expansion on congealing and to misfit with the substrate surface structure, furnish a prime cause of joints not reaching the levels of the wetting theory of adhesion. Another cause is heterogeneities, which no doubt play a vital role in adherance to solid surfaces. We shall examine these next.

Heterogeneities on Solid Surfaces

There are two regimes in the study of solid surfaces. One regime concerns itself with carefully prepared and idealized surfaces. They may be single crystals of metals or semiconductors, or a sheet of mica split open to yield fresh surface, all in an ultrahigh vacuum. The other regime deals with "real" surfaces made up of both geometrical and chemical heterogeneities.

It is the latter circumstance with which we shall deal here. Low-energy electron diffraction (LEED) and field emission and field ion microscopy are not of use for multigranular substrates like powders and sheet metals. Instead, spectroscopic methods—IR, NMR, and ESR—and direct adsorption and heat of immersion methods prove fruitful. The latter two methods enable us to determine site energy distributions. The actual spatial distribution of site energies from a particular site outward would be valuable information but it is not available. Nor in many cases can we ascribe site energies to specific defect centers and impurities.

Adsorption Isotherms and Heats of Immersion

Site energy distribution can be developed from adsorption isotherms. Generally, it may be stated that the fraction of the surface covered should be some function of E, P, and T, where E is the adsorption energy. The function $f(E)\,dE$ describes the probability of there being an adsorption energy between E and $E + dE$. The adsorption isotherm will be the sum of the adsorption on the different kinds of surface sites, and thus will be a function of P and T:

$$\theta(P, T) = \int_0^\infty \theta(E, P, T) f(E)\,dE \tag{20}$$

There is a corresponding integral distribution function, F, which gives the fraction of surface having adsorption energies greater than or equal to a given E:

$$\frac{dF}{dE} = f(E) \tag{21}$$

Hence:

$$\theta(P, T) = \int_0^1 \theta(E, P, T) f(E)\,dF \tag{22}$$

Some adsorption will occur on all portions of the surface, so that the heat liberated for dn moles adsorbed will be a weighted average; that is, dF will not be the same as $d\theta$. At least, only at $0°$K will the sites fill sequentially on sites of increasing E.

One approach is to assume functions for both $\theta(E, P, T)$ and $f(E)$ so that integration is possible. For example, for the first function the Langmuir equation or a two equation of state like the van der Waals' equation may be chosen; the $f(E)$ may be taken to be Gaussian.

Another approach is to use a transform procedure. A function for $\theta(E, P, T)$ is chosen so that the integral equation can be inverted to give $f(E)$ from the observed isotherm. This rather elegant approach still requires the choice of an analytical expression, and there are serious mathematical limitations to the choices that may be conveniently made. In addition, good low-pressure data are required to analyze the isotherm for the sites of highest energy. Very few data as good as those being developed by Hobson and Armstrong (3) now exist, and particularly absent are such data on carefully prepared surfaces. We shall now leave this subject of direct adsorption measurements and proceed to the heat of immersion technique. Without vapor adsorption measurements as well, however, heats of immersion lack supporting information which is almost always needed.

Heat of Immersion Technique To Determine Heterogeneities

Heat is evolved when a solid is immersed in a liquid, and the enthalpy change is always negative. The total heat of immersion, ΔH_I, the heat evolved on the immersion of a clean outgassed solid into a liquid, is the most straightforward quantity. Since specific surface areas of most solids can be established with reasonable certainty, the heat of immersion may be put on a unit area basis:

$$\frac{\Delta H_I}{\Sigma} = h_{SL} - h_{S^\circ} = h_{I(SL)} \cong e_{I(SL)} = e_{SL} - e_{S^\circ} \qquad (23)$$

where $h_{I(SL)}$ is the heat of immersion per unit area of the solid, ΔH_I the total heat of immersion per gram of solid, Σ the surface area of the solid per gram, h_{SL} the enthalpy of the solid–liquid interface, h_{S° the enthalpy of the solid–vacuum interface, $e_{I(SL)}$ the change in internal energy for the immersion process, e_{SL} the internal energy of the solid–liquid interface, and e_{S° the internal energy of the solid–vacuum interface. Since there is very little pressure-volume work during the immersional process, the enthalpy change is essentially the same as the change in internal energy. The total heat-of-immersion values are themselves useful in comparing different solid surfaces and in assessing their polarity. Variation of heat of immersion with activation temperature has been of great value in studying the nature of bound water on oxide surfaces. This aspect will be discussed later.

The immersional wetting can be carried out after the solid surface has been precovered with a known amount of the adsorbate from the vapor phase. This process is depicted in path 2 of Fig. 4. Path 1 represents the immersion of the bare solid. The heat of immersion can then be related to the heat of adsorption of the same molecular species. To make path 2 equivalent to path 1, the molecules for adsorption have to be evaporated from the liquid. Then the enthalpy changes for the two paths can be equated:

$$-h_{I(SL)} = -h_{A(SV)} - \Gamma h_L - h_{I(S_fL)} \cong e_A - e_L \qquad (24)$$

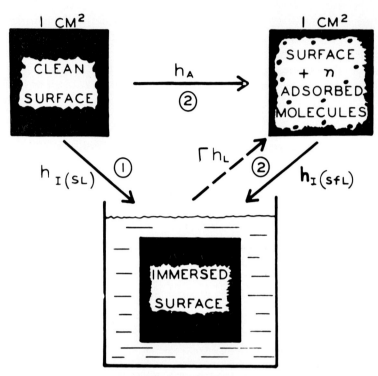

FIG. 4. Diagrammatic representation of the relation between adsorption and immersional processes (after Harkins). In path 1 the clean surface is immersed in the liquid. In path 2 the surface is first covered by N_A molecules of the adsorbate from the gas phase and subsequently immersed in the liquid. The relationship between the two processes is discussed in the text. Following this method the integral and differential heat of adsorption and site energy distribution on a solid surface can be determined.

where $h_{A(SV)}$ is the integral heat of adsorption of the coverage obtained at the vapor pressure represented by V, $\Gamma = (N_A/\Sigma)$ represents the molecules adsorbed per square centimeter during precoverage, h_L is the heat of liquefaction per molecule of the vapor of the wetting liquid, and $h_{I(S_fL)}$ is the heat liberated on immersion of the precovered solid. The negative signs are included to emphasize the exothermic character of the various heat effects. On rearrangement of equation 24, the integral heat of adsorption is given by

$$h_{A(SV)} = h_{I(SL)} - h_{I(S_fL)} - \Gamma h_L \tag{25}$$

Equation 25 clearly illustrates the relationship between the immersional and adsorption processes.

When the solid is precovered to the extent that the film is liquid-like (at least with reference to enthalpy), then $h_{I(SfL)}$ becomes $h_{LV°}$, the enthalpy of the liquid surface per unit area. Sometimes the situation develops when only one monolayer has been preadsorbed, as for water on some oxides. In other cases, additional layers are required. The net integral heat of adsorption is then obtained from equation 26:

$$h_{A(SV)} = h_{I(SL)} - h_{LV°} - \Gamma h_L \cong e_A - e_L \qquad (26)$$

Equation 26 has been used in some instances to calculate net integral heats of adsorption. It is not difficult to determine $h_{I(SfL)}$ experimentally and employ equation 25 to calculate the net integral heats of adsorption. The use of equation 26 may yield erroneous values of $h_{A(SV)}$, especially in the case of organic liquids interacting with polar surfaces.

Immersional calorimetry yields integral heats, since all the interaction up to the precoverage is included. The integral heat and the isosteric heat are related by

$$h_{I(SL)} - h_{I(SfL)} = \int_0^\Gamma q_{St}\, d\Gamma + \Gamma h_L = (e_A' - e_L)\Gamma \qquad (27)$$

Equation 27 relates the measured heat effects to the molar energies of the adsorbate solid system and that for the liquid. If perturbations of the solid surface may be neglected, e_A' may be regarded as the molar energy of the adsorbate itself. Adsorption indeed implies that the solid surface is perturbed, but the energy involved in the perturbation is likely to be small in the case of physical adsorption.

Chemisorption, on the other hand, implies severe surface perturbations and may lead to largely irreversible adsorption isotherms. Heats of adsorption cannot be calculated from isotherms which are not reversible. However, the direct determination of heats of immersion as a function of precoverage allows reasonably accurate evaluations of heats of adsorption from equation 25 even when rates of attainment of equilibrium are very slow.

Oxide Surfaces

One of the most interesting aspects of the surface chemistry of oxides is the presence of surface hydroxyls with their resulting influence on the adsorption of water and other molecules. Evidence for the presence of surface hydroxyls is derived from a variety of experimental techniques, such as direct dehydration (at temperatures above 180°C), reactions of active molecules with the OH's, and infrared and NMR spectroscopy (4, 5). The surface hydroxyls are rather resistant to dehydration at room temperature, although most of the hydroxyls undergo dehydration at temperatures of the order of 400° to 500°C. A few hydroxyls sometimes persist even after outgassing in a vacuum at temperatures of the order of 600° to 700°C (6).

The picture regarding hydroxyls is most clear in the case of silicas; variations exist, depending on the mode of preparation and the history of the sample. Silicas prepared by hydrolysis at ordinary temperatures possess fully hydroxylated surfaces. On the other hand, silicas prepared by the flame hydrolysis of silicon halides, sulfides, or organosilicon compounds are only partially hydroxylated. The surface concentration of hydroxyls is rather obscure in the case of titanium dioxide surfaces.

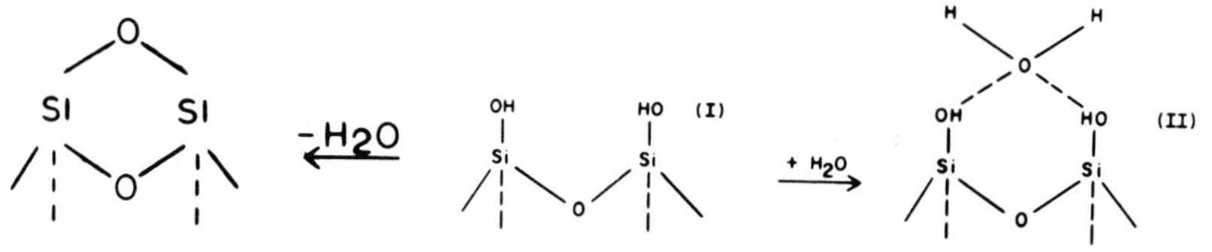

FIG. 5. Diagrammatic representation of a hydrated silica surface and one containing strongly bound adsorbed water (8). The stretching frequency for OH changes on the adsorption of the water. A dehydrated or siloxane surface is shown to the left.

Kiselev (7) proposed that water adsorption on oxides such as silica takes place entirely by hydrogen bonding between the oxygen of the water and hydrogens of two adjacent surface hydroxyl groups. The same mechanism is apparently prevalent in the adsorption of alcohols and other polar organic molecules. According to Kiselev, the physisorption of benzene can be explained in terms of the interactions between the π-electrons of the aromatic ring and the hydrogen of the surface hydroxyls. The benzene lies flat on the surface over the surface hydroxyl with which it interacts.

The above mechanisms for the adsorption of water and benzene are supported by a variety of experimental evidence. It is known that silica surfaces become progressively hydrophobic on heating at temperatures above 200° to 300°C. The hydroxyls dehydrated at temperatures below 300°C rehydrate readily and interact with water. Upon heating above 400°C, the rehydration is easily completed, and the interaction with water is often limited.

The most conclusive evidence for the mechanism proposed by Kiselev for the adsorption of water on silica surfaces has come from the work of Anderson and Wickersheim (8) on the infrared spectra of both dehydrated and rehydrated silicas. Samples of silica dehydrated at temperatures above 200°C show only a sharp OH(I) stretching frequency due to the silanol groups. The absorption due to the silanol groups decreases with progressive hydration and completely disappears in fully hydrated samples. The water adsorbed on the silanol groups masks this absorption. This process is accompanied by the appearance of a new band due to a different type of OH(II) group formed by the hydrogen bonding of water molecules to the silanol groups. At high levels of hydration the spectrum of the adsorbed water shows some of the characteristics of liquid water. It has been suggested that the water molecules form a hydrogen-bonded network on the surface. Surface hydroxyls and adsorbed water are depicted in Fig. 5. There are no doubt lone hydroxyls which can adsorb water and also geminal hydroxyls having two hydroxyls on a single Si.

Variation of Heat of Immersion with the Temperature of Activation

A surface interacting in a simple manner with water should show a continuous increase in heat of immersion with increasing activation temperature, and any peculiarities in the interaction of water will be reflected in the variation of heat of immersion with the activation temperature. The general types of heat-of-immersion curves observed in the case of oxides interacting with water are depicted in Fig. 6.

Silica surfaces generally show a variation of heat of immersion with activation temperature corresponding to curve I. Curve II is characteristic of alumina and thoria, and curve III is usually given by hydrous titanium oxide. These differences between the oxides will now be discussed in the light of present knowledge regarding surface hydroxyls.

In the case of a variety of quartz and silica samples, the heat of immersion initially increases with activation temperatures, passes through a well-defined maximum at about 350°C, and then decreases with higher activation temperatures (*9–14*). The temperature at which the maximum occurs varies and the absolute heats of immersion vary. These depend on the nature of the sample. A typical curve from the work of Young and Bursh (*11*) is reproduced in Fig. 7.

The initial flat portion of the curve up to 180°C is due to the removal of only physisorbed water. It should be noted, however, that the curve should normally show a slight rise as more water is removed with increase in activation temperature. Heterogeneous distribution of sites is possibly responsible for constancy in the heat of immersion. The desorption of this weakly bound water is essentially complete at 180°C (*8, 15*). The integral heat of adsorption corresponding to the flat portion of the curve, if we assume complete coverage of the surface and $h_{I(S_fL)} = h_{LV^\circ}$, is 12.0 kcal/mole of water. This value is reasonable for a polar surface interacting with water through hydrogen bonds.

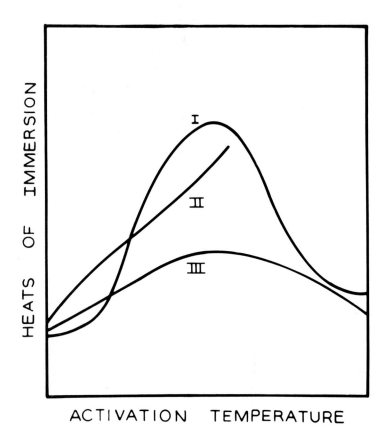

Fig. 6. General types of variation of heat of immersion with activation temperature. Silica, alumina, and hydrated titania surfaces exhibit variations of types I, II, and III, respectively. The temperature scale covers a range from 25° to 600°C. The absolute values of heats of immersion vary from sample to sample, and the curves do not necessarily indicate the magnitude of the heat of immersion.

At temperatures above 200°C the silanol groups start dehydrating to siloxane groups. The heat of immersion of samples activated above 200°C consists of the heat of rehydration of the siloxane groups plus the heat of physisorption of water. When the silica is heated above 400°C, the siloxane groups become relatively stable, and the rehydration of these groups is not complete within the length of time of the immersion experiment (*16*). The lesser degree of hydration and the fewer silanol sites for physisorption of water account for the decrease in the heat of immersion.

It is possible to calculate the heat of dehydration of the silanol groups from the concentration of the silanol groups at the various activation temperatures. On plotting the heat of immersion versus the number of hydroxyls in the temperature range 200° to 400°C, Young and Bursh (*11*) obtained a straight line. The slope of the plot gives 8.05 kcal/mole for the heat of dehydration of silanol groups, or -16.1 kcal/mole of water adsorbed, one molecule of water producing two surface hydroxyls. Brunauer (*17, 18*) reports a value of -4660 ± 230 cal/mole of water for the rehydration of siloxane groups on silica gel and tobermorite surfaces. The value of -16.1 kcal/mole extracted from the results of Young and Bursh comprises both the rehydration of siloxane groups and the heat of physisorption of water. On subtraction of -12.0 kcal/mole of physisorbed water, the heat of rehydration is found to be -4.1 kcal/mole. This value is in good agreement with that reported by Brunauer.

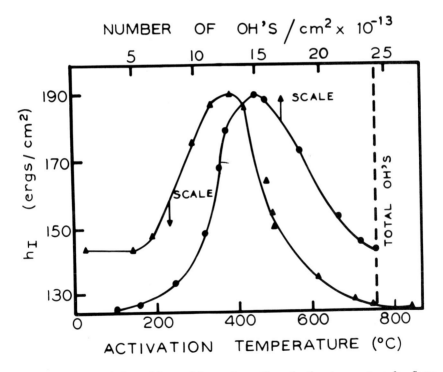

FIG. 7. The variation of heat of immersion with activation temperature for flame-hydrolyzed silica. The heat of immersion as a function of the number of surface hydroxyls is also shown.

Whalen (9) has extended the analysis to the variation of the heat of immersion with activation temperature for the descending portion of the curve. His values for the rehydration of siloxane groups are presented in Table VI. It is seen from the results of the table that the rehydration becomes progressively more difficult as the concentration of silanol on the surface decreases.

TABLE VI
VARIATION OF THE HEAT OF REHYDRATION OF SILANOL GROUPS WITH THE NUMBER OF SURFACE HYDROXYLS (9)

Sample	Silanol Content (OH groups/100 Å^2)	Rehydration Energy (kcal/mole of water)
Silica SB	11.4	−3.4
Silica FS	5.5	−4.4
Silica SL	5.0	−6.4

High-area titanium dioxide shows a variation of heat of immersion similar to that of silica, but the maximum is not sharp (19). Low-area rutile does not show a maximum. The nature of the surface hydroxyls and the mechanism of dehydration are not clear in the case of titania. No suitable explanation can be suggested for the difference in behavior of high-area and low-area titania samples.

Alumina shows a monotonic increase in heat of immersion with outgassing temperature at least up to 500°C. It is suggested that the rehydration of the alumina surface is rapid.

Heat of Immersion of Oxides in Organic Liquids

The characteristics of immersion of oxides into polar organic liquids, such as alcohols, are similar to those found with water. Maxima in the curves of heat of immersion versus activation temperature have been reported in the case of methyl alcohol (19). Oxides activated at temperatures above 400°C show a linear decrease in the heat of immersion into alcohols with decrease in surface hydroxyl content (20–22). The effect becomes less pronounced with increase in chain length of the alcohol. Shielding effects reduce the hydrogen-bonding tendency. These findings are consistent with the mechanism proposed by Kiselev (7) for the interaction of alcohols with oxide surfaces.

Benzene differs from other nonpolar hydrocarbons in its adsorption on silica surfaces. Kiselev (23) found that the differential heat of adsorption of benzene decreased drastically on dehydrating silica surfaces, whereas the adsorption of *n*-hexane was virtually unaffected by the hydroxyl content of the surface. Whalen (24) obtained values of 2.6×10^{-13}, 1.7×10^{-13}, and 0.4×10^{-13} erg/site for the interaction of benzene with hydroxyl groups for three silica samples. These values are in agreement with the value of 3.5×10^{-13} erg/site calculated by Kiselev and Poshkus (25) for the interaction of the π-electron cloud of benzene with the point charges of the hydroxyl dipole. Moreover, the Kiselev model requires the interaction of one benzene per hydroxyl site, and stearic factors limit this to the region where the number of hydroxyls is less than 3 per 100 $Å^2$. In agreement with this model, Whalen obtained a linear relationship between the heat of immersion and hydroxyl content at hydroxyl concentration of less than 3 per 100 $Å^2$.

Heats of Immersion Based on Water Areas

The energy of interaction of water with oxide surfaces depends strongly on the surface concentration of hydroxyls. The water molecules hydrogen-bond, one molecule to two surface hydroxyls, when the latter are in juxtaposition (7). When oxane groups are produced by high-temperature treatment or in production, for example, by flame hydrolysis, these interact only weakly, mostly through dispersion forces, with water molecules.

For example, a fully hydroxylated, amorphous silica surface (eight hydroxyls per 100 $Å^2$) yields a heat of immersion of 190 to 200 ergs/cm^2 (after outgassing at 100° to 180°C); a fully dehydroxylated surface of the same substance yields a heat of immersion of 120 to 130 ergs/cm^2.

The difference is of the order of 3 kcal/mole if all the interaction is ascribed to the first layer. The interaction of water is specific. Indeed, the water-vapor isotherm will possess an inflection or B point indicative of a lower area than the nitrogen area for many silica samples. In certain special cases, too, the water may penetrate pores which the nitrogen cannot; an example was furnished by Zettlemoyer and Iyengar (26) for silica-coated rutiles.

Consideration should therefore be given to the concept of putting heats of immersion in water on a water-area basis. If the nitrogen area is larger than the water area, a lower contribution can be ascribed to the difference between the two.

Heterogeneities from Adsorption and Heat-of-Immersion Measurements

A study of the heterogeneous distribution of sites on a solid surface is of fundamental importance to many surface phenomena. The specific interactions occurring at surface heterogeneities are important to such seemingly diverse phenomena as corrosion, lubrication, catalysis, and heterogeneous nucleation. The variation of heats of adsorption with coverage offers a means of studying such energetic heterogeneities. In addition, use of specific adsorbates allows the study of the nature of the heterogeneities—for example, hydrophilic sites on a largely hydrophobic matrix such as graphite using water as adsorbate and the acid sites on a cracking catalyst using an amine.

Many physical measurements of metals through a single crystal or over different lattice planes lead to the conclusion that properties vary significantly in different directions. For example, a voltage can be applied to the different planes of a metal crystal to measure the energy required to liberate electrons.

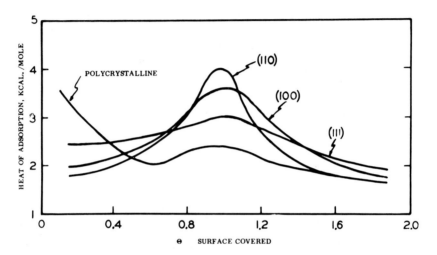

FIG. 8. Differential heats of adsorption for nitrogen on three single crystal faces of copper, and on polycrystalline copper. The large heat values at low coverages in the case of polycrystalline copper are indicative of surface heterogeneities. The low heat of adsorption on single crystal faces implies the absence of active heterogeneous sites.

The most reliable measurements are available for tungsten (27); here the electron work function, φ, varies from 4.3 to 5.5 electron volts. The work function increases with the packing density of the atoms in the various planes, and also the specific surface energy decreases linearly with increase in φ (28). These findings are expected, but more important for the present discussion are expected differences in adsorbate–adsorbent interactions on different planes.

One of the best examples of directional effects during the physical adsorption on a homogeneous, well-defined crystallographic surface was provided by Rhodin (29) on the (100), (110), and (111) faces of a single crystal of copper. Figure 8 shows that at low coverages the isosteric heats of adsorption of nitrogen on the single faces are constant but are higher for the planes with higher packing density. Note that the heat of adsorption rises to the highest maximum on the least densely packed (*110*) face. Here the dipoles induced in the nitrogen are lowest because the van der Waals' attractive forces more readily surmount the repulsive forces between the dipoles as the surface becomes covered with one layer. (The high initial heat on the polycrystalline copper which tends to erase the maximum at one layer is the result of intercrystalline sites; this type of heterogeneity will be discussed in more detail next.) Of the new tools, field emission microscopy provides decisive information on the effectiveness of different lattice planes.

As Rhodin's work showed, effects of intercrystalline sites on polycrystalline surfaces are readily revealed in adsorption studies. The evidence for this type of heterogeneity is the initially high heat of adsorption, followed by a gradual decrease (rather than stepwise) as coverage is increased. Along a similar vein, work in this laboratory (*30*) showed that the interaction of argon with polycrystalline molybdenum powder decreases as sintering increases.

The site energy distribution can be obtained by differentiating the curve of isosteric heat of adsorption of 0°K versus coverage (*31*). The distribution function is given by

$$\frac{dN_A}{dq_{st^\circ}} = g(q_{st^\circ}) \tag{28}$$

where q_{st° is the isosteric heat of adsorption at 0°K, and $g(q_{st^\circ})$ is the distribution function. The value of q_{st° can be deduced from

$$\frac{dq_{st^\circ}}{dT} = C_{PG} - \bar{C}_{N_A} \tag{29}$$

where C_{PG} is the molar heat capacity of the gas at constant pressure and $\bar{C}_{N_A}$ is the differential heat capacity of the adsorbed phase at constant amount adsorbed.

The product $g(q_{st^\circ}) dq_{st^\circ}$ gives the number of molecules that can be accommodated on sites with energy between q_{st° and $q_{st^\circ} + dq_{st^\circ}$. It is not always possible to obtain q_{st°, and hence q_{st} at the temperature of adsorption is used in calculating distribution functions. In view of this approximation and because of the double differentiation involved, the distribution curves must be considered as only rough approximations. At high coverages, lateral interaction between the adsorbed molecules becomes important, and the distribution function may not be sensitive to heterogeneities.

The topographical distribution of sites can be studied by employing adsorbates of different molecular sizes. As an example, the differential heat curve for the adsorption of butylamine on a kaolin-based cracking catalyst (*32*) showed a maximum of 0.4θ. Generally such maxima occur at $\theta = 1$.

The maximum was suspected to be caused by lateral interactions, even though $\theta = 1$ for butylamine (after 25°C outgassing) corresponded to 50% coverage. To explore this possibility, sterically hindered diethylamine and pyridine were employed as adsorbates. Only about 40% of the butylamine coverage was obtained in these cases, suggesting strongly that the acid sites were close neighbors and so adsorption occurred in patches.

Another interesting study of surface heterogeneities is provided by the work of Zettlemoyer et al. on the Graphon–water system (33). Graphon possesses a well-defined hydrophobic and almost homogeneous surface with a very small fraction of hydrophilic heterogeneities. The fraction of hydrophilic sites could be increased by the controlled oxidation of Graphon, with a consequent increase in water adsorption. The heat of immersion of Graphon increased almost linearly with increase in hydrophilicity. From the slope and intercept of the linear plot of heat of immersion versus hydrophilicity, Healey et al. (34) obtained a value of 31 ergs/cm² for the heat of wetting of the hydrophobic sites, and 730 ergs/cm² for the hydrophilic sites.

The isolated hydrophilic sites in a hydrophobic solid may be critical in its practical utility. It has been shown, for example, that the lubricity of graphite is mainly due to the water adsorbed on these isolated sites. Zettlemoyer and co-workers (35) have suggested that a critical hydrophobic-to-hydrophilic ratio (3 or 4:1) may be an important factor in the efficiency of silver iodide in ice nucleation. Indeed, several hydrophobed silicas having similar characteristics were found to be good nucleating agents (36).

It is to be expected, furthermore, that such sites on polymers and organic coatings (produced, for example, by surfactant molecules) could provide the seat of undesired attack by water and other ambients.

Electronic State and Adsorption

As mentioned before, the adsorption of water on hydroxylated oxide surfaces is believed to be donor fashion with the oxygen of the water interacting with two neighbor hydroxyls wherever possible. Oxides often possess defect structures so that they are n- or p-type semiconductors. The question arises as to whether the semiconductor properties can affect the adsorption as to both energetics and amounts.

A classic case, studied in our laboratory (37), is the adsorption of water on n- and p-type germanium samples (propanol adsorption was shown to be similar). Germanium single crystals were crushed and stored in air for a brief period before study; two or three layers of oxide formed on the surfaces. For the n-type samples (see Table VII), both the amounts and energetics of chemisorbed water increased with increase in bulk resistivity—that is, with availability of electrons to be trapped in the oxide film. For the *p*-type samples, the amounts increased with decreasing bulk resistivity. These results are in accord with the model of water adsorbing donor fashion. They also emphasize the electronic effect in adsorption.

TABLE VII
TOTAL, PHYSICAL, AND CHEMICAL ADSORPTION OF WATER AT 25°C ON *n*-TYPE GERMANIUM POWDERS AFTER ACTIVATION AT 300°C

Sample	H_2O (water area, m²/g)			$\Sigma H_2O/\Sigma A$		
	Total	Physical	Chemical	Total	Physical	Chemical
0.14 ohm-cm	0.107	0.072	0.035	0.62	0.42	0.20
4 ohm-cm	0.135	0.090	0.045	0.93	0.63	0.30
5–10 ohm-cm	0.161	0.096	0.065	0.94	0.56	0.38

Wetting and Adhesion to Oxide Surfaces

It is of interest to speculate on what these studies mean to the wetting and adhesion of polymers to oxide films. Obviously, we have oxide films on real metal surfaces and we have adsorbed water on these films under ordinary circumstances.

We have seen that dispersion forces are modest between water and organic molecules. Free energy calculations show that dispersion forces are only 22 out of 72 ergs/cm². On the other hand, the development of hydrogen bonding would raise the interaction enormously. Therefore, polymers containing hydroxyl or amine groups, for example, would be expected to develop much stronger interaction. In addition, the multiple linkages to the substrate would be most helpful in developing and maintaining adhesion.

The omnipresent water is a concern. For low coverages on hydroxylated surfaces, we can expect that each water molecule is hydrogen-bonded to two hydroxyls so that it is rather tightly held. Thus, adhesion should not be greatly affected at this level of coverage. When the surface water is at higher coverage than one layer, however, this loosely bonded water should adversely affect adhesion. Solubility of such water in coatings applied to oxide-wetted substrates should aid in developing better adhesion.

Summary

The Berthelot relation together with the additivity rule of Fowkes allows us to examine interfacial free energies in a systematic way. Dispersion forces only interact across an interface in many instances, but hydrogen bonding can enhance adhesion. The wetting theory is more than adequate to explain forces of adhesion realized in practice. Strain in dried coatings and lack of good wetting explained by slow kinetics and/or loose adsorbed films can be the cause of adhesion failures. Heterogeneities are often responsible for these effects; they can be physical or chemical in nature.

Adsorption and heat of immersion techniques are powerful tools for examining heterogeneities. Mostly, however, these techniques have been tools for studying heterogeneities on plane substrates. Oxide film structure and behavior as a function of gross size and as a function of minor additions to the metal have yet to be studied. Much progress can be expected even from electron microscopic studies of such films. The studies reported here point the way to results to be expected from new enterprising studies of real substrates.

References

1. F. M. FOWKES, in "Surfaces and Coatings Related to Paper and Wood," ed. by R. H. Marchessault and C. Skaar, Syracuse University Press, Syracuse, New York, 1967, Chapter 5.
2. H. Schonhorn, *Advan. Chem. Ser.*, **43**, 189 (1964).
3. J. P. Hobson and R. A. Armstrong, *J. Phys. Chem.*, **67**, 2000 (1963).
4. J. Bastick, *Bull. Soc. Chim. France*, **1953**, 437.
5. J. R. Zimmerman and W. E. Brittin, *J. Phys. Chem.*, **61**, 1328 (1957); J. R. Zimmerman, B. G. Holmes, and J. A. Lasater, *ibid.*, **60**, 1157 (1956).
6. W. K. Lower and E. C. Broge, *J. Phys. Chem.*, **65**, 16 (1961).
7. A. V. Kiselev, *Proc. Symp. Colston Res. Soc.*, **10**, 195 (1958).
8. J. H. Anderson and K. A. Wickersheim, "Solid Surfaces" (Proc. Intern. Congr. Phys. Chem. Solid Surfaces), North-Holland, Amsterdam, 1964, p. 252.
9. J. W. Whalen, *Advan. Chem. Ser.*, **33**, 281 (1961); *J. Phys. Chem.*, **65**, 1676 (1961).
10. A. C. Makrides and N. Hackerman, *J. Phys. Chem.*, **63**, 594 (1959).
11. G. J. Young and T. P. Bursh, *J. Colloid Sci.*, **15**, 361 (1960).
12. M. M. Egorov, K. G. Krasilnikov, and E. A. Sysoev, *Dokl. Akad. Nauk SSSR*, **108**, 103 (1956).
13. F. E. Bartell and R. M. Suggit, *J. Phys. Chem.*, **55**, 1456 (1951).
14. M. M. Egorov, V. F. Kiselev, K. G. Krasilnikov, and V. V. Murina, *Zh. Fiz. Khim.*, **33**, 65 (1959) [*C.A.* **53**, 2103e (1959)].
15. G. J. Young, *J. Colloid Sci.*, **13**, 67 (1958).
16. C. A. Guderjahn, D. A. Paynter, P. E. Berghausen, and R. J. Good, *J. Phys. Chem.*, **63**, 2066 (1959).
17. S. Brunauer, D. L. Kantro, and C. H. Weise, *Can. J. Chem.*, **34**, 1483 (1956).
18. S. Brunauer, D. L. Kantro, and C. H. Weise, *Portland Cement Assoc., Res. Develop. Lab., Develop. Dept. Bull.*, **105** (1959).
19. W. H. Wade and N. Hackerman, *Advan. Chem. Ser.*, **43**, 222 (1964).
20. A. V. Kiselev, *Dokl. Akad. Nauk SSSR*, **130**, 569 (1960).

21. A. K. Bonetskaya and K. G. Krasilnikov, *Dokl. Akad. Nauk SSSR*, **114**, 1257 (1957).
22. L. G. Ganichenko, V. F. Kiselev, K. G. Krasilnikov, and V. V. Murina, *Zh. Fiz. Khim.*, **35**, 844 (1961).
23. A. V. Kiselev, *Zh. Fiz. Khim.*, **35**, 233 (1961).
24. J. Whalen, *J. Phys. Chem.*, **66**, 511 (1962).
25. A. V. Kiselev and D. P. Poshkus, *Dokl. Akad. Nauk. SSSR*, **120**, 834 (1958).
26. A. C. Zettlemoyer and R. D. Iyengar, paper presented at the 149th American Chemistry Society Meeting, Detroit, April 1965; also A. C. Zettlemoyer and R. D. Iyengar, unpublished work.
27. I. Langmuir and K. H. Kingdon, *Science*, **57**, 58 (1923).
28. R. Suhrmann, *Advan. Catal.*, **7**, 320 (1955).
29. T. N. Rhodin, *J. Am. Chem. Soc.*, **72**, 5692 (1950).
30. F. H. Healey, J. J. Chessick, and A. C. Zettlemoyer, *J. Phys. Chem.*, **57**, 178 (1953).
31. L. E. Drain and J. A. Morrison, *Trans. Faraday Soc.*, **48**, 316 (1952).
32. A. C. Zettlemoyer and J. J. Chessick, *J. Phys. Chem.*, **64**, 1131 (1960).
33. G. J. Young, J. J. Chessick, F. H. Healey, and A. C. Zettlemoyer, *J. Phys. Chem.*, **58**, 313 (1954).
34. F. H. Healey, Y. F. Yu, and J. J. Chessick, *J. Phys. Chem.*, **59**, 399 (1955).
35. A. C. Zettlemoyer, N. Tcheurekdjian, and J. J. Chessick, *Nature*, **192**, 653 (1961).
36. A. C. Zettlemoyer, N. Tcheurekdjian, and C. L. Hosler, *Z. Angew. Math. Phys.*, **14**, 496 (1963).
37. G. Srinivasen, J. J. Chessick, and A. C. Zettlemoyer, *J. Phys. Chem.*, **66**, 1819 (1962).

Discussion

JELLINEK: You expressed surprise that the π-electrons apparently do not contribute to the heat of adsorption on Graphon. We have recently studied the adsorption of water vapor on polystyrene, which has some similarities with Graphon. We do not find any contribution of the π-electrons; nor any adsorption of water on polystyrene up to a relative pressure of about 0.95. About ten years ago, we measured the heat of immersion of polystyrene, and we always found a positive heat of adsorption. I think that the positive enthalpy is compensated by an entropy effect so that you still get the negative free energy.

ZETTLEMOYER: I did not say a positive heat of adsorption. Do you mean a net heat of immersion?

JELLINEK: Yes. We measured the heat of immersion of polystyrene in water and always found an absorption of heat instead of heat given out. It seems to me that this can be the case only if the entropy compensates for the heat of adsorption. We have also investigated other polymers. In some polymers, the water is more ordered, and there is some sort of ice structure on the polymer. For instance, with polymethylmethacrylate the adsorbed water is more ordered for at least three layers so that it is more ice-like than water-like. In the case of polyacrylonitrile, it is the other way around.

ZETTLEMOYER: In the example that I quoted of π-electron interaction with hydroxyls, which is exemplified by benzene against silica, the π-electron effect is present, so this makes it surprising that you do not find it with polystyrene. I wonder if it is due to how the polystyrene molecules are stacked?

JELLINEK: We tried all sorts of surface treatments and purifications, and it always came out the same.

MICHAELS: About Dr. Jellinek's observations on what appears to be positive heats of immersion of polystyrene, I tend to be a bit skeptical. Polystyrene has a fairly respectable sorptive capacity for water. In immersion experiments, there should be a fair amount of dissolution of water within the polymer; also, the solubility of water in polystyrene increases with the temperature. It is thus an endothermic sorptive process. I am wondering whether or not your immersion heats are being dominated by sorptive heat within the polymer. I find it hard to believe that an adsorptive energy should be opposite in sign to what you would expect.

JELLINEK: Thermodynamically it is quite possible that the enthalpy is positive because it can be compensated by the entropy.

MICHAELS: I do not deny that you have reconciled the measurements on thermodynamic grounds.

JELLINEK: I do not know about the solubility of water in polystyrene, but I do not think it is very soluble.

MICHAELS: I do not know what size particles you were working with, but the solubility is reported to be in the range of about 0.05% at atmospheric temperatures. So, it is not inconsequential.

JELLINEK: Maybe you had a very high molecular weight.

MICHAELS: That does not make any difference. The glassy polymer has appreciable water absorptive capacity.

GRISKEY: Harkins studied the adsorption of sodium dodecyl sulfate on polystyrene and got tremendously large areas for adsorbed molecules. He found polystyrene to be highly porous. Polystyrene will take materials, and the areas that he got were in no way consistent with other things that he measured, like graphite. So I think some of what Dr. Michaels says might be true; there might be something else happening in these systems and not just adsorption.

JELLINEK: If water is soluble in polystyrene, why don't you get any adsorption up to a relative pressure of 0.95? We could not find any water adsorption.

ZETTLEMOYER: Do you know the purity of the polystyrene?

JELLINEK: Yes. We purified it several times by solvent precipitation.

ZETTLEMOYER: Was there an inhibitor present in the monomer?

JELLINEK: Yes.

ZETTLEMOYER: This might reside in the surface of the polymer.

JELLINEK: We purified it. There is still the fact that water is not adsorbed on polystyrene up to a relative pressure of 0.95. This does not contradict the fact that polystyrene is permeable to water.

MICHAELS: I do not believe it. Polystyrene permeates water at a respectable rate, and it does so by dissolving it, and your data show no detectable uptake of water up to 95% humidity.

GRISKEY: To prepare polystyrene by thermal polymerization, you had to have a catalyst. So you had some kind of impurity in the system.

ONSAGER: We do not have quite the same type of hydroxyls on silica as we do in water. It is a weaker interaction, there, so it could show in one case and not in the other. On the silica, there is no opportunity for the hydroxyls to cooperate, in the first place, as in liquid water, where they are very much different. So it is a very different sort of competition. I have been wondering whether or not, once in a while, you should see some symptoms of π-electron interactions—for example, copper on nitrogen. This seems not to be evident, but it is one thing we might bear in mind as a possibility. We do know a few instances in which nitrogen tacks onto something else by π-electrons (for example, biochemical nitrogen fixation).

ZETTLEMOYER: I think the answer there is that for a *sp* metal like copper it does, but for a *d* metal it does not. I also think that the quadrupole moment of nitrogen is contributory.

ONSAGER: If, somehow, we take one electron off the copper and put it down in the *d* class, the surfaces differ in both. One interaction that can exist is the image force of, say, a water molecule and metal, like mercury, which I think is your example. When you get down to these distances, the dielectric constant is not entirely a local property, and the infinite dielectric constant of a metal is by no means a local property, either. It might be particularly nonlocal in mercury. Then, the solid particles are a factor in the theory of shielding of electrostatic forces in metals. It is further recognized that there is a great question of finite distances here and also, perhaps, an exaggerated finite distance from water. So, the full influence of a dielectric constant of a metal must be taken with a grain of salt. You must not deal with the dielectric constant but with the dielectric function. At the surface this is a particularly difficult task because you do not have the advantage of dealing with a homogeneous system in which the progressive wave functions, and so forth, are fairly amenable to analysis, and you have to bear in mind the shielding. One other booby trap in the theory of dielectrics is this idea that, say, you have a silica layer of hydroxyls and then one layer of water molecules all pointing one way, all with the dipoles in the same direction. This makes me a little skeptical. That is a maximum energy configuration, so that, although individual hydroxyls might tend to coordinate in this manner, by the time you get a lot of dipole water molecules coordinating with the dipoles pointing in the same direction, you impose a weird constraint on the water molecules. If you continue that, there is only one choice, and that is to make ferroelectric water.

ZETTLEMOYER: It does not work that way, because the surface is very rough on an atomic scale. Real surfaces are not smooth enough to develop the organization required for the ferroelectric effect.

ONSAGER: Even so, there is still a problem because it is still ferroelectric water, which, by the way, does not have a very high energy anymore. Of course, it has no entropy, so one has to be a little cautious in looking at these localized interactions. You have got to take a look at what is going to happen to that enormous dipole moment and then, perhaps, also speculate on some of the symptoms that seem to indicate that organized water might be ferroelectric. You might have a long, long look at dielectric interactions between molecules on the surface and within the molecules themselves.

BOLGER: Referring to Professor Onsager's comment that one might expect to have ferroelectric water if the dipoles were oriented as regularly as Professor Zettlemoyer has shown, one runs into a real problem in trying to draw a picture of an oxide surface and the first layers of water that are adsorbed. Silica, for example, calculates to about one hydroxyl per 60 Å^2 on the surface. It has a very rough surface on this type of scale, so it really gets difficult to picture, two-dimensionally, an array of hydroxyl groups on the surface. These groups are not close enough or regularly enough spaced for the next layers of dipoles to be in a regular enough array to worry about really parallel orientations. The hydroxyl population also varies per oxide. It is different on silica than on titania. I think we could be misled unless we recognize that there is quite a bit of randomness at the surface, and any pictorial representation looks more regular than it really is.

ZETTLEMOYER: You can monitor how close these hydroxyls are by using a sterically hindered molecule, like isopropanol, to compare the water uptake to the isopropanol uptake. It is very interesting because one observes quite different characteristics depending on the thermal history of the silicas compared. Sometimes we find that the hydroxyls lie in patches, and other times we find that they are isolated. But I was talking about a preponderance of the water going down one per two hydroxyls on a fully hydroxylated surface. I was not talking about all of the adsorbed water molecules.

MATIJEVIĆ: I wish to corroborate some of Dr. Zettlemoyer's data with some of our results obtained with a different system. We measured adsorption of hafnium ion on glass using tracers.

The interesting property of hafnium ion is that as the result of hydrolysis at pH > 4 the entire amount of hafnium in solution is in the form of soluble neutral $Hf(OH)_4$ species. When these neutral species adsorb, there is no lateral repulsion, and one can expect a tightly packed monolayer to be formed.

We used glass beads, the surface area of which was determined by the adsorption of water vapor. This is obviously the most appropriate procedure to obtain the surface area of an adsorbent onto which hydrolyzed species are to be adsorbed. We found indeed that a monolayer of adsorbed $Hf(OH)_4$ is eventually formed with one complex species per 25 $Å^2$. This is exactly in agreement with data given by Dr. Zettlemoyer: one water molecule per two hydroxyls, or eight hydroxyls per 100 $Å^2$ units. Thus, two independent techniques produced the same answer, which is rather encouraging.

Localized Oxidation Processes on Iron
A Contributed Discussion

E. A. GULBRANSEN

Westinghouse Research Laboratories
Pittsburgh, Pennsylvania

Introduction

The discovery of oxide nuclei (*1, 2*) and of oxide whiskers (*3, 4*) and platelets (*5*) in the oxidation of metals came as a surprise to many workers in the field. The then-accepted model of oxidation of metals (*6*) assumed the presence of a uniform, coherent oxide film with reaction occurring by the diffusion of oxygen or metal atoms and ions through lattice defects in the oxide. Dislocation and grain boundary diffusion processes and the influence of stress in the metal structure were not considered in the theory.

The formation of localized oxide growths suggests that the simple model of oxidation is not complete. In addition, the formation of these growths gives a new insight into the nature of oxide films and metals, stress corrosion cracking, and metal fatigue.

The availability of oxide nuclei and oxide whiskers and platelets for study by electron diffraction and by electron microscopy makes possible a determination of some of the structural and kinetic details of their growth mechanisms. Results from these studies may be carried over to a study of the general oxidation mechanism. Thus, the kinetics of metal and oxygen transport through oxide lattices and through well-defined crystal imperfections can be studied.

Oxide Nucleation

Thin Film Range—Normal Pressure

Microstructural studies of gas–metal reactions became possible with the development of the electron microscope and electron diffraction cameras in the late 1930's and early 1940's. The microstructure of oxide films and metals can be defined by the following factors: (1) crystal structure, (2) crystal size and habit, (3) crystal orientation relative to the metal, (4) presence of dislocations, twins, and other structural factors in the metal and oxide, and (5) presence of stress in the metal–oxide system.

Visual and optical microscopic observations of surface reactions are the oldest methods for the study of the chemical activity of surfaces. Catalysis and wet corrosion scientists have long recognized that certain sites were especially reactive. However, the nature of these reaction sites commonly remains undetermined because of lack of resolving power of light microscopic techniques and the difficulties inherent in defining the surface and the chemical environment at the surface.

In the field of gaseous oxidation of metals, conditions are more favorable for the study of localized chemical activity. Phelps *et al.* (6) reviewed the literature in 1946 on the microstructure of oxide films on the common metals. An extensive electron microscope and electron diffraction study was also made on the microstructure of oxide films less than 500 Å thick on a number of metals and alloys, including iron, nickel, iron-based alloys, and stainless steels.

The oxidation of a pure grade of iron at 250°C in dry oxygen for 30 minutes produced a continuous oxide film consisting of irregularly shaped oxide crystals 500 to 1500 Å in size. All parts of the metal grain were covered with oxide crystals.

This assembly of oxide crystals was partially oriented to the metal grain as shown by reflection electron diffraction patterns and by transmission electron micrographs of the oxide film removed from the metal. An analysis of the patterns showed that spinel Fe_3O_4 and hexagonal α-Fe_2O_3 were formed.

High-purity nickel oxidized at 500°C for 20 minutes yielded a microstructure similar to that found on iron at 250°C. The oxide film was NiO and was nearly randomly oriented to the metal.

Phelps *et al.* (6) concluded that most metals oxidized at temperatures of 200° to 500°C and 0.1 to 1 atmosphere of oxygen pressure formed granular oxide films up to 500 Å in thickness. Some of the films contained thicker and thinner sections indicating a multilayer structure. Oxide crystals were nucleated on the initial metal surface covered by the room temperature oxide film and on top of the first layer of oxide crystallites.

The oxide crystals 500 to 1800 Å in breadth are much smaller than the metal grains. The surface density of the oxide crystal is large, about 10^{10} cm^{-2}. The crystal size was a function of both time and temperature. Some evidence was found for preferred orientation with respect to the metal grain. Most oxide films were not uniform in thickness.

Thin Film Range—Low Oxygen Pressures

The nucleation of oxide crystals in the thin film range was found to be a function of both temperature and oxygen pressure in the early 1950's by

Bardolle and Bénard (1, 2). These workers studied the initial stage of oxide film formation on carefully annealed and electropolished ARMCO iron at temperatures between 650° and 850°C in vacuums of 10^{-1} to 10^{-3} torr. At 850°C and at pressures between 10^{-1} and 10^{-2} torr, many oriented oxide crystals were formed on each metal grain, while at pressures between 10^{-2} and 10^{-3} torr only a few oxide crystals were formed. None of the conditions studied produced oxide nuclei on the metal surface in the form of one single

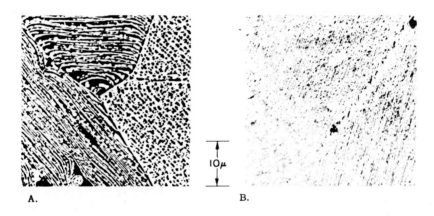

FIG. 1. Oxide film stripped from Puron annealed in hydrogen passed over hot copper. H_2, 850°C, 20 hours, cooled in H_2. (A) 752-Å-thick oxide, 850°C. (B) 322-Å-thick oxide, 750°C.

crystal. However, the ensemble of oxide crystals on each metal grain had a definite crystal pattern which suggested a definite orientation of the crystals to the metal grain. These oxide crystals were much thicker and larger than those studied by Phelps et al. (6) at lower temperatures.

The shapes of the individual oxide nuclei could be studied when low oxygen pressures of 10^{-3} to 10^{-2} torr were used. The crystal shape was dependent on the orientation of the metal grain. The space between the oxide crystallites was occupied by a thin film of randomly oriented oxide crystals.

Gulbransen et al. (7) extended the studies of Bardolle and Bénard (1, 2). Electron optical instead of light optical methods were used to study the oxide crystals formed in the oxidation process and electrochemically separated from the metal. High-purity iron was used, and the oxidation conditions were carefully controlled. A vacuum microbalance was used to determine the extent of reaction.

Figure 1 shows two electron micrographs of oxide films stripped from iron samples that were oxidized at 850°C and 750°C to weight gains of 10.73 $\mu g/cm^2$ and 4.59 $\mu g/cm^2$, respectively. Figure 1A, a micrograph of the sample reacted at an oxygen pressure of 2×10^{-3} torr, shows oxide crystals of about 1 micron in size and well oriented to the metal grain. The oxide crystals were formed in straight rows across the metal grain. The spacing between rows of crystallites was between 0.5 and 1.0 micron. Figure 1B shows the results at 750°C for an oxygen pressure of 1×10^{-3} torr and a total weight gain of 4.37 $\mu g/cm^2$. The size of the crystallites and the spacing between rows of oxide crystallites were smaller than that observed at 850°C. Except for the size of the crystallites and their spacing, the oxidation process was similar to that noted at 850°C.

At higher oxidation pressures and for large oxygen additions, the oxide crystals grew to a larger size and filled in the void between the rows of oxide crystallites. The density of oxide crystallites was of the order of 10^8 cm^{-2}.

Circular oxide growth patterns were found on some of the grains. This pattern may be due to the presence of inclusions in the metal grain. The oxide growth patterns decorate some metallurgical substructure in the metal. Further experiments (8) showed that the pretreatment procedures determined the pattern of oxide growth. Annealing pure iron in high vacuum gave random oxide growth patterns after oxidation, while hydrogen annealing led to line growth patterns. In both pretreatments, the oxide crystals were oriented to the metal grain.

Orientation of Oxide Nuclei Formed on Iron

The orientation of FeO on α-Fe was studied by means of x-ray diffraction methods by Mehl and McCandless (9) and by Bardolle and Bénard (1). Gulbransen and Ruka (10) studied the same problem, using reflection electron diffraction. In addition, they studied the orientation of Fe_3O_4 on α-Fe. All the results indicate that the cube plane of FeO and Fe_3O_4 grows on the cube plane of α-Fe, while the [100] direction of the oxide lies parallel to the [110] direction in the metal.

Model of Initial Oxide Formation

One concept (7, 8) of the initial stages of the oxidation of iron at 850°C and low pressures is shown in Fig. 2. The first stage involves a chemisorbed layer of oxygen atoms. This would be a very transitory stage at 850°C with the quantity of oxygen available. The second stage is the formation of a film of 100-Å crystallites which is of the order of the thickness of the oxide film. The third stage is the nucleation and growth of larger crystals along certain crystallographic directions in the metal grain. Along these growth sites, a lower-energy barrier exists for the formation and growth of new oxide crystals.

These linear growth sites may represent a substructure in the metal which is influenced by the pretreatment. Hydrogen and impurities in the hydrogen gas appear to order the sites for oxide nucleation on pure iron.

These large oxide crystals are well oriented to the metal grain. As oxidation continues, these crystals grow, and new ones are formed, completing the

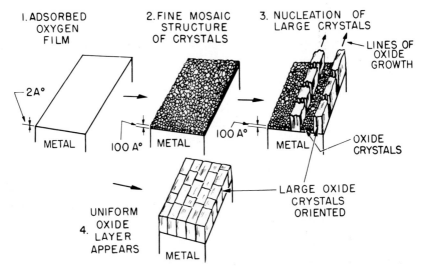

FIG. 2. Oxidation mechanism.

mosaic structure. This is shown in part 4 of Fig. 2. Finally, an oriented granular oxide film is formed. Although the oxide crystals are well oriented, there are many oxide crystals on each metal grain.

At 850°C iron atoms are assumed to diffuse through the oxide crystal by lattice diffusion. Under conditions where the oxide crystal is not restrained, growth can occur in all directions, laterally as well as normal to the surface. The oxide crystals in the original thin 100-Å film that are favorably oriented to the metal grain may grow in size, while those that are not favorably oriented remain small.

The detailed mechanism of nucleation and growth of oxide nuclei are not well understood. Surface defects, dislocations, and impurities have been suggested as sites for oxide nuclei formation. Bénard et al. (11) proposed an oxygen adsorption process on the surface to be rate-limiting in the growth of oxide nuclei. Rhead (12) has developed this idea and has proposed an equation relating the area of the nuclei to the surface diffusivity. One of the difficulties in applying quantitative relationships to the mechanism of nucleation and growth is that for most oxidation studies the gaseous environment during growth was not controlled and the gaseous species contributing to the reaction environment were not determined. Microbalance studies by Gulbransen et al. (7) on the low-pressure oxidation of iron at 2×10^{-3} torr and 850°C showed complete oxygen pickup by the sample within 1 to 2 minutes. This was the time to make the first reading on the microbalance. Kinetic theory suggests that, at 2×10^{-3} torr, 3.7×10^{17} oxygen molecules collide with 1 cm^2 of surface per second. For FeO this is equal to an outward growth rate of 1.5×10^3 Å/sec. With this potential growth rate, surface adsorption as a rate-limiting process [proposed by Bénard et al. (11) and Rhead (12)] is hard to visualize.

It is necessary to assume that some specially oriented crystals of the thin oxide film can grow with grain boundary diffusion processes contributing to the transfer of reacting species.

Oxide Whiskers and Platelets

Oxide Whiskers

The formation of long filaments of oxide whiskers during the oxidation of metals was first observed by Pfefferkorn (3) and Takagi (4). Both workers studied the edges of wires and of fine holes in metal sheets using transmission electron microscopy after oxidation. At this early date, the metals used and the reaction conditions were poorly defined. The oxide whiskers were formed after considerable oxidation of the metal had occurred. A study of oxide whiskers was begun in 1957 for the purpose of understanding localized corrosion and oxidation on iron and stainless steel. Five factors must be considered in discussing the growth of oxide whiskers and platelets. These are: (1) the crystal structure, (2) the crystal morphology and size, (3) the orientation of the oxide crystal with respect to the underlying interfaces, (4) the nature of the reaction site in the metal or on the oxide, and (5) the mechanism of growth.

Gulbransen and Copan (5, 13) made a careful study of the localized oxide growths formed on high-purity iron under carefully controlled oxidation conditions. In addition to the normal oxide film, fine oxide whiskers were formed on annealed high-purity iron in dry oxygen atmospheres at temperatures of 400° to 500°C. A few small fan-shaped oxide platelets were formed in the early stages of oxidation before the formation of oxide whiskers.

Figure 3 shows an electron micrograph of an oxidized disk of PURON grade of iron. The sample was reacted for 48 hours at 500°C in a dry oxygen atmosphere at 76 torr. The oxide whiskers were 150 Å or more in diameter, excluding contamination, and 15,000 Å in average length. Some grew to lengths of 75,000 Å. A surface density of about 10^8 cm^{-2} was estimated.

FIG. 3. Fine oxide whiskers formed on PURON at 500°C in dry oxygen for 48 hours. No applied stress. Micron length shown.

Transmission electron diffraction patterns of the oxidized wire gave a pattern of α-Fe_2O_3. A summary of the properties and conditions of formation of oxide whiskers from our studies is given in Table I. The conditions for formation of oxide whiskers are dry oxygen, pure metal, and no stresses. The diameters of the simplest whiskers were uniform along the length of the whisker.

Pointed Blade-Shaped Oxide Platelets

When annealed or cold-worked iron is reacted with water vapor atmospheres containing oxygen, pointed blade-shaped oxide platelets form. Figure 4 shows the genesis of blade-shaped oxide platelets when iron is reacted with a

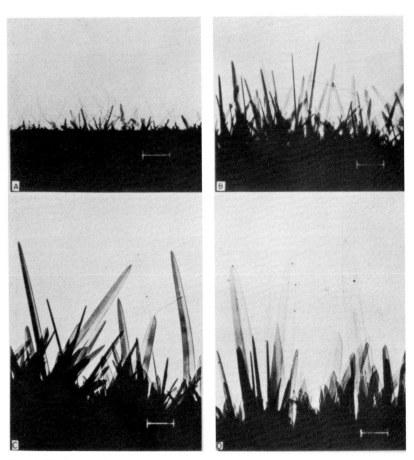

FIG. 4. Blade-shaped oxide platelets formed on annealed pure iron in 10% H_2O and 90% argon at 400°C. Micron length shown. (A) 2 hours. (B) 6 hours. (C) 12 hours. (D) 23 hours.

10% water vapor and 90% argon mixture. Traces of oxygen were present in the argon gas. Figure 4*A* shows that both oxide whiskers and blade-shaped platelets were formed after 2 hours of reaction. The whiskers were about 150 Å in diameter and averaged 10000 Å in length. The oxide platelets were about 150 Å thick, 1000 to 1500 Å wide, and grew to lengths up to 10000 Å. Figure 4*B* shows that most of the whiskers have disappeared

TABLE I
Summary of Data on Oxide Whiskers

1. Diameter $\sim$150 Å
2. Length up to 4×10^5 Å
3. Density up to 10^8 cm^{-2}
4. Weight of 10^5-Å whisker $= 9.25 \times 10^{-15}$ g
5. Crystal structure—hexagonal α-Fe$_2$O$_3$—single crystal
6. Site area—1.77×10^{-12} cm^2
7. Percentage of surface covered for density 10^8 cm^{-2} = 0.0177%
8. Conditions—dry oxygen, no stress, pure metal, random sites on oxide or metal
9. Elastic strain limit = 5% for 150-Å-diameter whisker bent to 1500-Å radius
10. Stress at root = 3.1 psi for 4×10^5-Å whisker, 150 Å in diameter
11. Thickness—nearly independent of time
12. Melting of tip—prevents further growth
13. Effect of H$_2$—oxide reduced to metal

after 6 hours and only oxide platelets have formed. Figures 4*C* and 4*D* show that the size of the oxide platelets increases with time. After 23 hours of reaction, the platelets were 2500 to 6000 Å wide and up to 70000 Å long, with a surface density of 10^8 cm^{-2}.

Table II shows a summary of the properties and conditions of formation of the blade-shaped oxide platelets of an iron specimen reacted with steam and argon at 400°C for 7 hours. The d_{hkl} values could be correlated exactly with the x-ray diffraction values for hexagonal α-Fe$_2$O$_3$.

The shape of the platelets suggests that localized oxidation starts at point sites on the surface. As oxidation continues, the point site grows to a line site to form a blade-shaped oxide platelet. Hydrogen atoms or ions act on the surface of the oxide or metal to enlarge the reaction area along certain directions. The thicknesses of the platelets are uniform and remain constant over the period of oxidation; they are nearly the same as the uncontaminated oxide core of whiskers formed in dry oxygen atmospheres. The platelet shape suggests that the width of the platelets is a function of time or extent of reaction.

Gulbransen and Copan (*14*) have shown that blade-shaped oxide platelets are the predominant localized oxide growths for a wide range of O_2–H_2O gas mixtures. To prevent platelet growth, the O_2–H_2O ratio must be greater than 30:1. Other studies (*15*) have shown that traces of oxygen must be present for growth of blade-shaped platelets. A comparison of Fig. 3 with Fig. 4 shows the oxide present in blade-shaped platelets to be over 250 times the oxide present as oxide whiskers in Fig. 3. Water vapor has a unique effect on the corrosion of iron.

TABLE II
SUMMARY OF DATA ON POINTED BLADE-SHAPED OXIDE PLATELETS

1. Thickness ~150 Å
2. Width up to 1.5×10^4 Å
3. Length up to 9×10^4 Å
4. Density up to 10^8 cm^{-2}
5. Weight = 5.3×10^{-13} g for platelet 9×10^4 Å long, 1.5×10^4 Å wide, and 150 Å thick
6. Crystal structure—hexagonal α-Fe_2O_3 (400°C temperature of formation)
7. Site area—2.3×10^{-10} cm^2 for platelet 9×10^4 Å long, 1.5×10^4 Å wide
8. Percentage of surface covered for 10^8 cm^{-2} = 2.25%
9. Conditions—$H_2O + O_2$, stressed or unstressed metal, probably oriented sites, pure metal
10. Reaction site—grows in width with time
11. Thickness—nearly independent of time

Rounded Oxide Platelets

Thin, rounded oxide platelets formed when cold-worked iron was reacted with dry oxygen. Figure 5 shows an electron micrograph of the crystal habit of the localized oxide growths formed after oxidation of cold-drawn iron wire at 400°C for 48 hours in 760 torr of dry oxygen.

Thin, rounded oxide platelets grow along the axis of the wire to a height of 40000 Å and to lengths up to 70000 Å. The thickness of the platelets was estimated to be of the order of 150 Å or the same as that found for the blade-shaped oxide platelets and oxide whiskers. The thickness appears uniform over the platelet and the same for a large number of platelets examined.

Table III shows a summary of the properties and conditions of formation of rounded oxide platelets.

Blunted Blade-Shaped Platelets

Reaction of high-purity annealed and etched iron wires in an atmosphere of 10% H_2O + 90% (He + 1000 ppm O_2) at 400°C for 4 hours has shown a

FIG. 5. Rounded oxide platelets formed on cold-worked pure iron in 1 atmosphere of dry oxygen at 400°C for 48 hours. Micron length shown.

new crystal habit (16). Instead of pointed blade-shaped oxide platelets, blunted blade-shaped plates are formed. The blunted end is characteristic of these platelets even from the early stages of reaction. Figure 6 shows a typical electron micrograph. The blunted end has a chisel-shaped appearance. However, the density of the material to passage of electrons is nearly constant.

TABLE III
SUMMARY OF DATA ON ROUNDED OXIDE PLATELETS

1. Thickness —150 Å
2. Width up to 2×10^5 Å
3. Length up to 1.5×10^5 Å
4. Density—10^6 cm^{-2}
5. Weight = 3.1×10^{12} g for platelet 2×10^5 Å long and 1.5×10^5 Å wide
6. Crystal structure—hexagonal α-Fe$_2$O$_3$
7. Site area—3×10^{-9} cm^2 for platelet 2×10^5 Å wide and 1.5×10^5 Å long
8. Percentage of surface covered for 10^6 cm^{-2} = 0.3%
9. Conditions—O$_2$, metal in stressed conditions, impurities in metal
10. Reaction site—line site on metal, grows with time
11. Thickness—nearly independent of time

This suggests that the main body of the platelet may be hollow and the end crimped closed.

The possibility that these platelets are not solid and uniform in cross section is shown in Fig. 7. Here the blunted blade-shaped platelets were tested in the electron beam. The helium gas trapped in the blade-shaped crystal expands and forms a gas bubble or ruptures the platelet, leaving the walls exposed. Many smaller bubbles also are formed on the flat side of other oxide platelets.

The shape of these bubbles and ruptured platelets indicates that the platelets have an internal surface between faces and that the oxide is ductile. We conclude that the platelet structure is partially revealed by the helium gas. Helium diffuses rapidly in most solids. In the oxidizing atmosphere, helium can be

Fig. 6. Blunted blade-shaped oxide platelets formed on annealed pure iron in 10% H_2O + (He + 1000 ppm O_2) at 400°C and 5 hours.

Fig. 7. Effect of electron beam heating on blunted blade-shaped oxide platelets. (A) Ruptured platelet. (B) Hollow bubble.

trapped in the structure between parallel crystals of the blade-shaped structure. If heated rapidly, bubbles form or the structure is ruptured by the high gas pressure.

Helium as a carrier gas has also revealed a new structure—that is, blunted blade-shaped platelets. The blunted shape must be related to the initial condition of the reacting interface. Helium acts in the presence of H_2O and O_2 to give initial line shaped reaction sites.

Figure 8 shows a schematic interpretation of the formation of localized oxide growths in oxygen and in water vapor plus oxygen mixtures. Figure 8a shows a nearly random pattern of nucleation sites. A metal having this arrangement of nucleation sites may react with oxygen to form long fine whiskers of oxide in addition to the normal oxide film.

Figure 8b shows a picture of the nucleation pattern of a cold-worked specimen of iron. An ordered arrangement of nucleation sites is formed in the metal by the cold-working treatment. Rounded oxide platelets of oxide are formed when a metal having this structure is reacted with oxygen. LINE growth of oxide nuclei are formed.

Figure 8c shows a picture of the nucleation pattern of cold-worked or annealed iron after reaction with water vapor plus oxygen. After reaction, pointed blade-shaped oxide platelets are formed. Hydrogen ions or atoms from the water vapor reaction may diffuse into the metal at 400°C and modify the impurity distribution or the nucleation sites. The water vapor generates new reaction sites as reaction occurs.

Model for Formation of Localized Crystal Growths

The experiments using helium as a carrier gas suggest that oxide platelets and whiskers have internal boundaries or axial screw dislocations (15). The essential feature of these internal boundaries and defects is to provide paths for rapid diffusion of iron at rates that are large when compared to normal lattice diffusion.

Let us first consider the model for whisker growth as shown in Fig. 9. The whisker contains an internal hole or defect where grain boundary or surface diffusion of iron occurs along a concentration gradient $\dfrac{(C_0 - C_{m_1})}{x_1}$

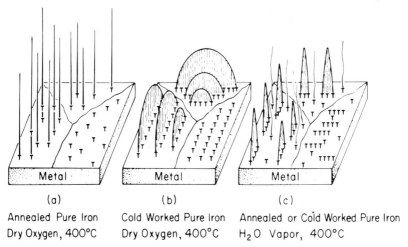

(a) Annealed Pure Iron Dry Oxygen, 400°C

(b) Cold Worked Pure Iron Dry Oxygen, 400°C

(c) Annealed or Cold Worked Pure Iron H_2O Vapor, 400°C

FIG. 8. Localized oxide growth habits in the oxidation of iron. (a, b) Oxygen reaction (c) Water vapor reaction.

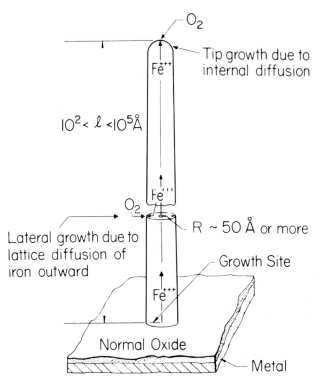

FIG. 9. Model of whisker growth during oxidation of iron, 400°C, dry O_2.

Here C_O is the concentration of iron in α-Fe_2O_3 in equilibrium with oxygen; C_{m_1} is the concentration of iron in Fe_3O_4 in equilibrium with metal; and x_1 is the length of diffusion path. The transfer of iron to the tip is given by $D_{GB\text{ or }S} \cdot (C_O - C_{m_1})/x_1$, where $D_{GB\text{ or }S}$ is the grain boundary or surface diffusion coefficient. Since diffusion of iron can occur through the α-Fe_2O_3 lattice also, the transfer of material through the wall is given by $D_L \cdot (C_O - C_{m_2})/x_2$. Here C_{m_2} is the concentration of iron in α-Fe_2O_3 in equilibrium with metal, D_L is the diffusion coefficient of iron in the lattice of α-Fe_2O_3, and x_2 is the whisker wall thickness.

The condition for growth of oxide whiskers is

$$D_{GB\text{ or }S} \cdot \frac{C_O - C_{m_1}}{x_1} > \frac{D_L \cdot (C_O - C_{m_2})}{x_2}$$

Most of the growth occurs by reaction with oxygen atoms or ions at the tip of the whisker.

Since the diameter is nearly uniform during the growth cycle, it follows that $D_L < D_{GB}$ or D_S. In dry oxygen atmosphere, the size is independent of extent of reaction.

The model for growth of blade-shaped oxide platelets is shown in Fig. 10. Again, the assumption is made that the platelet contains an internal surface or grain boundary where grain boundary or surface diffusion of iron ions or atoms can occur. The equations governing the transfer of iron to the tip or to the sides of the growing platelet are the same as for whisker growth. The major difference between whisker growth and platelet growth is the influence of water vapor on the growth site. Water vapor extends the point site into a line site. The line site grows as reaction proceeds, whereas the point site is independent of time. The growth of the oxide platelet in the edge direction must be related to stresses set up in the platelet at the line site.

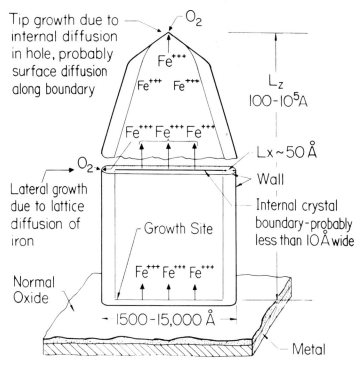

FIG. 10. Model of oxide platelet growth on iron, 400°C, $H_2O + O_2$.

Stress Corrosion

The role of localized oxidation in practical corrosion problems such as stress corrosion cracking was considered by Gulbransen and Copan (13). The effect of stress and water vapor and hydrogen chloride gas on localized corrosion of 304 stainless steel was to change the crystal habit from oxide whiskers and small platelets to large rounded platelets. The formation of rounded oxide platelets under stress and in water vapor atmospheres with

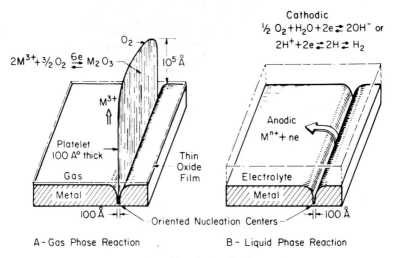

FIG. 11. Trenching by localized corrosion.

hydrochloric acid vapor present leads to the conclusion that a highly localized corrosion process exists. Since the rounded platelets are only 150 Å thick, the process is very localized.

A suggested mechanism of how localized chemical reaction can lead to the formation of submicroscopic trenches or cracks in the metal is depicted in Fig. 11. Figure 11A shows an oxide platelet formed as a result of gas-phase reaction of an ordered arrangement of nucleation sites. A deep trench may form at the root of the platelet. The formation of trenches in the metal a few hundred angstroms thick can lead to a high stress concentration and to a condition for propagation of the crack nuclei.

Figure 11B shows a schematic drawing of how localized chemical or electrochemical reaction can occur for liquid-phase reactions. As in Fig. 8b

and 8c, the presence of a set of aligned nucleation sites for liquid-phase reactions is assumed. The corrosion product may dissolve and as a result may not accumulate to form a coherent oxide structure. Again, a narrow trench may form which leads to a high stress concentration and conditions for the propagation of the crack nuclei.

The chemical and electrochemical conditions of the liquid-phase reaction suggest that more rapid localized corrosion may occur than in the gas-phase reactions. Some of these are as follows: (1) Diffusion of metal ion through the oxide platelet to the lip is not necessary, (2) local electrochemical cells may accelerate the rate of localized corrosion; (3) pH shifts may occur at the site of the reaction which could accelerate the rate of reaction; and (4) a sharper radius of curvature and larger stress concentration may exist at the root of the trench if a coherent metal oxide structure is absent.

Much work remains to be done to demonstrate the role of localized oxidation in stress corrosion cracking. Some empirical evidence supports the picture given. Attempts to form fan-shaped oxide growths on pure nickel and on 80 nickel–20 chromium alloys have been unsuccessful. These materials are reported to be less sensitive to stress corrosion cracking.

Summary

Electron optical studies of the oxidation of iron show that a wide variety of localized oxidation processes can occur depending on the environment, temperature, pressure, metal structure, and stress. Iron oxidized in dry oxygen at temperatures of 250°C to 300°C forms an oxide film consisting of nearly randomly oriented submicroscopic oxide crystals. At temperatures of 650°C to 850°C and oxygen pressures of 10^{-3} torr, small oxide nuclei form which are highly oriented to the metal grain. Metal substructure is important in determining the pattern of the oxide nuclei on the grain.

At temperatures of 400° to 500°C and near normal oxygen pressures, long oxide whiskers in addition to the normal oxide film are formed. In mixtures of water and oxygen, blade-shaped oxide platelets are formed. When stress and impurities are present in the metal, rounded oxide platelets are formed.

It is concluded that the microstructure of the oxide film is related to fundamental physical chemical activity of the underlying metal. Much can be learned about oxidation, general corrosion, stress corrosion, and fatigue of metals by microstructural studies of the chemical activity in gas–metal reactions.

References

1. J. Bardolle and J. Bénard, *Mem. Sci. Rev. Met.*, **49,** 613 (1952).
2. J. Bénard, "Oxydation des Metaux," Gauthiers-Villars, Paris, 1962.
3. G. Pfefferkorn, *Naturwissenschaften*, **40,** 551 (1953); *Z. Metallk.*, **46,** 204 (1955).
4. R. Takagi, *J. Phys. Soc. Japan*, **12,** 1212 (1957).
5. E. A. Gulbransen and T. P. Copan, *Discussions Faraday Soc.*, **28,** 229 (1959).
6. R. Phelps, E. A. Gulbransen, and J. Hickman, *Anal. Chem.*, **18,** 391 (1946).
7. E. A. Gulbransen, W. McMillan, and K. F. Andrew, *J. Metals*, **200,** 1027 (1954).
8. E. A. Gulbransen and K. Andrew, *J. Electrochem. Soc.*, **106,** 551 (1959).
9. R. F. Mehl and E. L. McCandless, *Trans. AIME*, **125,** 531 (1937).
10. E. A. Gulbransen and R. J. Ruka, *Ind. Eng. Chem.*, **43,** 697 (1951).
11. J. Bénard, F. Grönlund, J. Oudar, and M. Duret, *Z. Elektrochem.*, **63,** 799 (1959).
12. G. E. Rhead, *Trans. Faraday Soc.*, **61,** 797 (1965).
13. E. A. Gulbransen and T. P. Copan, "Physical Metallurgy of Stress Corrosion Fracture," ed. by T. Rhodin, Interscience, New York, 1959, p. 155.
14. E. A. Gulbransen and T. P. Copan, *Proc. European Regional Conf. Electron Microscopy 1960, Delft*, **1,** 225 (1961).
15. E. A. Gulbransen, *Mem. Sci. Rev. Met.*, **62,** 253 (1965).
16. E. A. Gulbransen, T. P. Copan, and W. M. Hickam, *Proc. 5th Intern. Congr. Electron Microscopy, Philadelphia*, **1,** 1 (1962).

Interface Conversion of Polymers by Excited Gases

R. H. HANSEN*

*Bell Telephone Laboratories, Incorporated
Murray Hill, New Jersey*

Many polymers require a surface treatment before they can be successfully printed on or bonded to other materials. Surface modification is usually accomplished by oxidative treatments. These may be relatively slow and at relatively low temperature as, for example, the use of glass cleaning solution, or they may be rapid and at very high temperature as, for example, flame treatment. To compare reactions which occur during one such oxidative surface process (exposure to atomic oxygen) with those which occur during simple thermal oxidation, let us consider both processes.

Simple Thermal Oxidation

Equations for the reactions which are believed to occur during simple thermal oxidation of a hydrocarbon polymer such as polyethylene are shown in Table I. Thermal oxidation takes place in three steps: initiation, propagation, and termination. Decomposition of hydroperoxide molecules into several active species which can attack unoxidized polymer molecules and result in their eventual oxidation is responsible for the autocatalytic nature of simple thermal oxidation. Autocatalysis of oxidation is shown in Fig. 1. It is seen that simple isothermal oxidation is initiated immediately on contact of the polymer with oxygen and that the reaction proceeds at an ever-increasing rate. Not shown here is the fact that the presence of antioxidants delays onset of oxidation. The amount of delay depends on the kind and amount of antioxidant used to protect the polymer. However, once the antioxidant is consumed, oxidation of the polymer occurs.

Simple thermal oxidation is a bulk reaction and occurs in accessible regions of the polymer. The reaction takes place as a continuously accelerating stage (because of autocatalysis) followed by a continuously decelerating stage as accessible material is progressively depleted. Equations for both of these stages follow the Arrhenius relationship and agree well with experimental points (Fig. 2). In general, the reaction is limited to noncrystalline regions of semicrystalline polymers (Figs. 3 and 4). As a matter of fact, the ratio of

* Present address: J. P. Stevens and Co., Inc., Garfield, New Jersey.

TABLE I
Oxidation of Polymers

Initiation:	$RH \rightarrow R\cdot + H\cdot$ or $R'\cdot + R''\cdot$
	$R\cdot + O_2 \rightarrow ROO\cdot$
	$ROOH \rightarrow RO\cdot + OH\cdot$
Propagation:	$ROO\cdot + R'H \rightarrow ROOH + R'\cdot$
	$RO\cdot + R'H \rightarrow ROH + R'\cdot$
	$OH\cdot + R'H \rightarrow H_2O + R'\cdot$
	$R'\cdot + O_2 \rightarrow R'OO\cdot$
	$ROOH \rightarrow RO\cdot + OH\cdot$
	$2ROOH \rightarrow RO\cdot + ROO\cdot + H_2O$
Termination:	Decrease in number of participating species by:
	(1) Action of antioxidant
	(a) $HA + ROO\cdot \rightarrow ROOH + A\cdot$, where $A\cdot$ is inactive
	(b) heterolytic hydroperoxide decomposition
	(2) Disproportionation
	(3) Combination
	(4) Depletion of accessible material

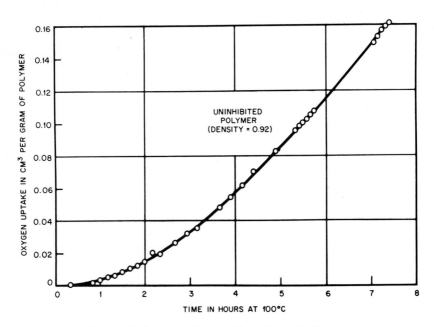

FIG. 1. Isothermal oxidation of branched polyethylene.

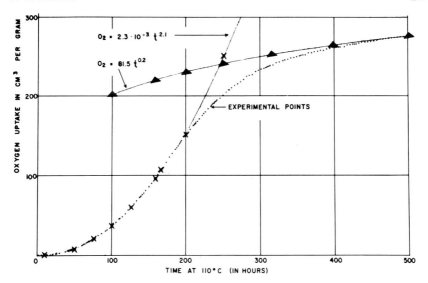

Fig. 2. Oxidation of polyethylene.

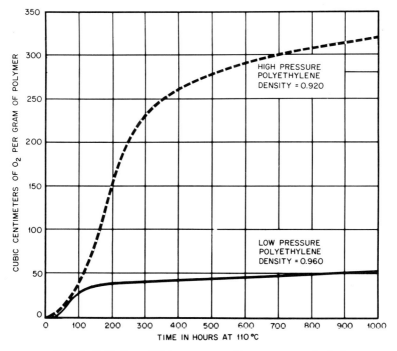

Fig. 3. Oxidation of high-pressure and low-pressure polyethylene.

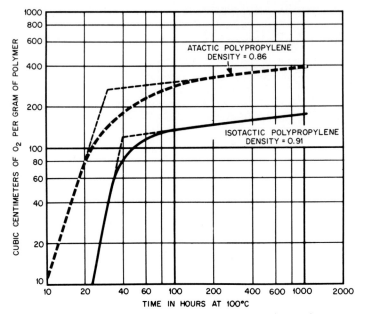

FIG. 4. Oxidation of atactic and isotactic polypropylene.

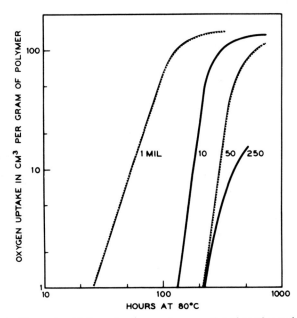

FIG. 5. Thermal oxidation of polypropylene as a function of sample thickness.

the amount of oxygen uptake observed during oxidation of a semicrystalline polymer to the oxygen uptake which would occur if the polymer was completely noncrystalline can be used to determine the crystalline content of the polymer.

Rate of simple thermal oxidation is affected by thickness of the specimen (Fig. 5). Although rates of oxidation for the various thicknesses of polypropylene shown in this figure are decreased by a factor of 10 in going from 1-mil to 50-mil specimens, the overall extent of reaction is not affected.

TABLE II
EFFECT OF POLYMER STRUCTURE ON EASE OF OXIDATION

Polymer	Induction Period, 110°C	t, 10 hours, 80°C
Atactic polypropylene	4.5	95
Isotactic polypropylene	7.5	130
Poly(vinylcyclohexane)	800	>500
Poly(allylcyclohexane)	35	700
Poly(allylcyclopentane)	1.5	
Polystyrene	>10000	—
Poly(3-phenylpropene-1)	1900	>10000
Poly(4-phenylbutene-1)	30	500
Poly(5-phenylpentene-1)*	23	360
Poly(6-phenylhexene-1)*	13	200
Commercial Polymers		
0.92 density polyethylene	45	800
0.96 density polyethylene	70	1300
Polypropylene	20	370

* Tacky.

Accessible portions of the polymer are eventually oxidized, but not before diffusion delays the oxidation of polymer beneath the surface.

Structure of the polymer exposed to simple thermal oxidation greatly affects ease of oxidation. Highly branched polymers oxidize at a much more rapid rate than a linear polymer such as polyethylene unless the branches are sufficiently bulky to sterically hinder oxidation. The data in Table II show that large substituents such as phenyl or cyclohexyl groups are able to hinder simple thermal oxidation even when they are removed from the main chain by one or two methylene groups as in poly(allylcyclohexane) or poly(3-phenylpropene-1). When the bulky group is further from the main

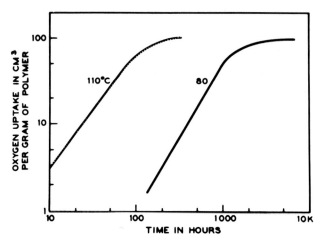

Fig. 6. Oxidation of poly(5-phenylpentene-1).

chain, oxidation occurs more rapidly than it does in relatively unbranched polyethylene. If the polymer contains both aliphatic and aromatic moieties, only the aliphatic portion is oxidized (Fig. 6). Poly(5-phenylpentene-1) is completely noncrystalline at temperatures above 80°C, and so, although the rate of oxidation is more rapid at 110°C than it is at 80°C, the overall extent of oxidation is the same at both temperatures and corresponds to exhaustive oxidation of accessible aliphatic portions of the polymer.

Oxidation with Atomic Oxygen (*1, 2*)

Equations for the reactions which are believed to occur when a polymer such as polyethylene is exposed to atomic oxygen are shown in Table III. Slow reactions which also occur during simple thermal oxidation can take place during oxidation with atomic oxygen, but these slow reactions are of

TABLE III
REACTION OF POLYETHYLENE WITH ATOMIC OXYGEN

Rapid:	$RH + O\cdot \rightarrow R'\cdot + RO\cdot$ or $R\cdot + OH\cdot$
Rapid:	$R\cdot + O\cdot \rightarrow RO\cdot$
Slow:	$RH \rightarrow R\cdot + H\cdot$ or $R\cdot + R\cdot$
Rapid:	$R\cdot + O_2 \rightarrow ROO\cdot$
	$ROO\cdot + R'H \rightarrow ROOH + R'\cdot$
Slow:	$ROOH \rightarrow RO\cdot + OH$

little importance because of the rather rapid oxidation which usually occurs. As shown in Fig. 7, reaction rates are linear, and there is not even a hint of autocatalysis. In further contrast to simple thermal oxidation, there appears to be little or no temperature coefficient, no induction period even in the presence of large amounts of antioxidants, and no effect due to crystalline content of the polymer or to thickness of the specimen exposed to the stream of atomic oxygen. Instead, there is a dependence on the surface area of the specimen and a dependence on the structure of the polymer. The data presented in Fig. 7 show that sulfide and formaldehyde polymers are readily attacked by atomic oxygen. As might be anticipated, polypropylene is more readily oxidized than polyethylene, while the branched but sterically hindered polystyrene is more resistant to oxidation by atomic oxygen. Also expected was the resistance of polytetrafluoroethylene toward attack by atomic oxygen, but it is not likely that anyone would have predicted the remarkable resistance

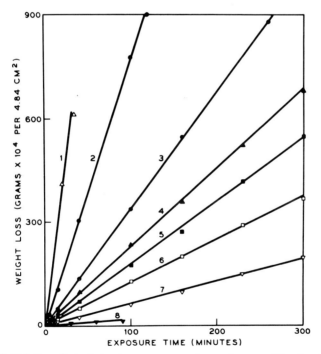

FIG. 7. Relative effect of atomic oxygen on various polymers: (1) polysulfide polymer; (2) formaldehyde polymer; (3) polypropylene; (4) low-density polyethylene; (5) polyethylene terephthalate; (6) polystyrene; (7) polytetrafluoroethylene; (8) sulfur-vulcanized natural rubber.

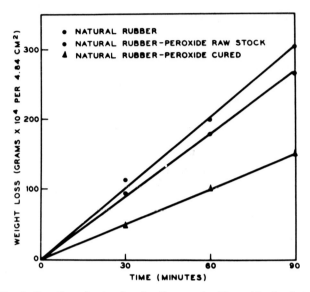

FIG 8. Reaction of natural and sulfur-cured rubber with atomic oxygen.

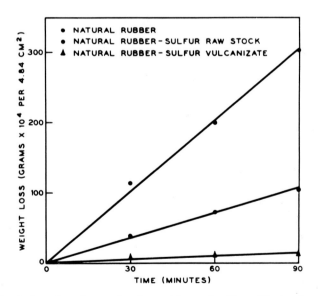

FIG. 9. Reaction of natural and peroxide-cured rubber with atomic oxygen.

of sulfur-vulcanized rubber (natural rubber or ethylene–propylene rubber in a variety of hardnesses) toward attack by atomic oxygen. The data in Fig. 8 show that natural rubber alone is about as readily oxidized as polypropylene, which is reasonable, since it has about the same number and kind of side chains. Simple incorporation of typical ingredients for a sulfur cure on a rubber mill imparts some resistance against attack by atomic oxygen, while complete vulcanization renders the rubber even more resistant to atomic oxygen than polytetrafluoroethylene. Peroxide-cured natural rubber also is more resistant to attack by atomic oxygen than uncured natural rubber (Fig. 9). Cross-linking makes polyethylene more susceptible to attack by atomic oxygen whether the cross-linking is accomplished by use of peroxides or by irradiation with high-energy electrons (Fig. 10). Since both the presence of sulfur and the presence of cross-links in a polymer appear to enhance its reactivity with atomic oxygen, it is difficult to reconcile the relative inertness of vulcanized rubbers with their structures. It may be possible that in rubber, instead of introducing a susceptible tertiary carbon atom, vulcanization apparently shields vulnerable sites, possibly by forming highly gelled structures.

Although its rate of reaction with atomic oxygen is apparently unchanged when vulcanized rubber is exposed while under stress, many of the desirable

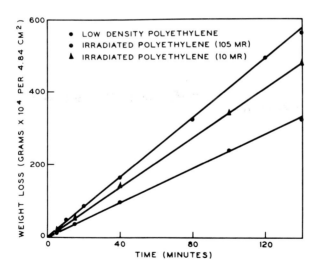

FIG. 10. Treatment of polyethylene with atomic oxygen after previous electron bombardment under high vacuum.

properties of the polymer are soon destroyed. The same is true for saturated hydrocarbon polymers. For example, if polyethylene under stress is exposed to atomic oxygen, the failure resembles that which occurs in rubber which is exposed under stress in an atmosphere containing ozone. The appearance of a stress-cracked specimen (ASTM D-1693) after short exposure to atomic oxygen while under stress is shown in Fig. 11. Much of the applied stress

FIG. 11. Micrograph of polyethylene after exposure to atomic oxygen while under stress.

has been relieved by the development of a plenitude of surface cracks even though the specimen has only lost about 5×10^{-4} gram of its original weight of 1.25 grams.

However, although polymers literally fall apart when exposed to atomic oxygen while under tension, the stress-crack resistance of polyethylene is somewhat increased after short exposure to atomic oxygen before stress is applied. It is believed that this improvement is a result of elimination of surface defects.

There are other, more important uses for the process. The most important application is in the preparation of polymer surfaces for printing and adhesive bonding. Although bulk properties of unstressed polymers are unchanged during attack by atomic oxygen, the surfaces of most polymers are affected.

The most obvious change is an oxidation of the surface which results in an increase in the wettability of the polymer. Polymers which showed a great decrease in contact angle with water after treatment with atomic oxygen included polyethylene, polypropylene, polybutene-1, polycarbonate, polyimide, and various types of nylon.

A potential laboratory application for the technique is to appraise the suitability of polymers for upper atmosphere applications where a plenitude of atomic and excited gases prevails. The equipment is relatively simple, as is the technique. Polymer samples about the size of microscope slides are placed in a glass chamber, and a stream of atomic oxygen, produced by passing oxygen at about 1 mm pressure through a radio-frequency coil, is passed over them. The equipment used in these and subsequent studies was the Tracerlab LTA-500A Asher.

Effect of Other Excited Gases (3, 4)

What happens when polymers are exposed to excited species of non-oxygen-containing gases (that is, not H_2O, CO, CO_2, etc.) under otherwise very similar conditions in the same equipment? As might be anticipated, bulk properties are unaffected, while the surfaces and regions below the surface are affected quite differently than when the same polymers are exposed to atomic oxygen or excited species of oxygen-containing gases. The infrared transmission spectra of a 9-mil polyethylene specimen before and after exposure to excited helium are identical (Fig. 12). Attenuated total internal reflection spectra of the same specimen showed that there was a great deal of *trans*-ethylenic unsaturation produced in the surface regions of polyethylene by this treatment (Fig. 13). When a powdered specimen of the same high-density polyethylene was treated with activated helium, its molecular weight was increased considerably (Table IV). Furthermore, mass spectrographic

TABLE IV
EFFECT OF ACTIVATED HELIUM ON MARLEX 5003

	Untreated	After 10 Minutes at $\sim$100 Watts
Melt index	0.15	0.11
M_W	86000	135000*
M_N	1800	4800*

* No fractions having lower molecular weights than those present in the untreated material were observed. The reported values do not include the effect of almost 1% of insoluble gel produced by CASING.

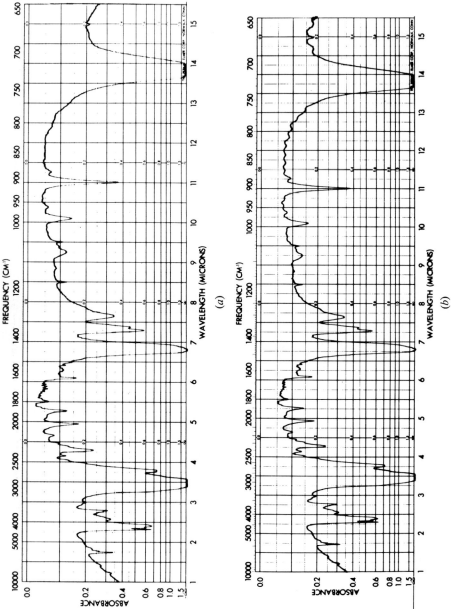

FIG. 12. Transmission infrared spectra of polyethylene: (a) Before treatment. (b) After treatment with excited helium.

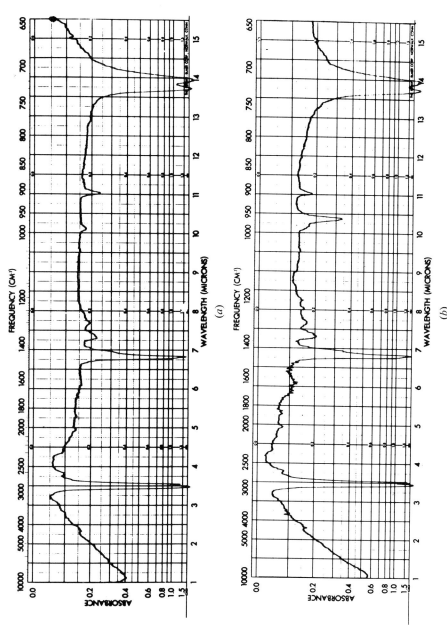

Fig. 13. Attenuated total internal reflection spectra of polyethylene: (a) Before treatment. (b) After treatment with excited helium.

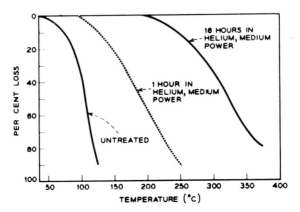

Fig. 14. Effect of treatment of perfluorokerosene with activated helium.

analysis of the effluent gases showed that only hydrogen was produced. Higher-molecular-weight species as well as insoluble gel and unsaturation were also observed when lower-molecular-weight materials were treated with activated species of helium. For example, when a volatile perfluorinated liquid was treated with activated helium for 1 hour, it was converted to a tacky solid. After 16 hours, it became a hard glassy solid. The effect of treatment with activated helium on the molecular weight of perfluorokerosene is shown by the thermograms in Fig. 14. In each case the specimens were heated in air from room temperature on a shallow pan at a rise in temperature of 2°C per minute. A similar treatment of the linear hydrocarbon *n*-octadecane is

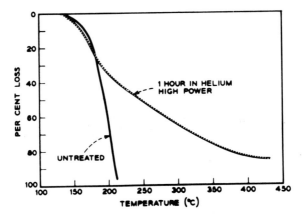

Fig. 15. Effect of treatment of *n*-octadecane with activated helium.

shown in Fig. 15. Part of the hydrocarbon (that at the surface exposed to activated helium) is converted to a cross-linked polymer resembling crosslinked polyethylene. The cross-linking reaction does not occur if the hydrocarbon being treated is crystalline; for example, n-hexatriacontane ($C_{36}H_{74}$) is apparently unaffected in the solid highly crystalline state. It is easily crosslinked when it is first melted and exposed while in the molten state to activated helium. In this respect, exposure of a hydrocarbon to activated helium resembles simple thermal oxidation of hydrocarbons; that is, reaction appears to be limited to noncrystalline regions in both cases. The effect of crystalline content on the simple thermal oxidation of hydrocarbon polymers is shown in Fig. 16. It is seen that the considerably more crystalline low-pressure polyethylene reacts autocatalytically with much less oxygen than does the less crystalline high-pressure polymer.

When a polymer such as polyethylene is exposed to excited species of helium, hydrogen is evolved and the polymer is seen to increase in gel content, molecular weight, and *trans*-ethylenic unsaturation. Reactions such as those presented in Table V could be invoked to interpret these results. For example, all the metastable and ionic species of helium possess more than

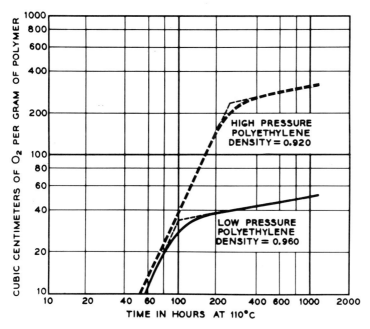

FIG. 16. Effect of crystalline content of polyethylene during thermal oxidation.

TABLE V
Formation of Higher-Molecular-Weight Species with Activated Helium

Excited Molecules
$R_1H + He^* \rightarrow R_1H^* + He$
$R_1H^* + R_2H \rightarrow R_1R_2 + H_2$
$R_1CH_2CH_2R_2^* \rightarrow R_1CH{=}CHR_2 + H_2$

Ion-Molecule Reactions
$He^+ + RH \rightarrow He + RH^+$
$CH_4^+ + CH_4 \rightarrow CH_5^+ + CH_3\cdot$
$CH_3^+ + CH_4 \rightarrow C_2H_5^+ + H_2$
$C_2H_5^+ + CH_4 \rightarrow C_3H_7^+ \rightarrow H_2$

enough energy to cause the formation of higher-molecular-weight species. However, since electron spin resonance studies of polymers such as polyethylene and polytetrafluoroethylene after bombardment with activated helium showed that radicals were present, the reactions in Table VI are considered to be those most likely to occur. Reactions such as those shown in Table VII apparently do not occur to any great extent because they would lead to lower-molecular weight species, and only higher-molecular-weight species and hydrogen were observed during the treatment of polyethylene with activated species of helium. For the case of polyethylene, the equations can be written as as shown in Table VIII. These reactions account for the formation of *trans*-vinyl unsaturation and for the increased molecular weight (and eventual gelation) of the polymer which occurs at the gas–solid interface.

This reaction is of considerable practical importance and is also theoretically significant, for it will be shown that bombarding polymers such as polytetrafluoroethylene and polyethylene with activated species of helium results in the formation of surfaces which are printable and which can be bonded to other materials. These useful surfaces are created without observed changes

TABLE VI
Possible Free Radical Reactions Resulting from Treatment with Activated Helium

$He^* + RH \rightarrow R\cdot + H\cdot + He$
$He^* + R_1R_2 \rightarrow R_1\cdot + R_2\cdot + He$
$H\cdot + RH \rightarrow H_2 + R\cdot$
$R\cdot + R_1\cdot \rightarrow RR_1$
$R\cdot + R_2\cdot \rightarrow RR_2$
$R\cdot + R\cdot \rightarrow RR$

TABLE VII
UNLIKELY REACTIONS FROM TREATMENT
WITH ACTIVATED HELIUM

$R_1 \cdot + RH \rightarrow R_1H + R \cdot$
$R_2 \cdot + RH \rightarrow R_2H + R \cdot$
$R_1 \cdot + H \cdot \rightarrow R_1H$
$R_2 \cdot + H \cdot \rightarrow R_2H$
$R_1CH_2CH_2 \cdot + R_2 \cdot \rightarrow R_1CH{=}CH_2 + R_2H$
---

in wettability. Improved adhesion is also observed when activated species of the other inert gases are used as well as when activated species of nitrogen and hydrogen are employed. Atomic hydrogen changes the wettability of polytetrafluoroethylene slightly so that after long exposure it begins to resemble polyethylene, while activated species of nitrogen cause some increase in the wettability of both polytetrafluoroethylene and polyethylene.

The effect of exposure of these polymers to activated species of helium on the strength of adhesive joints prepared from the polymers is shown in Fig. 17. Tensile shear strength experiments show that this new surface treatment increases the strength of joints prepared from polyethylene to nearly 3000 psi (as compared with a few hundred psi for untreated polymer); the tensile shear strength of joints prepared from treated polytetrafluoroethylene is over 1000 psi as compared with essentially no strength for untreated polymer.

These greatly improved joint strengths are achieved without change in wettability of the surface of the polymer. They are not a result of the *trans*-ethylenic unsaturation which is introduced by exposure to activated helium. The strength of joints prepared from freshly prepared samples (lower curves on Fig. 18, polyethylene on left and polytetrafluoroethylene on the right) was essentially the same as the strength of joints prepared from treated surfaces in which the unsaturation was removed by exposure to a solution of

TABLE VIII
TYPICAL REACTIONS OCCURRING DURING THE TREATMENT
OF POLYETHYLENE WITH ACTIVATED HELIUM

$RCH_2CH_2CH_2CH_3 + He^* \rightarrow R\overset{H\cdot}{\overset{\cdot}{C}}HCH_2CH_2CH_3 + He$

$R\overset{H\cdot}{\overset{\cdot}{C}}HCH_2CH_2CH_3 \rightarrow RCH{=}CHCH_2CH_3 + H_2$

or

$RCH_2CH_2CH_2CH_3 \quad\quad RCHCH_2CH_2CH_3$
$+ \quad\quad\quad\quad\quad\quad\quad \rightarrow \quad | \quad\quad\quad\quad + H_2$
$R\overset{H\cdot}{\overset{\cdot}{C}}HCH_2CH_2CH_3 \quad\quad RCHCH_2CH_2CH_3$

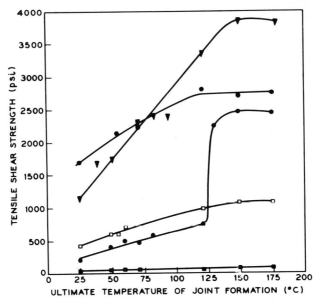

Fig. 17. Tensile shear strengths of lap shear composites: (■) untreated G-80 polytetrafluoroethylene film; (□) G-80 polytetrafluoroethylene film treated with activated helium; (●) untreated Marlex 5003 polyethylene film; (○) Marlex 5003 film treated with activated helium; (▼) aluminum composites.

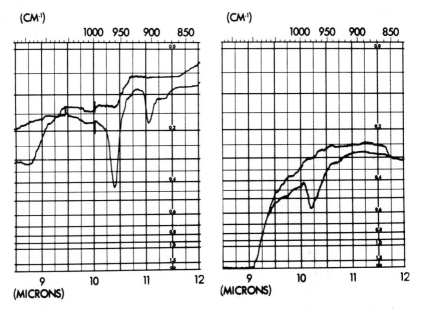

Fig. 18. ATR infrared spectra of Marlex 5003 polyethylene (left) and G-80 polytetrafluoroethylene (right).

bromine after bombardment with activated helium (upper ATR infrared curves on Fig. 18). Instead, the improvement in adhesive joint strength is due to the formation of a strong cross-linked layer at the surface. This "skin" does not swell in solvents, and so it is highly cross-linked. It can be separated from unreacted polymer beneath the surface regions by dissolving away unreacted polymer. The relationship between weight loss on bombardment of Marlex 5003 polyethylene with helium and the weight of residual gel obtained after extraction with xylene is shown in Fig. 19. Note that the reaction appears to be diffusion-controlled, as was the case for simple thermal oxidation. In this case, it is believed that the reason for the diffusion-controlled reaction is the diffusion of hydrogen atoms formed during collision of the activated helium particles with polymer molecules. As the hydrogen atoms diffuse away from the surface, they cause polymerization reactions to occur beneath the surface of the polymer. These reactions, as was the case for low-molecular-weight waxes, apparently occur only in noncrystalline regions of the polyethylene, for the melting characteristics of the extracted skin are indistinguishable from those of untreated polymer.

This surface skin may be of interest in applications where it is desirable to alter the permeability, abrasion resistance, coefficient of friction, and other mechanical properties of polymers such as polyethylene and polytetrafluoroethylene. It is of great importance in improving the printability and

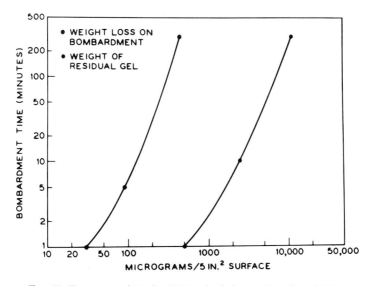

FIG. 19. Treatment of Marlex 5003 polyethylene with activated helium.

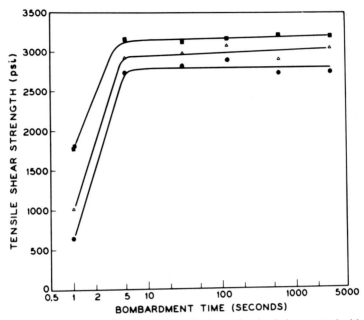

Fig. 20. Joint strength between epoxy adhesive and polyethylene treated with activated helium: (●) 60°C; (△) 82°C; (■) 104°C.

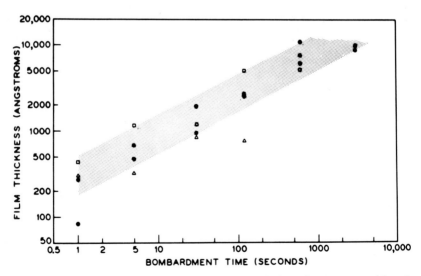

Fig. 21. Thickness of surface skin produced on polyethylene after treatment with various activated gases: (□) hydrogen (1 mm pressure); (●) helium (1 mm pressure); △ helium (0.4 mm pressure); (○) neon (1 mm pressure).

bondability of these and other low-surface-energy polymers. The process is simple and rapid. Adhesive joint strength which is attained between a conventional epoxy adhesive and untreated polyethylene is about 200 psi. It can be seen from Fig. 20 that maximum strength of about 3000 psi is developed between 1 and 5 seconds of bombardment with activated helium at rather low power levels (rated below 100 watts; developed temperature is less than 35°C). The three curves in Fig. 20 represent different cure temperatures for the epoxy adhesive. Similar families of curves result when the other rare gases or hydrogen or nitrogen are used to treat the polymers.

Surface skins produced by bombarding polyethylene over a range of times were extracted with xylene to constant weight. If a density of 1.0 is assumed (a value in good agreement with an experimental value of 1.04 measured for a very thick skin obtained by long exposure to activated helium), it is possible to convert weight of the skin to thickness. Values for the thickness of the skins produced have been plotted against bombardment time in Fig. 21. A comparison between Fig. 20 and Fig. 21 shows that maximum adhesive joint strength is observed when the cross-linked skin produced is about 300 to 1000 Å thick. This apparently is the thickness of the weak boundary layer present in polyethylene as it is normally prepared. Maximum adhesive joint strength is achieved when this weak boundary layer is strengthened by cross-linking, and no further increases in adhesive joint strength are obtained by further treatment and hence further penetration.

CASING (Cross-linking with Activated Species of INert Gases) is effective in increasing the joint strength obtained with adhesives and other polymers. For example, even though nylon is more polar and more readily wetted than polytetrafluoroethylene or polyethylene, tensile shear strength of joints prepared with untreated nylon average less than one-sixth that of nylon which has been treated by CASING, as shown in Fig. 22. It is also effective for neoprene and other halogenated polymers such as poly(vinyl fluoride) (Fig. 23). It is not an effective surface treatment (Fig. 24) for poly(vinylidene fluoride). This material forms strong adhesive joints even without surface treatment and apparently has no weak boundary layer as it is ordinarily prepared. That CASING improves printability as well as adhesive joint strength is shown in Fig. 25. Two polytetrafluoroethylene strips, one untreated and one exposed to activated species of helium, were coated with marking ink. The ends of the strips were not coated, so it can be seen that CASING does not cause discoloration of the polymer. Transparent cellophane tape was applied to the inked area of both strips and peeled away. Ink was completely stripped from the untreated specimen, while excellent retention was observed for the specimen which had been exposed to activated helium.

278 INTERFACE CONVERSION BY EXCITED GASES

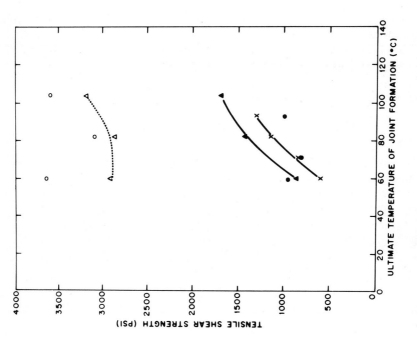

FIG. 22. Effect of activated gas treatment of nylon: untreated (lower curves); treated (upper curves).

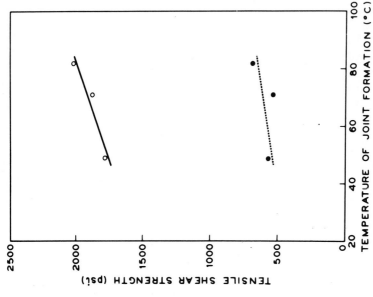

FIG. 23. Effect of activated gas treatment of poly(vinyl fluoride) on tensile shear strength: (●) untreated poly(vinyl fluoride); (○) poly(vinyl fluoride) treated with activated 1:1 mixture of H_2–He for 1 hour at medium power and 1 mm pressure.

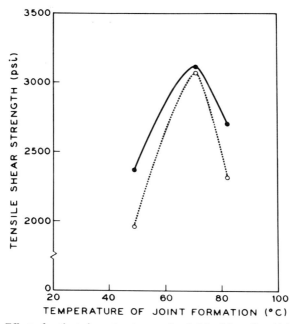

Fig. 24. Effect of activated gas treatment of poly(vinylidene fluoride): (●) untreated; (○) treated with activated helium.

Fig. 25. Effect of activated helium on printability of polytetrafluoroethylene.

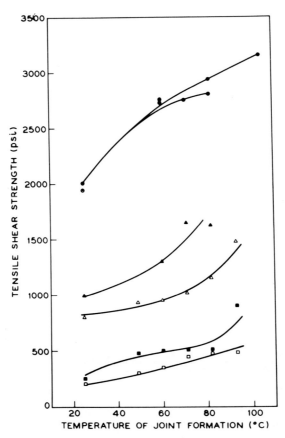

Fig. 26. Composites of Marlex 5003 polyethylene after a variety of treatments: (□) polyethylene (no surface treatment); (■) crystallized Marlex 5003 from xylene (no surface treatment); (△) polyethylene exposed to vapors of boiling 1:1 hexane–heptane mixture for 5 minutes; (▲) polyethylene irradiated with a Van der Graaf generator (dose 10 Mrads); (○) polyethylene treated with glass cleaning solution at 80°C for 4 minutes; (●) polyethylene treated with activated helium at 1 mm pressure for 5 seconds.

Results of this and other investigations are summarized in Fig. 26. Plotted against temperature of joint formation with a typical epoxy adhesive are tensile shear strengths obtained from composites prepared from Marlex 5003 polyethylene after a variety of treatments. Other adhesives could have been used, but the results would have been the same. All the curves point upward toward the right. This is because the strength of the epoxy adhesive increases as cure temperature is increased. The bottom curve in Fig. 26 is based on joint strength data obtained from untreated polyethylene film. All experiments were performed at temperatures below 100°C, since if the melting range of the polymer is exceeded it will spread on the cured epoxy adhesive surface, and in this process will preclude the weak boundary layer and result in strong joints. The curve just above that for untreated polyethylene is that for single crystals of well-characterized high-molecular-weight polyethylene which were remolded at 160°C. It appears either that the weak boundary layer which is responsible for the production of weak adhesive joints in untreated polyethylene is not due to the presence of low-molecular-weight polymer molecules, or that these low-molecular-weight species are regenerated during the remolding of the fractionated single crystals and forced to the surface during recrystalization of the melt, because no significant increase in adhesive joint strength was observed.

To avoid possible recrystallization from the melt, a solvent extraction method was employed. By proper choice of solvents to avoid swelling of the polymer, a two- to threefold increase in joint strength could be obtained (third curve from the bottom, Fig. 26).

Similar results were obtained when the weak boundary layer was eliminated by cross-linking the polymer with high-energy electrons. Again, a two- to threefold increase in adhesive joint strength was observed, but, since high-energy electrons are highly penetrating as well as nondiscriminating, the bulk properties of the polymer were greatly affected before adhesive joint strength was affected.

The uppermost curves in this figure are based on results obtained with surface treatment techniques which yield strong adhesive joints without affecting bulk properties of the polymer. High joint strengths were obtained both after CASING (5 seconds in activated helium at 35°C) and after etching with glass cleaning solution (4 minutes at 80°C), which, like all oxidative surface treatments, also causes ablation of the polymer and changes its wettability. What is generally not appreciated is that glass cleaning solution, like activated helium, also results in cross-linking and strengthening of the weak boundary layer at the surface of the polymer. Because ablation occurs concurrently with cross-linking during oxidative surface treatments, a thick skin is never created or observed, whereas cross-linked skins several mils thick can be produced by long periods of CASING.

It has been shown that adhesive joint strengths obtained by CASING are comparable to those obtained by oxidative treatments even though there is no change in wettability during CASING, while contact angle with water for polyethylene etched with glass cleaning solution is approximately 30°. It has also been shown that even a relatively readily wetted polymer such as

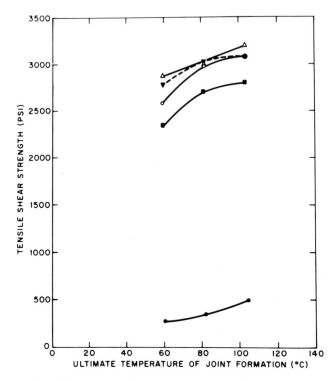

FIG. 27. Effect of treatment of polyethylene with mixtures of argon and fluorine and pure fluorine: untreated (lower curve); treated (upper curves).

nylon requires a surface treatment such as CASING in order for maximum adhesive joint strength to be obtained. It can also be shown that high adhesive joint strength is obtained by exposing polyethylene to elemental fluorine in the dark at ambient temperature (5). The values of tensile shear strength of adhesive joints prepared from untreated polyethylene are compared in Fig. 27 with those obtained from polyethylene which had been exposed to mixtures of argon and fluorine and to pure fluorine. Because treatment with fluorine causes radical reactions to occur, it is reasonable to assume that cross-linking occurs as it does during CASING. Thus, even though wettability is greatly decreased by fluorination of the surface, strong adhesive joints are obtained. Wettability apparently is less important than has commonly been believed. Fluorination is interesting but expensive and difficult to control. The most effective surface treatments are CASING and ablative oxidative techniques such as the use of atomic oxygen.

Conclusions

It can be said that there is no single general technique for the surface preparation of polymers for adhesive bonding and for printing. For example, oxidative techniques fail for polytetrafluoroethylene and other relatively inert polymers and cannot be used where samples are under stress or where changes in other surface properties are undesirable. CASING is not effective for treating highly branched polymers such as polypropylene because in this case, apparently, more chain scission than cross-linking is observed, and a strong surface region is not produced. CASING, however, can be used to successfully treat many polymers.

The CASING technique dramatically illustrates the importance of increasing the mechanical strength of weak boundary layers (in polymers where this phenomenon is observed) in order to prepare strong adhesive joints. By use of this technique, strong joints may be formed without change in wettability or electrical characteristics of the polymer surface. Polymers such as nylon, which are more readily wetted than polytetrafluoroethylene or polyethylene, also require a surface treatment such as CASING to promote maximum adhesive joint strength. Fluorination of the polyethylene surface also results in the formation of strong adhesive joints even though wettability of the surface is drastically decreased. Therefore, wettability is considerably less important than is commonly considered, and techniques for interface conversion of polyethylene and other low-surface-energy polymers in order to improve adhesive joint strength all must have in common the ability to strengthen mechanical properties of the surface region of the polymer.

Acknowledgments

I wish to acknowledge the collaboration of Dr. Peter Rentzepis and Dr. Harold Schonhorn during parts of the investigation reported here and express my appreciation also to William M. Martin, Thomas DeBenedictis, Joseph V. Pascale, Frank Ryan, Dr. Patrick K. Gallagher, and Frank Schrey for their contributions to this work.

References

1. R. H. Hansen, W. M. Martin, and T. DeBenedictis, *Trans. Inst. Rubber Ind.*, **39**, T301 (1963).
2. R. H. Hansen, J. V. Pascale, T. DeBenedictis, and P. M. Rentzepis, *J. Polymer Sci.*, **A3**, 2205 (1965).
3. R. H. Hansen and H. Schonhorn, *Polymer Letters*, **4**, 203 (1966).
4. H. Schonhorn and R. H. Hansen, *J. Appl. Polymer Sci.*, **11**, 1461 (1967).
5. H. Schonhorn and R. H. Hansen, *J. Appl. Polymer Sci.*, in preparation.

Discussion

MICHAELS: A process for making particulate reinforced polyolefins with significantly improved physical properties has been developed which involves the introduction between the filler and the resin phase of a boundary zone with a graded elastic modulus. We attribute the rather marked increase in impact strength observed with these composites to the fact that this zone delocalizes the stress which normally builds up at the interface between two mismatched modulus materials. What I am wondering is whether your improved adhesive bonds which you get in this system are not so much due to the increase in the strength of the material near the surface but rather due to the reduction in the stress at the polymer–metal surface interface because you have increased the modulus of the polymer by cross-linking.

HANSEN: If one puts a tab of adhesive on Teflon, for example, and pulls the tab off and measures wettability upon the tab, the surface of the tab resembles Teflon. In effect, you put Teflon onto the surface of the tab itself. However, if you treat the Teflon with activated helium, you cannot do this.

MICHAELS: You mean you do not remove the Teflon from the treated surface when you make the epoxy tab on it, but yet it could be attributable to the fact that you reduce the localization of the stress at the surface and the primary failure might occur either in the epoxy or in the underlying bulk of the material. Where does the failure occur?

HANSEN: In the bulk material, not primarily at the interface.

MICHAELS: This again could be attributed to stress delocalization at the interface.

ZIMMT: When you ran your molecular weight distribution, was this on a film, and how thick was it?

HANSEN: No. This was on Marlex crumbs (powder).

ZIMMT: Since it was on crumbs, only the surface was actually an interface. This means that the surface is of high molecular weight.

HANSEN: Yes. As a matter of fact, about 1% of the treated powder was converted to gel. I don't know what percentage of the crumb is actually surface, but it would be interesting to calculate.

H. MARK: I don't know how vulcanized the rubber was which you used, but if it did contain, let us say, something like 2% sulfur, it is necessary to think beyond just cross-linking. In reality, there is a great deal of cyclization n vulcanized rubber. Actually, under certain conditions, 75% of all the sulfur of vulcanization does not make cross-links but produces cyclic structures, and it might be that the rings are more stable than the open chains.

CUTHRELL: Did you try CASING on epoxy polymers?

HANSEN: I don't think CASING would be of value in epoxies because you can alter epoxies to make them cross-linked, if you wish, by chemical means.

CUTHRELL: Well, once you have done this, though, and still have possibly low-molecular-weight fractions, do you think you could increase the molecular weight of the epoxy in a similar manner?

HANSEN: The process works primarily with linear polymers and with chlorinated polymers such as neoprene and poly(vinyl chloride). I really don't know whether it works with epoxy polymers. I would suspect that it would work with them if they contained sufficiently long, linear aliphatic sequences.

GRISKEY: You irradiated some polyethylenes, and I think what it showed is that, with the radiation, susceptibility toward atomic oxygen got worse instead of better. With certain polymers, you get cross-linking with radiation. Did you evaluate your samples to see whether you got cross-linking with radiation? Was this the only one you checked with radiation? Did you check its effect on adhesion?

HANSEN: This is the only one we checked with radiation. We also studied chemically cross-linked polyethylene, and we observed pretty much the same thing. As we increased cross-linking, it became more reactive toward atomic oxygen. Cross-linking by irradiation (electron bombardment) does favorably affect ease of bonding, but it also changes bulk properties of the polymer. We considered taking Teflon and bombarding it with hydrogen to make a polyethylene-like surface and then subjecting this to atomic oxygen. This would create a Teflon with a very wettable surface. We have not tried this experiment yet.

ASBECK: Where does this layer come from? Presumably, you have a bulk system. You cut this, and all of a sudden you have the weak layer. Would it be possible to cross-link the system in bulk and then cut it? If everything is cross-linked, it can't exude. Do you then get good adhesion?

HANSEN: If you had a three-dimensional, cross-linked polymer, I think you would probably get away with this. In any cutting operation you're going to break molecules as well, and if you happen to break off short segments you would be in trouble again as you would with untreated samples. Schonhorn actually fractionated high-molecular-weight polyethylene and remolded it into a mat and then tried to form an adhesive joint. Apparently, the remolding process causes reorganization of this material, and joint strengths were very little better than they were with untreated polyethylene.

ASBECK: I'm just wondering where the weak layer really comes from. Obviously, it exudes somewhere from the inside, but what is the mechanism of this?

HANSEN: We're not even certain what it is yet. We have a number of theories. I don't really think it is low-molecular-weight material, although this would be one possibility. I do know that untreated polyethylene and Teflon feel waxy, and after treatment they no longer feel waxy.

FOWKES: When mercury is put on a Teflon surface, the surface tension of the mercury drops about 100 dynes per centimeter. This is presumed to be due to material on the surface of the Teflon spreading onto the mercury drop, so it would mean that this was fairly liquid-like material. When Gray tried a similar experiment and obtained comparable results, he didn't measure the surface tension but determined the contact angles on both polyethylene and Teflon. This might be an interesting method to use to examine this material.

HANSEN: This is possible, but we have some results which tend to refute this somewhat. It occurred to me that it would be possible to study the weak boundary layer by removing it with adhesive and then destroying the adhesive. The very first thing we tried as an adhesive was water, but this crystallizes too well. Glycerine, however, does not crystallize readily. It supercools very easily, and so we were able to bond two samples of Teflon to each other with glycerine at very low temperatures. We pulled them apart and then dissolved the glycerine in water, and we actually observed minute particulate pieces of Teflon and not a liquid-like material at all. We are trying to do the same thing with polyethylene, but so far we have isolated more dust than anything else. It is a very difficult process.

LEE: In improving the adhesive joint strength, one cannot neglect the importance of adhesion. Those who are talking about wettability are thinking predominantly about adhesion. In considering adhesive joint strength, we are thinking about stresses at the interface and mechanical interlocking. Undoubtedly you have found a new way of making adhesive joints, but I don't think your work can be used to discredit the work that has been done on wettability and adhesion. I also think your technique here is changing the rheology and the modulus of the polymer at the interface. In addition, you are making something such as a pressure-sensitive adhesive tape at the interface. Pressure-sensitive adhesives, most of the time, are made of two phases. You are creating a system of the cross-linked polyethylene dispersed in polyethylene at the interface. Therefore, I suggest that you have made a two-phase reinforced system, and this may be responsible for the increased adhesive joint strength.

HANSEN: If this were true, wouldn't you expect that with longer treatment you would see further changes? Yet, we see the most change between 1 second and 5 seconds in which we produce a cross-linked layer between

200 and 1000 Å thick. If you go to 100000 Å, you don't see anything different.

LEE: You might change the modulus of the thin interface adversely upon further treatment. There are two things here. Rheology is very important at the thin interface and the stress, too. The adhesive joint strength depends greatly on this interface.

MICHAELS: Our evidence is that you can delocalize the stress with an extremely thin layer of a few hundred angstroms; and this, incidently, would be consistent with your observation that if you overbombard with activated gases you do not get any further change. There is a critical thickness where you get the maximum benefit.

PETERLIN: I think the data that you have shown with polyethylene very closely relate with the special role of morphology. You only mentioned that high-density and low-density polyethylenes show a different weight loss when exposed to atomic oxygen. But the weak point in polyethylene is just the layers at the interfaces between the crystals where they contain plenty of folds. Cross-linking occurs mainly just in these folds, and for every such addition and connection between the crystals, you increase the mechanical strength. Isn't that what you have shown in some of your results? The amount of weight loss is not simply a question of crystallinity but also of the special morphology of the sample, which means how the crystals are folded in their relative thickness, lateral expansion, etc. So, the morphological components or the morphological characteristics of the sample are certainly very important factors just as in the case of polyethylene. On the other side, you see that this type of attack is particularly suited for this morphology study. If you attack by helium ions, gaseous oxygen, or fuming nitric acid, you can get rid of the amorphous regions, and what remains are the crystals. All these things can, to a large extent, be explained on this basis.

HANSEN: As a matter of fact, this was one of the first things we looked at when we obtained this apparatus—that is, to see if we could etch away amorphous material and study crystalline areas. It turns out that in the case of attack by atomic oxygen, this is not the case; crystallinity is not as important as a branched structure.

PETERLIN: Here I think there is a difference between branched and unbranched polyethylene. But if you take the same sample of unbranched polyethylene of high molecular weight, you can follow pretty well these changes in morphology by treating the samples with certain agents usually used in morphology studies of polyethylene, particularly fuming nitric acid.

HANSEN: I am familiar with this work, but we have not observed etching. We have not observed anything except hydrogen in our effluent gases, and it is difficult to see how one could etch away amorphous material without removing something besides hydrogen unless you cause crystallization to occur. Calorimetric studies of the skin produced during bombardment with helium, for example, show that crystalline content is the same as it was before bombardment.

Dynamics of Ionic Adsorption on Polymers

E. G. BOBALEK

Department of Chemical Engineering
The University of Maine
Orono, Maine

Introduction: Adsorption Theories

Monolayer Theory

A substantial part of the theory of ionic adsorption on polymers involves the assumption that an orderly adsorption of ions progresses to some saturation limit at polymer–water interfaces. This process deposits more or less condensed monolayers. Moreover, the density of these two-dimensional monolayers is some predictable function of the concentration (or activity) of the ions in solution. If this monolayer concept is accepted, a number of useful conclusions can be deduced from experiments. For example, adsorption isotherms can be constructed by a variety of curve-fitting techniques which establish correlations between ionic concentration in the serum and the quantity of ion adsorbed per unit mass of a solid. Such curves show a decline of adsorption potential to some static limit as the available surface area on the solid decreases. These isotherms give two important parameters. First, the slopes of the adsorption curves are steeper when adsorption tendency is high. Hence, where the solid is the same, and a number of adsorbates are compared, then the relative steepness of the slopes provides a measure of the relative affinity of different adsorbates for the same solid surface. Second, if the adsorbate ion and adsorption characteristics per unit area of the solid are the same, the apparent location of the saturation point on the concentration axis can discriminate differences in relative solid surface area. Conversely, if surface areas have been determined for the solid by other independent measurements, as, for example, particle size measurements, then it is possible to calculate an apparent molecular area per adsorbed molecule, especially at that level of adsorption where a saturated monolayer exists. In general, these relations assume an equilibrium distribution of ions between the solid and liquid phases. This equilibrium depends on the surface area of the solid and on the capability of the surface for collecting limited amount of ion per unit area in competition with the solvency power of the liquid (serum). A comprehensive demonstration of the value of these classical concepts is illustrated by the development of soap titration techniques for measuring surface areas of latex particles and of adsorbed soaps (*1*).

Such measurements have had great value in solving problems of colloid stability and in estimating the relative efficiency of surfactants. In general, the adsorbed monolayer theory has been very useful because the theory led to many practical applications.

Active Site Theory

Although there are areas where the monolayer theory of adsorption can be used to design and interpret experiments, it is not applicable to the surface chemistry of catalysts nor to the kinetics and mechanisms of galvanic corrosion of metals. In these areas, it is well known that only a small fraction of the surface participates in these reactions. This research led to the concept of adsorption occurring on selected areas of the surface which were called active sites. Thus, the concept of adsorbed material existing as compressible monolayers on homogeneous surfaces was replaced by considering either the dropwise condensation or crystallization of the adsorbate onto these nucleating active sites. With active site adsorption mechanisms, one is concerned less with monolayer properties such as saturation limits and equilibrium states. Instead, the number, size, and adsorptive capacity of active sites and the rates at which the interface is being converted are important.

In the field of polymer science, McKelvey (*2*) proposed the active site hypothesis to correlate peel adhesion tests with the extent of surface treatment of polyolefin substrates. This paper will illustrate the application of the active site hypothesis in the study of composites of cellulose and vinyl polymers.

Experimental Procedure

Nature of Cellulose

Cellulosic polymers, particularly natural fibers, present an interesting example of the morphological complexities seen in many other polymers. A substantial part of the cellulosic polymer of natural fibers is crystalline. The macroscopic fiber is made up of microfibrils, which in turn are made up of protofibers. Genetic effects determine the chemical morphology during biosynthesis when the crystalline microfibrils are assembled into complex filament-wound structures of the cell wall. This composite structure is cemented by interfibrillar adhesive materials of many chemical types (*3*). In general, cellulosic fibers of vegetable origin, after even separation from much extraneous embedment material, still have a complicated morphology on the microscopic and ultramicroscopic scale. The basic polymer making up the complex fiber composite contains both crystalline and noncrystalline (or amorphous) constituents. Such fibers can be made into paper, or they can be woven into yarns and textile fibers. Both paper and textiles are very strong, very light in weight, and very porous. One might say that, in one sense, they are fiber–air composites.

One special goal in new product design with papers and textile materials is to fill the pores of these network or oriented fiber structures with polymers, hoping thereby to obtain a material which retains some of the advantages of fibrous arrays but is expanded by a plastic or rubbery continuum. One obvious way to do this would be to evacuate or force out the air and impregnate the structure with something like polymeric hot melts. This is not easy to do. Impregnation processes produce thermal and mechanical stresses which can damage the fiber array of papers or fabrics. Also, the pores can resist filling so as to make impregnation very irregular and thus create defects which degrade many properties.

An alternative process would be to add the polymer as a deposit, impregnant, or coat on each individual fiber unit before the fibers are formed into paper or textile structures. This would have a commercial advantage in that processing equipment familiar to, and in the possession of, the paper or textile industry could be used to manufacture new products from these microscopic composites of polymer and fiber.

Preparation of Composites

The following general statements describe the process (4):

1. The cellulosic fiber was composited with synthetic vinyl polymer suspended in water at relatively low concentration (less than 4% solids). Polymer was added to the fiber by polymerizing vinyl monomers which were dissolved in aqueous solution or emulsified in the water in which the fiber was suspended.

2. The polymerization reaction was confined substantially to the region of the fibers; that is, most of the polymer which formed was deposited within the lumen, within the cell wall, or on the surface of the pulp fiber. All the vinyl polymer which formed was associated with or deposited inside the fiber or on the surface walls of the fiber.

3. The localization of the reaction to the fiber was controlled by use of a two-component initiator—a redox system. The initiator components must react to initiate free radicals on the fibers at some temperature which is inadequate for initiation of polymerization by other mechanisms elsewhere in the aqueous suspension system. The concept has been described by Bridgeford (4). For example, when the redox system was the ferrous ion plus hydrogen peroxide (Fe^{++}–H_2O_2), the iron was deposited by ion exchange or by some

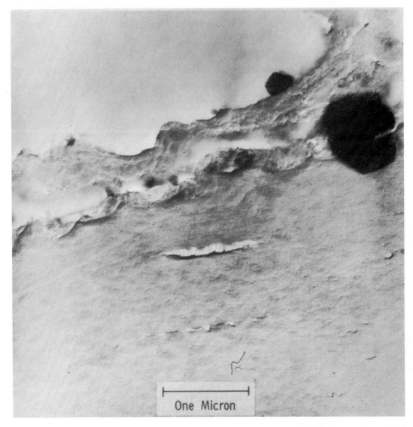

FIG. 1A. Microtome section of fragment of polystyrene-modified fiber cut parallel to long axis.

other mechanism of nucleated adsorption onto the cellulose fiber. After draining, washing, and redispersion in water, negligible quantities of iron remained dissolved in solution.

The monomer and the hydrogen peroxide were dissolved or suspended in the same water medium in which the fibers were suspended. The reaction proceeded at the correct low temperature only when both the vinyl monomer and the hydrogen peroxide diffused to meet the ferrous ion which was bonded onto the fiber. The reaction was then initiated, and polymerization continued at a rate governed by the diffusion rate of the monomer and peroxide from the serum into the domain of an active free radical on the fiber. Figures 1A through 1E are micrographs showing the end result of these reactions.

Figure 1A is a microtome section, less than 1 micron thick, of a cross section of a fragment of polystyrene-modified fiber, embedded in acrylic polymer matrix, and cut parallel to the long axis of the fiber. The dark spots are particles of polystyrene, on the surface and inside the fiber wall, when polymerization is initiated on sites activated by ferrous ions.

Figure 1B is a fragment of pulp fiber modified by polymerization add-on of about 30% of polystyrene. The electron beam fuses the polystyrene into a continuous film over the cellulose fiber surface. After prolonged exposure to the electron beam, the polymer decomposes to leave behind an opaque residue. The polystyrene acts like an *in situ* staining agent, developed by the electron beam. Here the decomposition residues of vinyl polymer outline

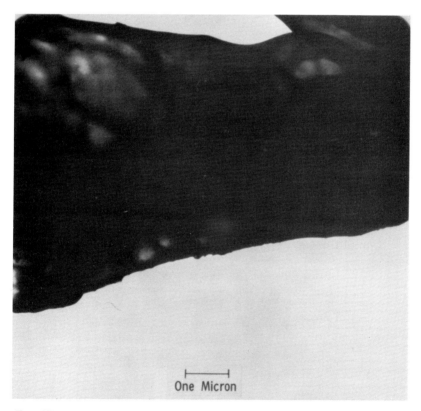

FIG. 1B. Fragment of pulp fiber modified by polymerization add-on of about 30% polystyrene.

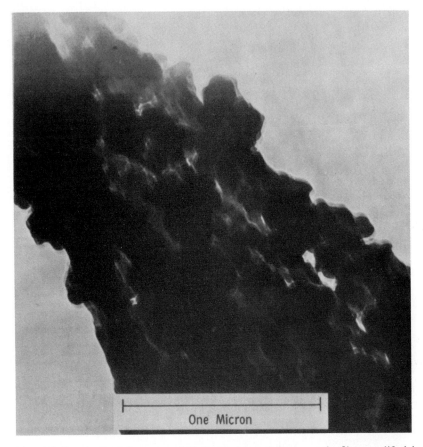

Fig. 1C. Longitudinal fragment of fibril delaminated from a pulp fiber modified by acrylonitrile.

the original vinyl polymer–cellulose fiber microstructure. If exposure to the electron beam is prolonged, only the staining residues remain. In early exposure stages, part of the photo shows residues of cellulose fibers, as in Fig. 1A. The comments on Figs. 1A and 1B apply to interpretation of other photos of this series in Fig. 1.

Figure 1C is a longitudinal fragment of fibril delaminated from a pulp fiber in a pulp which has been modified by add-on polymerization of acrylonitrile (about 150 parts of polyacrylonitrile per 100 parts of bone-dry fiber). The photograph is taken from a smear of the sample on the sample holder screen.

Figure 1*D* is a microtome cross section, less than 1 micron thick, normal to the longitudinal axis of the fiber, with fiber embedded in acrylic polymer matrix. This shows cross sections of fiber fragments from the same sample of polyacrylonitrile-modified fiber from which sample 1*C* was obtained.

Figure 1*E* is a cross section such as is shown in Fig. 1*D*, magnified to illustrate more detail regarding distribution of add-on polymer between fiber lumen and concentric cell wall. The figure shows mainly the residues of electron decomposition of polyacrylonitrile tagging the original positions of the polymer in the geometric matrix of the cellulose fiber.

An interesting hypothesis is supported by the work of de Mendoza (5). He claims that the number of polymer add-on units which can be accommodated on the fiber is limited. Once activated, the limited number of

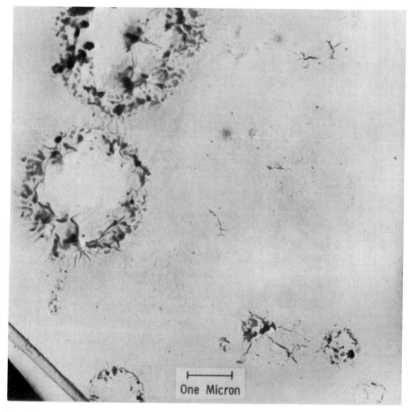

FIG. 1*D*. Microtome cross section normal to longitudinal axis of fiber modified by acrylonitrile.

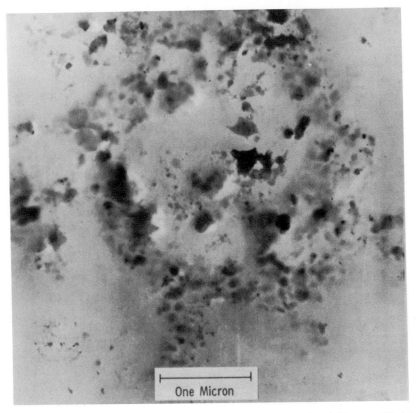

FIG. 1*E*. Microtome cross section normal to longitudinal axis of fiber modified by acrylonitrile.

polymerization unit cells can be grown to larger dendrites or globules. Their number does not increase as the quantity of added polymer increases to add-on weights which can exceed the weight of the fiber host to the polymerization reaction. It appears that the number of active sites generated by adsorption or absorption of the ferrous ion which creates catalytic unit elements on the surface for free radical initiation is very limited. Free radicals do not escape far from where they are generated at active sites. The resemblance to the proposed mechanism of emulsion polymerization (6) is obvious. It appears that nucleation and particle growth are separate process steps. In this case, however, the latex particle is bonded into the volume or onto the surface of another polymeric phase (the complex cellulosic fiber).

This presents some interesting problems. How can we control nucleation in quantity and place at ultramicroscopic dimensions within fibers having dimensions of fractions of a millimeter? How to control nucleation through control of ionic adsorption on cellulose is the subject we shall now develop in detail. If this specific chemisorption cannot be effected precisely, then much of what we hope to do in controlling the polymerization in other details will be very difficult. Although this example will describe some studies of the interaction of cellulosic fibers with the ferrous ions, this is only one of a large number of two-component redox systems which are possible. Bridgeford, who first proposed the concept which emphasizes the dynamics of ionic chemisorption, has suggested a substantial number of possibilities for converting a large variety of host substrates to a catalytic condition through adsorption of a variety of ionic initiator components (4).

The Bridgeford Concept

In general, the basic assumption which dominates the Bridgeford theory is that the ion exchange capacity of the host polymeric phase is critically important. An ion exchange capacity of 0.1 milliequivalent per 100 grams of cellulose is significant. For wood pulp fiber, delignified and bleached by multistage bleaching processes, an ion exchange capacity of 1 milliequivalent per 100 grams of dry pulp fiber is a high level, with reference to the catalytic conversion goal we contemplate. Levels of 10 milliequivalents per 100 grams of water-swollen polymer are very high and yet attainable. From the viewpoint of the active site concept, an ion exchange capacity of 1 milliequivalent per 100 grams of dry pulp represents more than 10^{18} active sites per cubic centimeter of the water-swollen and fibrous cellulosic phase. Microscopic and other evidence indicates that most of this ion exchange capacity, represented mainly by carboxyl or other acidic functions, is confined to the outer surface layers of the fiber, which has dimensions of fractions of a millimeter. While originally tubular in configuration, after extraction from the parent structure, wood, cotton, etc., the pulp fibers are more or less flattened. After high shear refining, the primary fibers become more or less delaminated into its constituent microfibrils and more readily swollen or hydrated by water. Figure 2 is an electron microscope photograph of a fragment of defibrillated pulp fiber. Cellulose decomposes and volatilizes in the electron beam without leaving behind on the specimen screen any residue which is opaque to electrons. Photographs like this are rare because focus and photography must be done in early stages of exposure, before the image disappears by decomposition of the specimen.

The fiber to be converted into a catalytic matrix has anionic sites which may be neutralized by some cation which is exchangeable by a ferrous ion. Most often, the ionic interchange associates the ferrous ion in a sparingly soluble salt configuration with the anionic functional group, and releases a hydrated

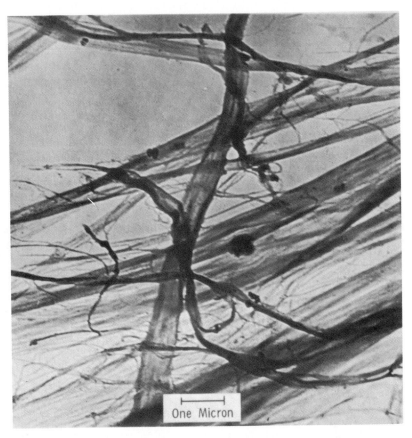

FIG. 2. Electron microscope photograph of fragment of defibrillated pulp fiber.

proton, or other ion, like sodium, into aqueous solution. In general, some anionic sites available to monovalent ions are not available to polyvalent ions like iron or aluminum ions. However, the capacity to accumulate polyvalent ions is proportional to capability for accumulating sodium ions, presumably by ion exchange with protons at high pH levels.

Conditions for Ion Exchange

For the special case of pulp fibers, Marshall (7) defined the following conditions as important for obtaining the maximum capability of the pulp fiber substrate for accumulating ferrous ion by ion exchange.

1. The hydronium ion concentration of the aqueous phase should be between pH 3.0 and 6. At a pH which is too low, ion exchange will not occur at all. At a pH near 6, the ferrous ion precipitates as a hydrous oxide, usually as a gelatinous precipitate which coats the surface of the pulp fibers (Fig. 3).

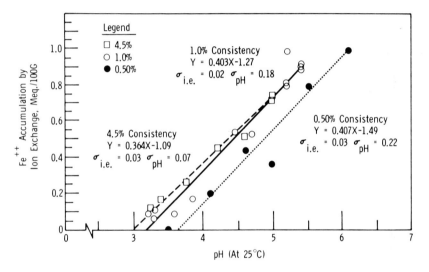

FIG. 3. Effect of pulp consistency (solids content in suspension) and pH on ion exchange of ferrous ions from 2.5% ferrous ammonium sulfate.

2. The maximum capacity of the fiber for accumulating ferrous ion is not independent of the concentration of ferrous ion in the solution (Fig. 4). The function is nonlinear, and maximum efficiency is realized if the pulp is mixed with a solution containing about 2% of ferrous salt. At 10% concentration of ferrous salt, or greater, irrespective of the pH, something has happened which prevents the accumulation of iron into the fiber matrix.

3. Hydration and delamination of the fiber through mechanical stressing of the pulp by refining processes do not significantly increase or decrease the capacity of the fiber to accumulate ferrous ion, but they may change the rate at which accumulation occurs (Fig. 5).

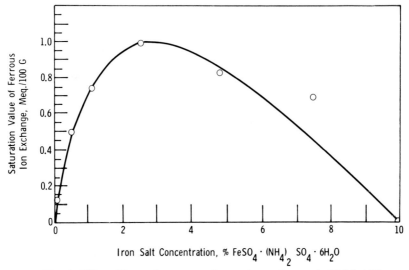

FIG. 4. Effect of iron salt concentration on ion exchange at pH 5.5 ±0.1.

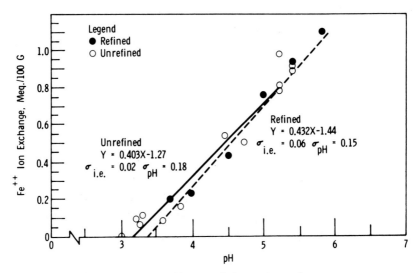

FIG. 5. Effect of beater refining on ion exchange.

4. In general, the effect of increasing temperature is to narrow the pH range where exchange occurs. For example, at 2°C, the tolerable pH range is 3.00 to 5.80. At 75°C, the acceptable pH range narrows to 3.65 to 4.60 (7).

5. Intensive washing of the pulp with acidic water before exchange with ferrous ion contributes nothing to increasing the ferrous ion accumulation capacity of commercial pulps when pH values are subsequently adjusted to near 6. The ferrous ion can displace not only protons, but also many of the usual ionic contaminants which contribute to the ash content of commercial

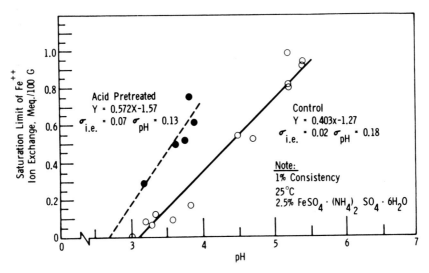

FIG. 6. Effect of acid pretreatment of pulp on ion exchange.

pulp. Acid prewashing, however, does accelerate the rate of ferrous ion adsorption, and it does improve adsorption capacity at lower pH values (Fig. 6).

6. The accumulation capacity for ferrous ion does not match the measurable ion exchange capacity of fibers for monovalent ions. For example, the maximum realizable capacity, under most favorable conditions, for exchange with ferrous ion may be 0.5 to 1.5 milliequivalents per 100 grams when the actual carboxyl content of the fiber is as high as 3.5 milliequivalents per 100 grams of bone-dry pulp. Apparently, not all acidic functions are available for exchange with the ferrous ion.

7. The optimum conditions for maximizing the capacity and rate of ferrous ion add-on (without superficial deposition of hydrous oxide) are: solution

of about 2.5% to 5.0% ferrous ammonium sulfate, refined pulp suspended at 2.5% consistency (solids content, 2.5 parts of bone-dry pulp per 97.5 parts of aqueous solution), a pH in the range of 5.0 to 5.8, and a temperature of less than 25°C.

In the design of processes to treat fiber with ferrous ion, it is important to recognize that the maximum ion accumulation capacity for ferrous ion may not be attainable at all for some process conditions, and that rates for achievement of preferred accumulation may be greatly retarded if controlling variables are not carefully adjusted. The important variables are salt concentration, fiber concentration in suspension, pH, and temperature.

While mathematical models familiar to chemical kinetics can be made to describe the rate processes and steady states achievable, it must be recognized that the most significant rate-limiting factors are diffusional phenomena. Neither the substrate nor the serum has physical properties sufficiently stable to allow for definition of unambiguous rate constants or equilibrium constants over wide ranges of reaction, pH, or salt concentration. The decline of adsorption potential as we approach saturation of available anionic sites is not the only factor that retards rate. The electrical environment of both serum and substrate changes during the process of the ion transfer between phases. Also, while the exchange of ions between solid and liquid phases is reversible, the rates of forward and back reactions are not the same at corresponding serum concentrations of both ferrous ion and solvated proton. A first step to managing control of the process requires a specific study of the kinetics of ion transfer between serum and the fiber in suspension.

Kinetics of Ion Exchange Reactions

The rate expressions to correlate data such as those shown in Fig. 7 will use as a point of reference the apparent saturation value of ion accumulation in fiber achievable after prolonged exposure under each different condition of the serum variables. It is clear that this quasi-steady-state saturation limit of iron in the fiber, C_e, depends on the pH, the initial iron concentration in solution, and the pulp–water ratio in the suspension. For fixed levels of these factors, each separate combination of these variables determines a limiting value of C_e. A rate equation with only one constant may therefore be written as

$$dC_x/dt = b(C_e - C_x) \tag{1}$$

where C_x is level of ion exchange realized (in milliequivalents of ferrous ion per 100 grams of bone-dry pulp) at time t in minutes, and b is the velocity coefficient. Rearranging equation 1 to

$$dC_x/(C_e - C_x) = b\, dt \qquad (2)$$

and integrating between the limits of $C_x = C_x$ and $C_x = 0$, and $t = t$ and $t = 0$, we obtain the integral equation:

$$\ln(C_e - C_x)/C_e = -bt \qquad (3)$$

This equation shows how fast the rate of further iron take-up by the fiber declines when the ferrous ion accumulation on the fiber substrate approaches

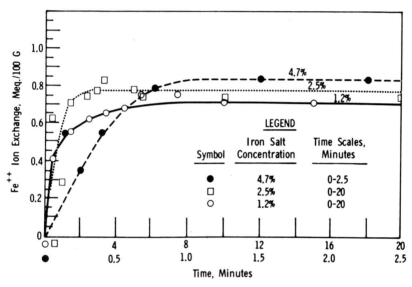

FIG. 7. Ferrous ion absorption rate in solutions of variable iron salt concentration at 25°C and pH 5.5 ±0.1.

a measurable and characteristic saturation limit, C_e. Rates under different conditions are compared graphically in Fig. 8 for reactions initiated at varying initial ferrous salt concentrations in the solution. Figure 8 uses three different time scales for the time coordinate. Apparently, the time to reach equilibrium increases with decreasing ion concentration. Here, the ion concentrations used are all near the optimum for maximum adsorption (Fig. 9).

Washing of Ferrous Pulps Containing Ferrous Ion

In general, if the pulp is filtered to remove the suspending solution containing soluble ferrous ions, and then reslurried before being filtered again, the amount of ferrous ion retained on active sites will depend on the pH of the suspending wash water. If the pH is near 6.0, the residual iron in the pulp after several stages of extensive washing remains at about the saturation value. If the pH of the wash water is less than 6, the residual retained in the pulp declines because ferrous ion is lost to the wash water. If extensive washing is done at a pH near 2.5 to 3.0, all the iron will be extracted from the pulp. There will be no residual iron (Fig. 10).

The interesting point shown by Marshall's data (7) in Fig. 10 is that extensive times and volumes of washing in several stages will fix a new saturation value which depends only on the pH of the wash water and not on the

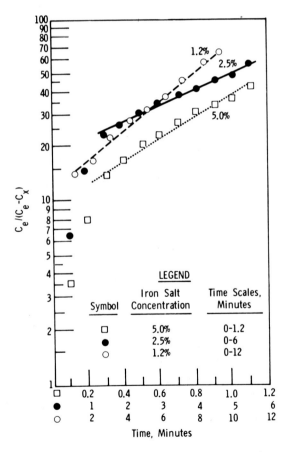

FIG. 8. Semilog plot of $C_e/(C_e - C_x)$ versus time.

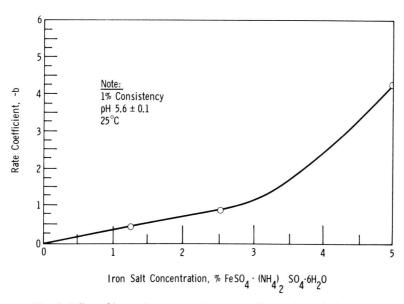

FIG. 9. Effect of iron salt concentration on specific reaction velocity coefficient.

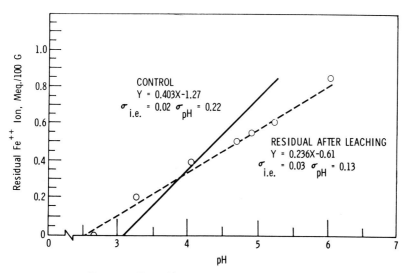

FIG. 10. Effect of initial slurry pH on iron salt leach-off.

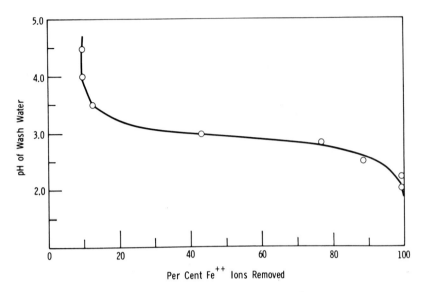

Fig. 11. Percentage of ferrous ions removed versus pH.

duration or number of stages of washing. The pH dependence of the residual is illustrated in Fig. 11, citing Jaeger's data (8).

This suggests an interesting possibility for controlling the number and perhaps also the distribution of active sites between surface and inner layers of the fiber structure. If this could be controlled, we might achieve ways to have more control over the morphology of the vinyl cellulose graft polymer produced when we activate vinyl polymerization at those active sites which are residual after washing, and which are distributed in different proportions between the surface and the interior of the fiber structure. Because we can visualize the possibility of eliminating ferrous ion from the surface while retaining such active polymerization sites within the fiber, we should explore the kinetics of the reverse exchange of ferrous ion by the solvated proton in more detail as a function of pH of the wash water. If we can clean the surface of the fiber of ferrous ion, and retain some inside, then the subsequent polymerization will deposit polymer inside the fiber, leaving the outside uncontaminated with respect to retaining interfiber adhesion by hydrogen bonding (5).

Kinetics of the Reverse Reaction—Displacement of Ferrous Ions by Solvated Protons

The pH controls not only the residual iron which can be retained after extensive washing, but also the rate of desorption. Some rate data are shown in Figs. 12 and 13. An attempt is made in Figs. 14 and 15 to represent the data by a first-order integral model, namely:

$$\ln C_x/C_0 = k\,dt \tag{4}$$

where C_x is the residual ferrous ion concentration on the washed fiber at washing time, t, and C_0 is the ferrous ion content of the fiber immediately before the 0.5% consistency pulp slurry is adjusted to some pH level less than 6, which induces the proton exchange process that returns ferrous ion into the wash water. The pH adjustment was made after the fiber was suspended in the wash water either by addition of dilute sulfuric acid (Figs. 11 though 15) (8), or by addition of acetic acid (Fig. 10) (7).

The curves of Figs. 14 and 15 are not linear with a constant slope. The maximum curvature appears to be caused by an intersection of two straight lines in this semilog plot with the intersections located at some critical point on the time–concentration coordinates. It is reasonable to assume that the initial fast rate, which follows immediately after the pH drops (shown by extending the line for the initial slope in Figs. 14 and 15), represents extraction of the easily accessible ferrous ions, which are near the exposed surface of the fiber. When these easily accessible ferrous ions are removed, a diffusion-controlled resistance to further extraction is encountered which retards the rate of removal of the remainder of the iron embedded deep in the fiber.

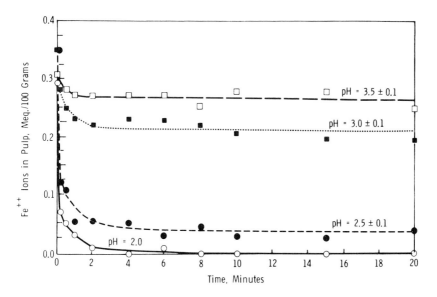

FIG. 12. Rate of ferrous ion removal by extensive washing at four pH levels.

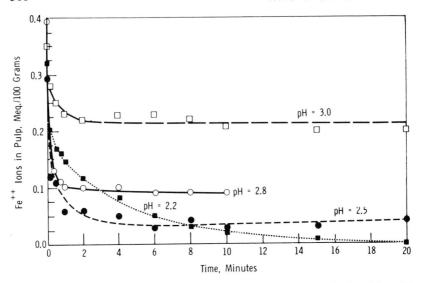

FIG. 13. Rate of ferrous ion removal at pH levels in the range of 2.2 to 3.0 ± 0.1.

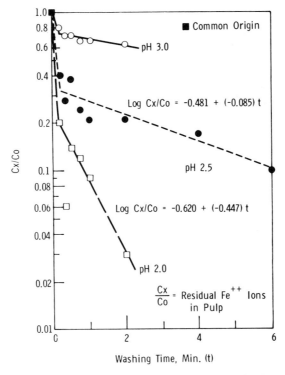

FIG. 14. Semilog plots of ferrous ion leaching rates at pH levels of 2.0, 2.5, and 3.0.

What this nonextractible residual iron will be as a function of pH is shown in Fig. 16 for three different types of pulp fiber. For all fibers, at a pH of 3.5, no more than about 10% of the bound iron can be removed even by extensive washing. At a pH of 2.0, all is removed very quickly.

The various data suggest four possible approaches for controlling chemisorbed iron at desired levels. These are:

1. Control the pH very closely in a favorable range of about 2 to 3% salt concentration, and allow sufficient time to reach a saturation, C_e, at a pH value which determines a saturation value near the iron content desired (see solid line of Fig. 6).

2. Add excess iron to the fiber at an elevated pH, and extract to the desired residual by decreasing the pH of the wash water to some fixed value in the range of 2.0 to 3.5 during washing. Not all the iron will be extracted. A predictable residual amount will remain, depending on the pH of the wash water.

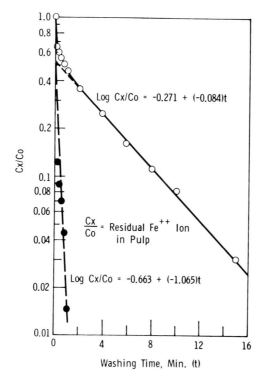

FIG. 15. Semilog plots of ferrous ion leaching rates at pH of 2.2.

3. Saturate with ferrous ion to a value greater than desired, drop the pH to the acidic side, and quench the extraction by raising the pH or by dewatering the pulp at some correct time after the extraction is initiated. Use pH to control leaching rate, but interrupt the process at the correct and convenient time.

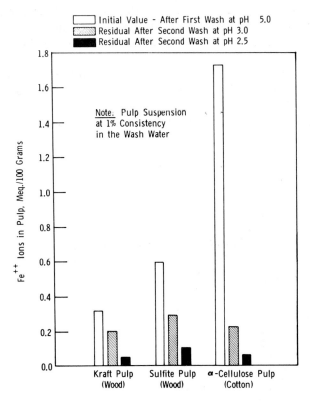

FIG. 16. Quantity of residual ferrous ions after washing at three initial pH levels.

4. Obviously, iron add-on or extraction at any reference pH level can also be interrupted short of completion if rates are known at some convenient time. These last two methods may actually be preferred to saturation limit techniques if the control of iron concentration ion in the fiber needs to be adjusted to a time scale convenient for execution of the process.

It is gratifying that so much control is possible over the process for introducing polymerization-initiating sites on colloidal suspensions of pulp fibers.

This control is manageable at any stage of mechanical refining of the fibers and also is independent of the extent of fiber flocculation, both of which change the apparent surface area of the cellulose–water interface.

The question of heterophase polymerization chemistry involved in modifying cellulose by addition polymerization of vinyl monomers on active sites is a subject of great complexity (4, 5, 7, 9). Here we are involved mainly with showing how cellulose fiber can be activated to provide a catalytic substrate which will accept localized deposition of polymer on activated sites. To show that this has been accomplished, let us consider an example.

An Example of Polymerization

To show the effectiveness of this mode of preparing the cellulose host polymer for localized acceptance of polymer by *in situ* polymerization, Marshall (7) prepared 1% pulp slurries from both unmodified and ferrous ion-modified pulp (bleached, soft-wood Kraft fiber). The activated pulp contained 0.83 milliequivalent of ferrous ions per 100 grams of bone-dry fiber. The aqueous solution in both instances contained also 5% acrylonitrile and 0.03% of hydrogen peroxide. After final mixing, the suspensions were held at 70°C, under reflux, with mild agitation. After one hour, the suspension was filtered and washed with several 1-liter volumes of cold water. The filtrate and wash water contained no discernible polymer, but did contain residual monomer. The ferrous ion-modified fiber accumulated 103% of its initial weight of polymer. The blank added only 8%. Several vinyl polymerizations were studied also by de Mendoza (5) (Table I). In most instances, electron microscopic

TABLE I
POLYMERIZATION OF VINYL MONOMERS ONTO PULP FIBERS IN AQUEOUS SUSPENSION

Monomer	Temperature (°C)	Polymer Add-on (% weight gain of fiber)
Styrene	25	15
Styrene	55	31
Methyl methacrylate	27	10
α-Methyl styrene	93	13
Methylacrylamide	70	13
Methylacrylate	85	20
Acrylic acid	103	11
Methacrylic acid	103	22
Acrylonitrile	80	151

Reaction continued for one hour, under reflux, in a nitrogen atmosphere. Bleached soft-wood Kraft fiber in 1% suspension, with 4% of monomer and 0.02% of hydrogen peroxide. The activated pulp contained 0.006 milliequivalent of ferrous ion per gram of bone-dry cellulose.

observation of the modified fiber showed that nearly all the polymer added to the cellulose was on the surface or within the fibers. Also, if the temperature of polymerization did not exceed 70°C, and if the acidity of the aqueous phases was maintained at a pH higher than 4.5, then negligible amounts of vinyl polymer formed in the aqueous phase apart from what was added to the fiber. This was true whether the vinyl monomer was water-soluble, like acrylonitrile, or was in emulsion, like styrene.

Many other examples of host polymer phases, with many different monomers and with several two-component, initiator systems, are described by Bridgeford (4). A great variety of polymer compositing techniques by *in situ* polymerization have been demonstrated in principle on the laboratory scale. A practical reduction of this process requires precision control of all steps. Above all, the most important control problem is associated with this first step of proper ionic accumulation on or within the host polymer phase.

Summary

In the localized and controlled polymerization of olefinic monomers onto cellulosic wood pulp fibers, the fibers first must be chemically treated to make them catalytically active. This chemical treatment generates polymerization nucleating active sites in the cellulose polymer. The fibers in aqueous suspension are activated to promote a free radical polymerization of vinyl monomers, which is initiated by free radicals generated by the ferrous ion (Fe^{++})–hydrogen peroxide (H_2O_2) redox couple. If the ferrous ion is bonded to the cellulose and excluded from solution in the aqueous phase (which contains monomer and hydrogen peroxide), then, at some favorable low temperature, diffusion of monomer and peroxide to the sites activated by the ferrous ion will sustain a polymerization. The polymer will accumulate as bonded microscopic particles only at these active sites on the surface or within the wall of the cellulosic fiber.

The principal variables which control efficiency of the subsequent chemical treatment processes of ferrous ion adsorption and washing are the bonded acidic functionality of the cellulose polymer, the concentration of ferrous and hydronium ions in solution, and the mixing conditions, which are affected mainly by the solids content of fiber in suspension.

Rate functions are illustrated for the ion exchange processes attending ferrous ion adsorption and also for desorption which can occur during subsequent washing of activated pulp. The rate data suggest that several ways can be used to control precisely the processes for activating the pulp to accept localized polymerization reactions. Important variables for all control processes are the pulp slurry consistency, the pH and ferrous salt concentration during the ion accumulation stage, and the pH during the washing stages which clean the activated pulp before it is suspended again for the polymerization process.

References

1. S. H. Maron, I. N. Ulevitch, and M. E. Elder, *Anal. Chem.*, **21**, 691 (1949); S. H. Maron, B. P. Madow, and I. M. Krieger, *J. Colloid Sci.*, **6**, 584 (1951); S. H. Maron, E. G. Bobalek, and S. M. Fok, *ibid.*, **11**, 21 (1956).
2. J. M. McKelvey, "Polymer Processing," Wiley, New York, 1962, Chapter 6, Section 6-3, pp. 161–171.
3. W. A. Cote, Jr., Editor, "Cellular Ultrastructure of Woody Plants," Syracuse University Press, Syracuse, New York, 1965.
4. D. J. Bridgeford, U.S. Patent 3,083,118 (March 26, 1963); *TAPPI*, **46**, 670 (1963); *Ind. Eng. Chem. Prod. Res. Develop.*, **1**, 45 (1962); British Patent 818,412 (Aug. 19, 1959).
5. A. P. de Mendoza, "Paper Formation from Pulp Fibers Composited with Vinyl Polymers," Ph.D. Thesis, University of Maine, Orono, Maine, 1965.
6. D. J. Williams and E. G. Bobalek, *J. Polymer Sci.*, **A1, 4**, 3065 (1966).
7. S. N. Marshall, "An Ion Exchange Process for Modification of Wood Pulp To Promote Localized Vinyl Polymerization within the Fiber," M.S. Thesis, University of Maine, Orono, Maine, 1964.
8. E. L. Jaeger, "Kinetic Study of Ferrous Ion Exchanges on Pulp," M.S. Thesis, University of Maine, Orono, Maine, 1967.
9. F. G. Lafratta, "A Study of Monomer Reactivity Ratios for the Heterophase Copolymerization of Vinyl Monomers onto Cellulose," M.S. Thesis, University of Maine, Orono, Maine, 1967.

Discussion

MATIJEVIĆ: How do you measure the ion exchange? Do you have any indication for ion exchange, or do you just measure the uptake of the species?

BOBALEK: Most of the data represent averages of four or five analyses, with the analyses taking different directions. The first of these was to measure the uptake by washed pulp; that is, we washed the pulp and measured the residual iron by colorimetric techniques. Also, this was checked with a material balance on the serum—that is, the depletion of serum concentration after filtration. There is no direct evidence that this process is ion exchange except that the pattern of add-on and extraction behavior, and in particular the pH dependence, so much resembles what is common for ion exchange phenomena. In addition, the diffusion limitation patterns are analogous to ordinary ion exchange behavior, except that we are dealing here with rather small quantities of exchangeable ion.

MATIJEVIĆ: I think you have no evidence for ion exchange at all. What you really did was to determine the difference in concentration of ferrous ions in the serum or on the surface. You should consider the adsorption of ionic species on the surface. Apparently, the soluble, ferrous hydroxide species adsorbs on the surface of the fiber. This is well corroborated by your experiments because of the increase in adsorption with pH. At higher pH the hydrolysis of ferrous ion is more complete and the adsorbed quantity of the hydrolyzed species is higher. Your hysteresis by washing also favors this explanation because, when you wash with a solution of pH 6, you maintain the hydrolyzed species and they stick to the surface. When you lower the pH, the ferrous ions are dehydrolyzed and they desorb.

BOBALEK: One thing that we have done in some instances was to produce conditions of more or less carboxyl content on different kinds of fiber. In the alpha cellulose, for example, we have actually increased the normal levels of carboxyl content by treatment with oxidative enzymes. Measurements using the TAPPI method give another measure of the carboxyl content of the fibers which is independent of iron-uptake measurements. Nearly all the limits that we found on how much ferrous ion you can put in and how much you can take out are related to changes of total carboxyl equivalents in the fiber. In general, less than half of the carboxyl equivalents are able to capture ferrous ion. Nevertheless, cellulose of higher carboxyl content can capture more ferrous ion than if carboxyl content is low. Whether the iron chemisorbed is there as ferrous hydroxide, or whether it is there as carboxyl-bound ferrous ion, or whether some other inorganic moiety can exist at that point, I am not sure.

Our micrographs show that polymerization developed globules of a size of 400 to 800 Å at the initation sites when total vinyl polymer add-on was less than about 10%. Obviously, many initiation and termination reactions must have occurred at each activation site. Once a particle was nucleated, monomer diffusion to a reacting site could take place much more readily. Until the polymer particle had grown to a size of at least 800 Å, we found that the copolymer ratios deduced from analytical data were characteristic of what is common for grafting of vinyl polymers onto cellulose. As add-on of polymers continued beyond about 20%, and growing separate particles coalesced into a sheet, the copolymer ratio of classical copolymerization theory drifted toward values more consistent for what is reported for emulsion polymerizations in the absence of cellulose.

Where is the catalytic feed coming from? It is too much to expect that, if you initiate one free radical from junction of one ferrous ion with peroxide, this is going to continue on as a growing free radical, having a life expectancy of an hour or more. I do not know what the catalytic system is or how it is sustained by the mobility and migration of the ferrous ions in the substrate to the active polymerization zones. At the maximum of iron add-on, we can have about 10^{18} ferrous ions per cubic centimeter of the swollen cellulose polymer. Obviously, we don't use any large fraction of these sites for the first initiation of the polymerization process. The problem is still open to discussion and speculation as to how the iron is bound in the substrate and how it migrates to reacting sites to sustain the polymerization situation which grows the particle.

Are you presuming, perhaps, that you get a crystallization of ferrous hydroxide at a site nucleated at some carboxyls?

MATIJEVIĆ: No. That is not what I presume. I presume that the soluble hydroxylated species adsorb via the hydroxyl group. Without this group, there is no adsorption. As a matter of fact, if you take, for example, rubber latex, add 0.001 M alum, and adjust the pH to 5, the rubber particles become strongly positively charged. You can dip a sheet of paper into such a latex and it will clear the latex solution within a short time. Now, if you have a mixture of two systems such as cellulose and rubber latex, alum will reverse the charge of the rubber latex (at pH $>$ 4) much before the paper or the cellulose fiber becomes positive because the adsorptivity of hydrolyzed metal ions on a lyophobic surface is much greater than on a "semilyophilic" surface. Therefore, what you considered as an isoelectric coagulation with aluminum is, in my opinion, not actually isoelectric. Instead we deal here with the mutual flocculation of oppositely charged species in a system in which the electrolyte recharges one kind of particle but not the other kind. I think that this explains many of the phenomena in the paper industry.

VALENTINE: Many years ago, I did some work on protein fibers, polymerizing acrylonitrile in an aqueous system, using persulfate as an initiator We were trying to adjust conditions so that virtually all the polymerization would take place inside the wool fiber, and the kinetics quite clearly showed that it was a diffusion-controlled process. You could get in this case up to 30% or so acrylonitrile in wool, but it followed a square root of time dependence throughout (that is, diffusion-controlled). I wonder whether you have been able to pick up in your studies any distinction of kinetics as to whether you are getting pretty much all polymerization within the fiber and/or polymerization on the outside.

BOBALEK: During the early stages of the polymerization process itself, where there is approximately 8 to 12% add-on, considerable evidence is apparent that a diffusion limitation is important. If you presume a quasi-first-order situation for both the early and late stages of polymerization, then the first-order time constant will indicate that the early stages are five or six times slower than are the late stages where the reaction speeds up. For the later stages we observe microscopically a dendrite growth away from the wall of the host polymer fiber.

If you go too far and if mastication of the slurry suspension is too severe, dendrites of vinyl polymer break away from the fiber, and cellulose-free polymer fragments collect in suspension in the serum. However, even at vinyl polymer add-ons of up to 400%, if you are very careful about agitation, you can drain the serum through a relatively coarse filter and find less than 1% cellulose-free vinyl polymer in the serum filtrate. We could detect in the filtrate as little as 1% of free polymer when we added it as latex. We could not detect even this much in the serum out of the properly controlled systems where polymer add-on to fiber was executed under conditions we here describe, up to add-ons exceeding 100%.

ONSAGER: Ferrous hydroxide does not precipitate at pH 5; it takes a higher pH. On the other hand, ferrous ion would complex quite readily with carboxylate groups. This is believable. Finally, the coordinating groups around a ferrous ion might very well affect the catalytic activity quite considerably and, for all we know, find a favorable situation in a particular type of coordination available in the internal pores on the cellulose.

KUMINS: In my experience with polymer films, I have yet to find a polymer which did not exhibit ion exchange properties. True, the capacity of these membranes was very low, but there was unequivocal evidence that these were ion exchange materials. The data you presented do not indicate whether you do have ion exchange, or whether the ion exchange phenomena supercede the adsorptive phenomena. I might also suggest an experiment to indicate this, in which the pulp is used as a chromatographic material, and then a salt is passed through it at very low concentration, which is more simple than ferrous ammonium sulfate (for example, potassium chloride). See whether you do get an ion exchange. I presume it is a cation exchanger, and hence potassium would be the critical ion.

BOBALEK: It is a cation exchanger.

KUMINS: By increasing the concentration of the solution, a point is reached where the membrane (or pulp in this case), becomes "leaky" and permits the passage of the counter ion. This is one of the criteria you could use to differentiate between the two phenomena. In regard to nonadsorption, what was the pH of the 10% ferrous ammonium sulfate solution?

BOBALEK: Five and one-half.

KUMINS: Since it was above 3, how do you know that you were not exchanging ammonium ions instead of the ferrous?

BOBALEK: I don't know what the answer is, but these exchange phenomena depend on ionic strengths. They depend on the pH and on the effect of the ionic environment on your polymer host substrate. At substantial ionic strengths, cellulose changes rather dramatically in its volume, density, and other properties. So, it may well be that just running up to too high an ionic strength, for example, could have just made the host phase not as good a host; that is, diffusion or other limitations were introduced which we have not as yet defined.

HELLER: Zeta potential measurements would help a great deal in pinning down what actually is going on in your system.

BOBALEK: That's a good suggestion.

SESSION IV

Chairman
H. BURRELL

Polymer Adsorption on Substrates*

R. R. STROMBERG

Institute for Materials Research
National Bureau of Standards
Washington, D.C.

Introduction

The adsorption of polymer molecules on surfaces is of both technological and scientific importance. It is of primary importance to the problems of adhesion, coatings, rubber–carbon black reinforcement, biological interactions, dispersion stabilization, etc. The many possibilities available to a polymer molecule on adsorption compared to that of a small molecule makes the determination of the configuration of an adsorbed molecule a scientifically interesting problem. The need for knowledge about the number and strength of attachments to the surface and the availability of portions of the molecule for interaction with other nonattached molecules makes information concerning the configuration important to the technological problems.

The process of polymer adsorption is relatively complicated. One of the principal problems of polymer adsorption to be resolved is the configuration of the polymer molecule on the surface. This was recognized by earlier workers who found that the amounts adsorbed would correspond to multilayers if all segments of the adsorbed polymer chains were attached to the surface. Consequently, Jenckel and Rumbach (*1*) proposed a model in which each polymer molecule is attached in sequences of segments adsorbed to the surface separated by loops of segments which extend into the solution. This would permit "monolayer" adsorption with adsorbance values far in excess of those that would be obtained for monolayer adsorption of monomer molecules. This concept of an adsorbed polymer chain is generally accepted today. Of considerable interest is a quantitative description of the arrangement of the trains of adsorbed segments and loops of unattached segments extending into the solution.

Theoretical Treatments

No single treatment has been generally accepted, partly because quantitative experimental verification is extremely difficult. Much of the theory has been

---

* Some of the work reported in this paper was supported by the Army Research Office—Durham.

restricted to negligible interactions between adsorbed polymer molecules. This requires extremely dilute solutions, and it is difficult to make accurate measurements under such conditions. The extension of these theoretical treatments to higher surface and solution concentrations has in some cases led to predictions of the necessity of very long times for equilibrium, again making experimental verification difficult. Thus far, because of these difficulties and also because of the large number of parameters involved in the theoretical equations for an adsorption isotherm, it has been possible to obtain only qualitative agreement or disagreement with the several published theoretical treatments.

The first quantitative equilibrium theory of polymer adsorption was derived by Simha, Frisch, and Eirich (SFE) (2) by a statistical mechanical treatment. The adsorbed polymer chains were assumed to be characterized by a Gaussian distribution of end-to-end distances. The surface was regarded as a reflecting wall, and a new distribution of the end-to-end distance was determined.

A major theoretical treatment was advanced by Silberberg (3). He attempted to overcome what he considered a fundamental weakness in the SFE treatment; that is, the shape of the molecule was determined with zero attractive energy, and this shape was not considered as a variable in the treatment of the thermodynamic equilibrium between the surface phase and the solution. Silberberg assumed the alternate sequences of loops and adsorbed trains to be essentially independent of each other. He evaluated the partition function as determined by the forces between the surface and the adsorbed units for the surface layer containing the units, and the internal partition function for the segments in the bulk for the units in the loops.

His treatment predicts that the sizes of the loops and the stretches along the surface are independent of the molecular weight, in conflict with the SFE theory. For a flexible molecule, the loops are small, and the molecule is close to the surface, if all surface sites are adsorbing and all polymer segments are adsorbable. At interaction energies about kT, more than 70% of the segments are in contact with the surface. The adsorbed molecule, consequently, is essentially two-dimensional, with a segment concentration which is a maximum at the surface and decreases rapidly with distance from the surface. Structural restrictions increase the size of the loops and increase the energy requirements. However, the energy requirements for adsorption are not very large, even for sterically unfavorable cases. Further, Silberberg predicts an adsorption isotherm in which a plateau does not exist at low molecular weights and the number of polymer segments adsorbed per site increases with increasing solution concentration. However, plateaus do occur for the higher-molecular-weight materials.

Hoeve et al. (4) reported that the assumption by Silberberg of uniform-size loops and uniform-size adsorbed segment stretches was too restrictive and led to some incorrect conclusions. For the case of the isolated molecule, in disagreement with Silberberg, for flexible molecules they predict a distribution of loop sizes—very large loops for very small adsorption energies, and small loops and larger fractions of segments attached as the energy increases. In agreement with Silberberg, they found that large energies are not required for large fractions of segments attached and small loop sizes. This is in disagreement with the SFE theory, which predicts large loops at relatively small energies. Results similar to those obtained by Hoeve et al. were also obtained by Roe (5).

Hoeve (6) has derived the density distribution for the adsorbed layer normal to the surface and found that, except for a surface layer at a very small distance from the surface, the distribution is well represented by an exponential function. For adsorption from a theta solvent, he calculated that the root-mean-square distance of segments from the interface should vary approximately proportionally to the square root of the molecular weight. Further, the amount of adsorbed polymer was determined to increase without limit as the molecular weight or solution concentration increased. However, although the amount of polymer adsorbed might be high, the polymer concentration immediately adjacent to the surface remained low.

DiMarzio and McCrackin (7) studied the conformations of a one-dimensional polymer molecule attached to a surface by means of theoretical and Monte Carlo calculations. This was also extended by McCrackin (8) to three-dimensional polymers with self-excluded volume. A transition was found at an attractive energy that depends on the assumed lattice. For large adsorption energies the molecule would lie relatively flat on the surface, with its dimensions independent of molecular weight. For attractive energies less than the transition value, the average number of contacts with the surface approaches a finite value as the molecular weight increases. In this region the end-to-end length increases with the square root of the molecular weight.

It is apparent from this short description of the theoretical treatments of polymer adsorption that one of the most useful and most interesting aspects of polymer adsorption is a determination of the configuration of the adsorbed polymer molecule. This can be attacked by measurements of the extension of the molecule normal to the surface and measurement of the fraction, p, of chain segments attached to a surface. In the remainder of this paper we shall describe some of our experimental approaches to this problem and discuss the results with respect to polymer configuration.

Adsorbed Film Thickness Studied by Ellipsometry

The model of the adsorbed polymer molecule that appears to be used by all theoretical treatments, then, is that of a molecule attached to a surface at trains of segments separated by loops extending into the solution. It is obvious, therefore, that measurements of the extension of the adsorbed chain normal to the surface will be extremely useful in determining which of the different theoretical treatments is more nearly correct. Furthermore, it is obvious that measurements must be made *in situ*, with the surface immersed in the polymer solution, to prevent the loops from collapsing.

One technique that we have found very useful for this measurement is ellipsometry. The technique (9) is based on the measurement of changes in the state of polarization of light on reflection from a surface. For a bare metallic surface two parameters are obtained from the instrument readings. These are tan ψ, the ratio of the magnitude of the reflection coefficient for light polarized normal to the plane of incidence, and Δ, the relative phase difference for these two polarizations. If the surface is now covered with a thin film, new values of ψ and Δ will be measured. From these new values and the previously determined values of the optical constants of the surface, the thickness and refractive index of the film can be calculated.

The model used for ellipsometric calculations is that of a homogeneous film with discrete boundaries. The adsorbed polymer film deviates from this model in that the film is not homogeneous. The calculations will, therefore, yield an "average" thickness value, and it is necessary to relate this calculated average for a homogeneous film model to the heterogeneous adsorbed polymer film. This has been done by McCrackin and Colson (10).

The changes in the state of polarization of reflected light are a function of the difference between the refractive index of the film and the index of the solution $(n_f - n_s)$. The values for the thickness, t_h, and the refractive index, n_h, for an equivalent homogeneous film are obtained by integrating this refractive index difference over the whole film:

$$\int_0^\infty (n_f - n_s)\, dt = (n_h - n_s) t_h \tag{1}$$

It is not possible to determine the distribution function by ellipsometry. However, the relationship between t_h, obtained directly from the measurements, and other averages can be obtained. The average that we have used is the root-mean-square thickness, which is given as

$$t_{rms}^2 = \frac{\int_0^\infty (n_f - n_s) t^2\, dt}{\int_0^\infty (n_f - n_s)\, dt} \tag{2}$$

McCrackin and Colson calculated values of Δ and ψ for given distributions, from which they then calculated the thickness and refractive index of an equivalent homogeneous film using the usual Drude equations. It was then possible to compare the thickness obtained from the homogeneous film model (equation 1) with the root-mean-square average (equation 2). They showed

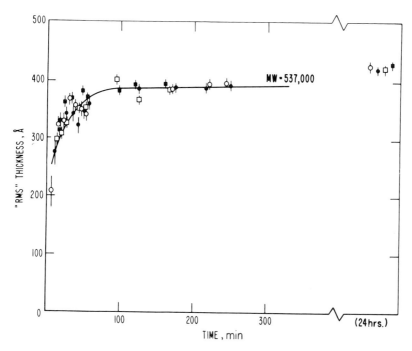

FIG. 1. Root-mean-square extension of adsorbed polystyrene (molecular weight 537,000) versus time. Adsorbed from cyclohexane solution (concentration 1.0 mg/ml) at 34°C on chrome surface. Different symbols represent separate runs.

that the value for the equivalent homogeneous film, t_h, was larger than $(t_{rms}^2)^{1/2}$ by a factor of 1.5 for an exponential distribution.

This technique was used to study the extension of the polystyrene molecule adsorbed near the theta temperature onto a number of polycrystalline metallic surfaces (*11*) and onto a liquid mercury surface under the same conditions (*12*). It has also been used to study the extension of a much more polar, but smaller polymer molecule—for example, a polyester on chrome surfaces (*13*) and coupling agents on glass surfaces (*14*).

The root-mean-square (rms) thickness of polystyrene adsorbed from cyclohexane near the theta temperature on a chrome surface is shown in Fig. 1 (*11*).

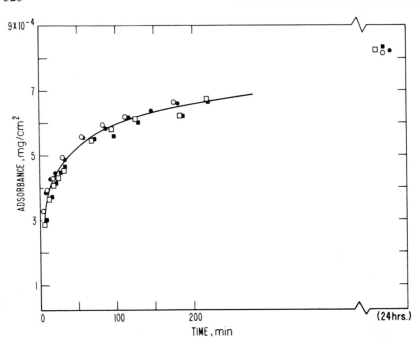

Fig. 2. Adsorbance of polystyrene (molecular weight 1,370,000) on chrome surface versus time. Adsorbed from cyclohexane solution (concentration 0.55 mg/ml). Different symbols represent separate runs.

The vertical lines for each point represent the uncertainty of the measurement for that point. The uncertainty is a function of film thickness, especially if the refractive index of the film is close to that of the immersion medium (polymer solution). The increase in thickness with time, shown in Fig. 1, is typical for the behavior of polystyrene adsorbed near the theta temperature on polycrystalline metallic surfaces. The time required to attain an "equilibrium" value is dependent on the solution concentration and molecular weight of the polymer. No further changes in thickness have been observed after a 24-hour period at the solution concentrations studied. This change in extension occurs during the adsorption period. From the value of the refractive index of the adsorbing film, which is obtained simultaneously with the thickness, and of the thickness, the amount of polymer adsorbed per unit area (adsorbance) can be calculated. A typical curve is shown in Fig. 2. We have interpreted these results to indicate that the polymer molecule is initially adsorbed in a relatively flattened configuration. As additional polymer molecules are adsorbed, some of the attached segments of the previously adsorbed molecules desorb, resulting in larger loops and increased rms thickness of the film. This process continues until an equilibrium thickness is established.

The film thickness was also found to increase with solution concentration, until again a plateau value was obtained from curves of thickness versus time and adsorbance versus time (Fig. 3) for polystyrene on a chrome surface. The interpretation given to this behavior is similar to that given for the change in extension with time. The polymer molecule appears to be spread out in a relatively flat configuration at low solution concentrations. As the solution concentration increases, the loop size extending into the solution increases. It can be observed in Fig. 3 that the plateau region for the thickness of the adsorbed polymer is attained at approximately the same concentration that the adsorbance curve, also shown in Fig. 3, attains a plateau.

If the polymer chain changes its configuration with increasing solution

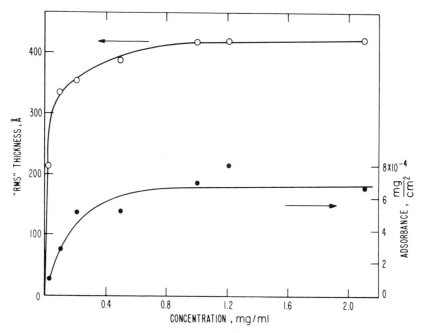

FIG. 3. Root-mean-square thickness and adsorbance of polystyrene (molecular weight 537,000) versus solution concentration. Adsorbed from cyclohexane at 34°C on chrome surface.

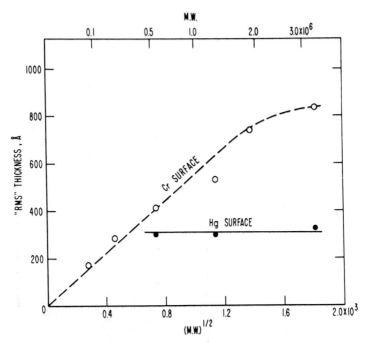

FIG. 4. Root-mean-square thickness of polystyrene versus the square root of the molecular weight. Adsorbed from cyclohexane at 34°C. The adsorbent surface is shown on the figure.

concentration, then the number of contacts of the chain with the surface should decrease with increasing solution concentration, and should also approach a plateau value. A study of the fraction of segments attached, p, by an infrared technique that will be described later in this paper, has been reported by Thies et al. (15) for polystyrene. They found a decrease in the value of p with increasing adsorbance, until a low value in the plateau region was attained. This supports the interpretation given above for the ellipsometric results.

The thickness of the adsorbed layer at the plateau region for the polycrystalline chrome surface is shown in Fig. 4. The thickness is proportional to the square root of the molecular weight except for the sample with a molecular weight of 3,300,000. A similar type of behavior for polystyrene adsorbed at the theta temperature from cyclohexane was reported by Rowland and Eirich (16). They studied the thickness of adsorbed polymer films on glass by an entirely different technique—the reduction in effective cross section in a capillary caused by polymer adsorption. In addition, they also reported a value for the thickness of polystyrene in close agreement with the value we obtained for a sample of approximately the same molecular weight.

Ellipsometric measurements similar to those on the polycrystalline metallic surfaces were also carried out on liquid mercury (12). In this case the behavior was remarkably different. The extension of the polymer molecule remained constant during the adsorption time period. This is shown in Fig. 5. Although the extension remained constant, adsorption did occur during this time period, and the adsorbance curve exhibited the same behavior shown in Fig. 2 for the polycrystalline surfaces. Most interesting, however, was the fact that the thickness of the adsorbed layer on mercury was independent of molecular weight. This is shown in Fig. 4. For liquid mercury it appears that polystyrene attains its final configuration relatively early in the adsorption period and retains this configuration as additional adsorption occurs. This configuration is relatively "flat," and the early arrivals do not change into more extended configurations as the surface population increases. Rather, "holes" in the adsorbed layer are filled with later arrivals, and the concentration of polymer in the adsorbed layer increases with time, until an equilibrium value is attained. It also appears, then, that the value of p remains constant and independent of the molecular weight as well as surface population. For the polycrystalline chrome surfaces, the value of p must decrease with increasing molecular weight if the extension is to be proportional to the square root of the molecular weight.

A number of explanations may be given for the difference in the results

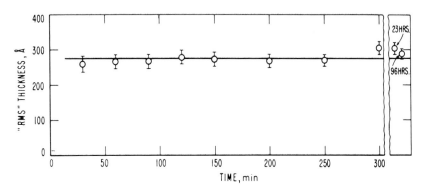

FIG. 5. Root-mean-square extension of adsorbed polystyrene (molecular weight 1,300,000) normal to mercury surface versus time. Adsorbed from cyclohexane solution (concentration 2.26 mg/ml).

between the solid polycrystalline surfaces, primarily chrome, and the liquid mercury surface. It has been pointed out by Fowkes (17) that hydrocarbons, including aromatics, have only dispersion force interactions with mercury. Therefore, it appears reasonable to use the value of the contribution of the London dispersion forces, γ^d, as a relative measure of the interaction energy between polystyrene and the various metallic surfaces, if we assume that small induced dipole interactions occur with the other metals. The values of γ^d for most of the solid metals that were studied were near 100

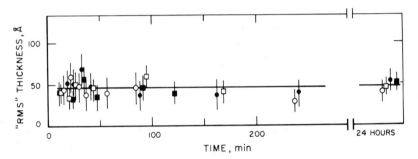

FIG. 6. Root-mean-square extension of adsorbed poly(ethylene o-phthalate) (molecular weight 7400) versus time. Adsorbed from ethyl acetate solution (concentration 0.28 mg/ml). Different symbols represent separate runs.

dynes/cm. The value of the γ^d of mercury is approximately 200. It is reasonable, then, that as a result of the higher interaction energy the value of p is increased and the loop size is decreased. In this case the value of p would be approximately independent of molecular weight, and the size of the loops would be unaffected by the coil dimensions in the solution. As a result of this higher interaction energy, any change in the extension of the film normal to the mercury surface during the adsorption period would be expected to be small and perhaps beyond experimental detection in this system.

The independence of extension with molecular weight is not necessarily in conflict with Hoeve's findings (6). The example that he discusses corresponds to a limiting case of weak interactions and large loop sizes, whereas in the case of mercury the interaction is probably strong. It is important to note, however, that fundamental differences between liquid and solid surfaces may exist, and this problem has not yet been resolved.

Ellipsometric measurements have been carried out on a system for which the interaction energies between a solid polycrystalline metallic surface and the polymer would be expected to be relatively high—a polar polyester, poly-(ethylene o-phthalate), on chrome and steel (13). The molecular weight of the polymer was relatively low, 7400. The extension as a function of time is shown in Fig. 6 for the polyester adsorbed from ethyl acetate on a chrome surface. There is no observable change in extension with time during the adsorption period, which is similar to the type of behavior observed for the polystyrene on mercury (Fig. 5). As in the case of adsorption on mercury, adsorption did occur during the time period shown in Fig. 6, the adsorbance curve exhibiting the same type of behavior shown in Fig. 2 for the adsorption of polystyrene on the chrome surface.

Fraction of Segments Attached Studied by Infrared Techniques

As was stated earlier, we would expect the value of p to remain relatively unchanged as the amount of polymer on the surface increases, if the interaction energy is sufficiently high. Measurements of the value of p for this polyester showed this to be the case. The technique used was similar to that described by Fontana and Thomas (18) and modified by Peyser and Ullman (19). It is dependent on the shift in the infrared absorption peak caused by the attachment of the absorbing group to a surface. If the original and shifted peaks are well separated, the value of p can be determined from the spectrum of the adsorbed molecule. However, in the case of the polyester adsorption, the peaks were not well separated, and it was necessary to resort to a differential spectral analysis. A differential spectrum is obtained between a cell containing only the original polymer solution and a cell containing the polymer adsorbed on silica suspended in polymer solution. The value of p is directly related to the optical density of the unbonded peak and the quantity of polymer adsorbed.

The results of this study are shown in Fig. 7 (13). The value of p is approximately 0.34 at all locations on the isotherm. It is also approximately independent of molecular weight for the two molecular weight samples studied. This value of 0.34 is relatively high and indicates that the molecule is adsorbed in a flattened configuration. As noted above, this is markedly different from the behavior exhibited by polystyrene. In that case (15), for adsorption on the same surface, silica, the value of p decreased as the amount of polymer on the surface increased.

Adsorbed Film Thickness Studied by Attenuated Total Reflection

Another technique that permits measurement of the extension and polymer concentration of an adsorbed film is internal reflection spectroscopy (20).

In this application of attenuated total reflection (ATR), polymer is adsorbed on a prism in contact with the polymer solution. Light transmitted through the prism will be totally reflected at the boundary between the prism and the solution if the angle of incidence is greater than the critical angle. If the polymer absorbs light at the wavelength used, some of the incident light will be absorbed by the adsorbed layer. Measurement of this attenuation is

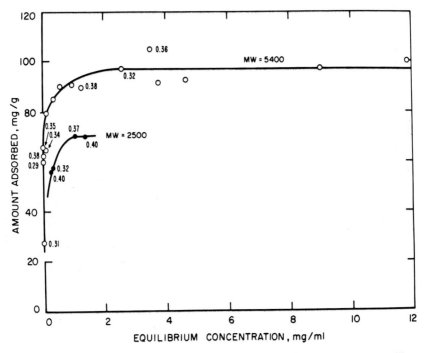

FIG. 7. Adsorption isotherm of poly(ethylene o-phthalate) from chloroform on silica at 35°C. The numbers on the figure represent the value p obtained at that point.

related to the thickness of the adsorbed film if the entire film is penetrated. In many respects the ATR method is similar to the technique of ellipsometry. With ATR, the absolute values of the two reflection coefficients are measured, whereas in ellipsometry a ratio of reflection coefficients and a phase shift of the light are measured. As is the case for ellipsometry, the sensitivity and utility of the ATR technique are dependent on the differences in refractive index among film, substrate, and surrounding medium. Also, as is true for ellipsometry, both the thickness and the concentration of polymer in the film can be simultaneously determined.

This method of ATR was used (20) to measure the thickness and concentration of an adsorbed film of polystyrene with a molecular weight of 76000. The polymer was adsorbed from cyclohexane on a quartz prism at the theta temperature, and the absorption was measured in the ultraviolet region after fifteen to sixteen reflections in the prism. The relationships between the parameters of the system including the thickness and refractive index of the film are given by the same Drude equations as were used in the ellipsometric calculations. In this case, however, the difference in absorption between polymer film and solution is related to the imaginary component of the refractive index. The thickness of the adsorbed polymer film was determined by this method to be 240 Å, and the film concentration 34%. This extension compares very favorably with the value obtained by ellipsometry in our laboratory and with the value obtained by Rowland and Eirich (16) by viscosity. The value for the concentration of polymer in the adsorbed layer was, however, somewhat higher than that obtained by ellipsometry.

In the study discussed above, the penetration of the light was much greater than that of the adsorbed polymer film, and only an "average" thickness value could be attained. However, in principle, the segment distribution normal to the surface for the adsorbed polymer layer can be determined by ATR. This is the most significant information that could result from ATR studies of polymer adsorption. It would require infrared-ray penetration depths of the order of the extension of the adsorbed polymer molecules normal to the surface. For sufficient sensitivity it would also be necessary to have large extinction coefficients. The ultraviolet region most closely meets these requirements, since the depth of penetration decreases with decreasing wavelength and many polymers have large extinction coefficients in this region.

Rates of Adsorption

Studies (21) of the rates of adsorption of tagged polystyrene adsorbed on chrome surfaces from cyclohexane at approximately the theta temperature and from benzene have shown the rates of adsorption to be slow. For solution concentrations of 10^{-1} and 10^{-2} mg/ml, the rates of adsorption were found to be approximately independent of solution concentration. This appears to support the concept of a rearrangement occurring on the surface during the adsorption period. Rates of desorption were found to proceed rather slowly, with approximately 40 to 60% remaining on the surface after a three-week period of desorption into solvent.

Reversibility of Adsorption

Although not large, there is a significant temperature dependence for the adsorption of polyester on glass. The effect of temperature cycling on the location of the adsorption isotherm was studied as a test of the reversibility of adsorption (22). This is, of course, also a test of the equilibrium of adsorption. The adsorption was carried out on glass powder, and the quantities adsorbed were studied by measuring changes in the concentration of the polymer solution, using the carbonyl absorption peak at 5.8 nm in the infrared region. The results are shown in Fig. 8. Essentially complete reversibility

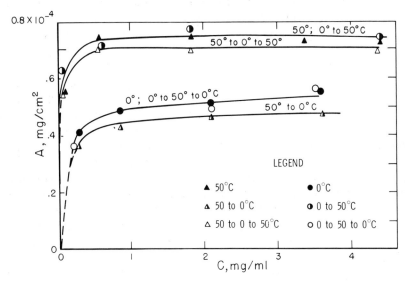

FIG. 8. Effect of temperature cycle on adsorption isotherms of poly(ethylene o-phthalate) (molecular weight 6250) adsorbed from chloroform solution on glass powder.

occurs when the temperature is cycled. When adsorption is carried out at 50°C initially and then at 0°C, the isotherm shifts from that of 50°C to that of 0°C. When the temperature is again changed back to 50°C, the isotherm again shifts back to the original isotherm of 50°C. The same reversibility occurs when 0°C is the initial adsorption temperature. This system indicated that equilibrium had been established. It should be mentioned that the curves shown in Fig. 8 represent isotherms on glass that had been exposed to the atmosphere prior to immersion in the polymer solution. If the glass is heated in vacuum, and the polymer solution added without prior exposure to the atmosphere, the results are not as well defined, and complete reversibility does not occur. This may be caused by a somewhat different surface structure due to the pretreatment.

Summary

Both theoretical and experimental polymer adsorption studies were primarily concerned with changes in the configuration of the molecule on adsorption and with the nature of the interaction with the surface. The experimental methods included studies of the extension of the adsorbed polymer molecule normal to the surface and the concentration of polymer in the film. The measurement techniques were ellipsometry and internal reflection spectroscopy. Also discussed were studies of the fraction of segments of the polymer chain attached to a surface. This quantity was determined by measuring a shift of an absorption peak in the infrared spectra that was caused by the attachment. Reversibility of adsorption and rates of adsorption and desorption, measured by using radioactive tracer techniques, were also briefly mentioned.

References

1. E. Jenckel and B. Rumbach, *Z. Elektrochem.*, **55**, 612 (1951).
2. R. Simha, H. L. Frisch, and F. R. Eirich, *J. Phys. Chem.*, **57**, 584 (1953); H. L. Frisch, R. Simha, and F. R. Eirich, *J. Chem. Phys.*, **21**, 365 (1953); H. L. Frisch and R. Simha, *ibid.*, **27**, 702 (1957); *J. Phys. Chem.*, **58**, 507 (1954); H. L. Frisch, *ibid.*, **59**, 633 (1955).
3. A. Silberberg, *J. Phys. Chem.*, **66**, 1872 (1962); *ibid.*, **66**, 1884 (1962); *J. Chem. Phys.*, **46**, 1105 (1967).
4. C. A. J. Hoeve, E. A. DiMarzio, and P. Peyser, *J. Chem. Phys.*, **42**, 2558 (1965).
5. R. J. Roe, *Proc. Natl. Acad. Sci. U.S.*, **53**, 50 (1965).
6. C. A. J. Hoeve, *J. Chem. Phys.*, **43**, 3007 (1965); *ibid.*, **44**, 1505 (1966).
7. E. A. DiMarzio and F. L. McCrackin, *J. Chem. Phys.* **43**, 539 (1965).
8. F. L. McCrackin, *J. Chem. Phys.* in press.
9. For descriptions of experimental techniques and calculations, see papers in "Ellipsometry in the Measurement of Surfaces and Thin Films," ed. by E. Passaglia, R. R. Stromberg, and J. Kruger, *Natl. Bur. Std. (U.S.) Misc. Publ.* **256** (1964).
10. F. L. McCrackin and J. P. Colson, see ref. 9, p. 61.
11. R. R. Stromberg, E. Passaglia, and D. J. Tutas, *J. Res. Natl. Bur. Std.*, **67A**, 431 (1963); R. R. Stromberg, D. J. Tutas, and E. Passaglia, *J. Phys. Chem.*, **69**, 3955 (1965).
12. R. R. Stromberg and L. E. Smith, *J. Phys. Chem.*, **71**, 2470 (1967).
13. P. Peyser, D. J. Tutas, and R. R. Stromberg, *J. Polymer Sci.*, **A1**, **5**, 651 (1967).
14. D. J. Tutas, R. R. Stromberg, and E. Passaglia, *Soc. Plastics Eng. Trans.*, **4**, 256 (1964).
15. C. Thies, P. Peyser, and R. Ullman, *Proc. 4th Intern. Congr. Surface Activity*, Brussels, 1964.

16. F. W. Rowland and F. R. Eirich, *J. Polymer Sci.*, **A1**, 2401 (1966).
17. F. M. Fowkes, *J. Phys. Chem.*, **67**, 2538 (1963); *Ind. Eng. Chem.*, **56**, 40 (1964).
18. B. J. Fontana and J. R. Thomas, *J. Phys. Chem.*, **65**, 480 (1961).
19. P. Peyser, Ph.D. Dissertation, Polytechnic Institute of Brooklyn, 1964.
20. P. Peyser and R. R. Stromberg, *J. Phys. Chem.*, **71**, 2066 (1967).
21. R. R. Stromberg, W. H. Grant, and E. Passaglia, *J. Res. Natl. Bur. Std.*, **68A**, 391 (1964); W. H. Grant and R. R. Stromberg, Abstracts American Chemistry Society, Meeting, Colloid and Surface Chemistry Division, Miami Beach Meeting, April 1967.
22. R. R. Stromberg and W. H. Grant, *J. Res. Natl. Bur. Std.*, **67A**, 601 (1963).

Discussion

MICHAELS: What is the volume fraction occupied by polymer in the adsorbed layer? How much solvent?

STROMBERG: The adsorbed layer of polystyrene on a chrome surface consists of approximately 10% polymer and 90% solvent.

HERMANS: If I remember correctly, you said that you have the surface covered with solvent first and then replace the solvent by the solution. If this is correct, under such circumstances, you may well have a layer of solvent on the surface. For a high-molecular-weight material, it might take something like 20 to 30 minutes for the polymer just to complete the diffusion to the surface, so that the first part of the process might well be, in large measure, diffusion-controlled. In a subsequent slide, the S-shaped rate curves indicated that it was a very slow process. What exactly is the difference in conditions shown on the slides?

STROMBERG: The solution concentrations used for the ellipsometry measurements were very different from those used for the rate studies. The concentrations used for the ellipsometry studies were of the order of 1 to 2 mg/ml. For the rate studies, they were 10^{-1} to 10^{-2} mg/ml, which would account for the slower rates. Thus, the conditions were very different in terms of solution concentrations. In all our polymer adsorption measurements we have always assumed that the surface is first covered with a layer of solvent. This would be true for both the rate and the ellipsometry studies. Thus, polymer adsorption occurs either by replacement of solvent molecules or by attachment across a preadsorbed solvent layer. We cannot state which occurs. With respect to diffusion, the rate studies did show that at short times the rates depended on concentration, but at long times they were relatively independent of solution concentration. Therefore, the rates showed a dependence on molecular rearrangement at long time intervals.

SCHWARTZ: In desorption into polymer solution, rather than pure solvent, what was the concentration of polymer in that solution?

STROMBERG: Desorption into polymer solutions was done at the same concentrations as the adsorption. This was 10^{-1} or 10^{-2} mg/ml, depending on the specific experiment, but always at the same concentration at which adsorption had been carried out so that the total amount of polymer on the surface would not be changed.

HELLER: One can, by the method of Heller and Tanaka [*Phys. Rev.*, **82,** 302 (1951)], actually fractionate polymers very well. If you adsorb from a mixture of molecular weights, then you get, on subsequent elution, only the lower-molecular-weight material. The latter, though less well adsorbed, is adsorbed much faster so that, in a nonequilibrium system, most of the high-molecular-weight material passes the column without effective adsorption compared to the oligomers.

STROMBERG: Yeh and Frisch [*J. Polymer Sci.*, **27,** 149 (1958)] fractionated polystyrene by elution with a series of solvents. It was an adsorption–desorption phenomenon that occurred, and it was followed by measuring the viscosities of the various molecular weight fractions.

HELLER: One does not need to use a different solvent; it can be done from the same solvent.

STROMBERG: In their case, they did not get complete elution with the same solvent, and, therefore, they used a series of solvents.

JELLINEK: Suppose you adsorb the same polymer, the same amount, on the same substrate, once from a good solvent, once from a poor solvent. How does the thickness of the layer compare in these two cases?

STROMBERG: I would assume that you would get a thicker layer when you adsorb from a good solvent. I think that Professor Eirich's work shows that. In our ellipsometric studies, refractive index differences must be sufficiently large to make measurements. We attempted to study the adsorption of polystyrene from methyl ethyl ketone, a better solvent than cyclohexane. Apparently, the adsorbed molecules were so swollen that the refractive index of the adsorbed layer was not different enough from that of the solution to enable us to make a measurement.

EIRICH: We made quite a study of this, and it turns out that when you go from a fluid to a solvent, you adsorb more, but the thickness becomes less. In other words, you get packing which becomes denser and denser, and eventually you get the densest packing in the lowest layer. Then, it collapses.

Adsorption of Surfactants at Polymer Interfaces
A Contributed Discussion

R. G. GRISKEY*

Chemical Engineering Department
University of Denver
Denver, Colorado

Introduction

The adsorption of surfactants at polymer interfaces is an important factor in many industrial processes, as, for example, the adsorption of surfactants by droplets or globules in emulsion and suspension polymerizations. The purpose of this paper is to discuss surfactant adsorption by polymer interfaces with one exception—adsorption on proteins and animal or vegetable fibers. These latter cases are excluded because they involve specific chemical interaction (1, 2). It should also be noted that this paper considers only the surfactant adsorption and not the chemical kinetics of emulsion and suspension polymerizations.

A number of investigations have considered the effect of surfactant on emulsion polymerizations. Many of these studies have, however, for the most part considered only the effect on the overall chemical kinetics and have not specifically studied the adsorption of surfactant in such systems. Studies in this category include the work of Baxendale and colleagues (3, 4), Evans (5), Bovey and Kolthoff (6), Bito and Yamakita (7), Khomikovskii and colleagues (8), McCoy (9), Yurzhenko and Vilshanskii (10), Motoyama and Mitsuoka (11), Chujo and colleagues (12), and Muroi (13).

Another group of investigators concerned themselves with micelle structure and the loci of emulsion polymerization. These included the studies of Dogadkin and co-workers (14, 15), Fikentscher (16), Gee and co-workers (17), Hohenstein and Mark (18), Fryling and Harrington (19), Hughes and co-workers (20), Harkins (21–24), Harkins and Stearns (25), McCoy (9), and Nestler (26). Studies concerned mainly with surfactant adsorption at polymer interfaces were those of Harkins and co-workers (1) and Griskey and Woodward (27).

* Present address: Chemical Engineering Department, Newark College of Engineering, Newark, New Jersey.

Experimental Methods

A number of techniques have been used to study the adsorption of surfactants. One method involved the use of surface tension measurements (23). The principle underlying the technique is that as surfactant is adsorbed the surface tension rises. The surface tension is usually measured by the du Nouy method (28). An inherent defect of this method is the difficulty in determining the exact surfactant concentration in dilute solutions.

The disappearance or take-up of surfactant can also be measured by a dye method (23). This method depends on the principle that certain dyes in solution fluoresce only when micelles are present. Therefore, the disappearance of detergent can be followed by measuring the quenching of fluorescence.

A radioactive tracer technique (27) can also be used to study detergent adsorption. This method involves preparing a radioactively tagged detergent and then counting detergent samples with a Geiger counter. This yields relative detergent concentration data which give a measure of detergent adsorption. This technique has the advantage that it gives a specific measurement rather than a gross measurement which can be clouded by other factors.

Experimental Results and Discussion

Harkins (23) first studied the adsorption of detergent as part of an investigation directed toward developing a mechanism for emulsion polymerization.

TABLE I

Disappearance of Sodium Oleate from the Aqueous Phase for an Isoprene–Styrene Emulsion Polymerization (23)

(Initial sodium oleate concentration 0.091 M or 2.8 wt%)

Time (hours)	Copolymer Yield (%)	Temperature (°C)	Soap Concentration (wt%)
0	0	20	2.8
3	8.1	24.8	—
5	13.5	26.1	0.15
5.5	17.7	35.2	0.05
6	21.1	36.2	0.04
6.5	20.9	37.4	0.04
8	28.5	42.0	0.025
11	32.5	45.9	0.02
12	37.6	45.8	0.015
14	46.8	50.0	0.015

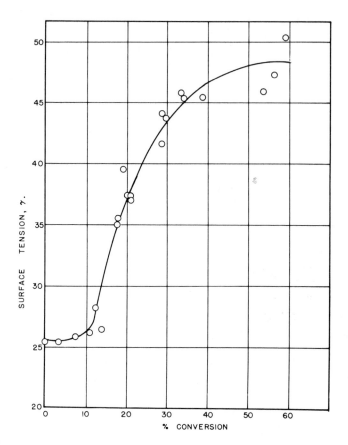

FIG. 1. Variation of surface tension with yield of isoprene–styrene copolymer in sodium oleate solutions (23).

He reported data relating changes in detergent concentration to percentage of conversion for a polymerization reaction. Such data for adsorption of sodium oleate in a 75:25 isoprene–styrene copolymer system are shown in Fig. 1 and Table I. The plot shows the behavior of surface tension with increasing percentage of conversion. As can be seen, the curve breaks rather sharply at about 15% conversion and then levels off from 50 to 60%. The data of Table I taken with the plot show that the break occurs because the bulk of the detergent (approximately 90% of the initial charge) has been adsorbed at 15% conversion. Furthermore, the leveling off at 50 to 60% conversion results because essentially no further adsorption takes place.

Additional data determined by Harkins (23) for the adsorption of potassium laurate and potassium myristate in a styrene polymerization system are shown in Fig. 2. Here again the adsorption of detergent is rather rapid during the early stages of conversion, although not as rapid as with the data of Fig. 1 and Table I. Figure 2 also shows that increasing the initial detergent concentration actually decreased the percentage of the initial charge adsorbed. This can be seen in Fig. 2 at about 23% conversion, since 73.5% of the original detergent was adsorbed for 3 wt% of potassium laurate charged, while only 61.5% of the original charge was adsorbed for a 5.2 wt% charge of potassium laurate.

It should be noted that the data of Harkins can be misleading if certain factors are not taken into account. First, the adsorption of detergent actually

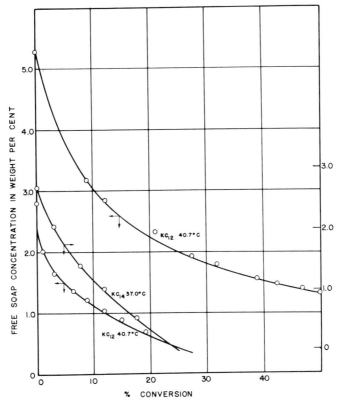

FIG. 2. Disappearance of detergent during polymerization of styrene. Yield is percentage of polystyrene. KC_{12} is potassium laurate and KC_{14} is potassium myristate (23).

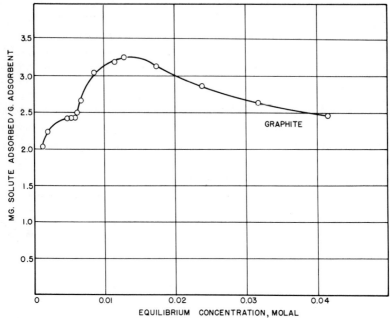

FIG. 3. Sodium dodecyl sulfate adsorbed as a function of initial detergent concentration (*1*).

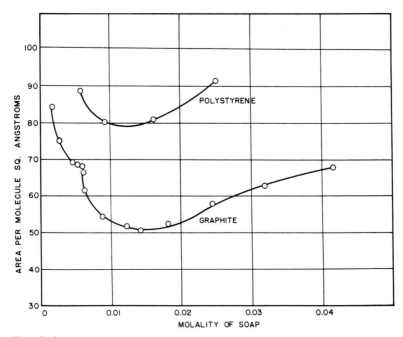

FIG. 4. Area occupied per adsorbed molecule (or ion) for sodium dodecyl sulfate on graphite and polystyrene (*1*).

took place on interfaces composed of multicomponent systems. In the isoprene–styrene copolymer there were three components (isoprene, styrene, and copolymer), while for styrene there were two components (styrene and polystyrene). Furthermore, composition of these multicomponent systems was different for each percent of conversion. This meant that Harkins' adsorption data represented information taken for a continually changing system of adsorbents. In addition the data of Table I and Fig. 1 are for a nonisothermal system, which further clouds their meaning.

Hence, the data of Figs. 1 and 2 and Table I are interesting in delineating the disappearance of detergent during emulsion polymerizations, but they are of limited applicability in determining the behavior of detergent adsorption, since they represent data taken in continually changing adsorbent systems.

In a later paper Harkins and co-workers (*1*) studied adsorption of sodium dodecyl sulfate onto solid polystyrene as well as graphite. Figure 3 is an adsorption isotherm for sodium dodecyl sulfate on graphite at 25°C. No direct data for polystyrene were given. However, the authors did indicate that the adsorption isotherm for sodium dodecyl sulfate on polystyrene was of the same shape as that for graphite and that the maximum for polystyrene occurred at approximately the same equilibrium detergent concentration as for graphite.

Figure 4 shows a plot presented in the same paper (*1*) which gives the area occupied per adsorbed molecule (or ion) for sodium dodecyl sulfate. These data show the occurrence of minima which correspond directly to the maxima of the adsorption isotherms. The data for solid polystyrene, although limited, are useful, since the system (polystyrene) is essentially a single adsorbent.

The most recent work on surfactant adsorption in polymeric systems was that of Griskey and Woodward (*27*), who studied adsorption of sodium dodecyl sulfate on polystyrene–styrene interfaces. This research differed from earlier work in that a constant composition of polystyrene–styrene was maintained for the adsorption tests.

The results of the study are shown in Figs. 5 through 9. Figure 5 gives adsorption data for a 20% polystyrene interface at 25°C with initial sodium dodecyl sulfate concentrations of 0.0004 and 0.004 mole/liter, and Fig. 6 gives similar data for a 60% polystyrene interface. Figures 7 and 8 show the effect of changing the polymer interface while keeping the initial detergent concentration constant.

At first glance the data of Figs. 5 through 8 appear to be anomalous. In Fig. 5 the data for both initial detergent concentrations and a 20% polystyrene interface seem to follow a common behavior. However, in Fig. 6 the data for the 60% interface show considerably different behavior for the two

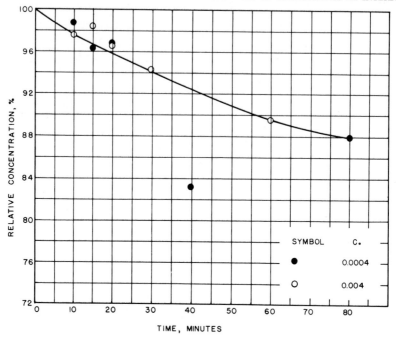

FIG. 5. Adsorption of sodium dodecyl sulfate with time for a 20% polystyrene interface at 25°C (27).

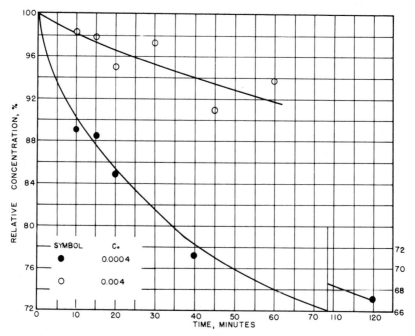

FIG. 6. Adsorption of sodium dodecyl sulfate with time for a 60% polystyrene interface at 25°C (27).

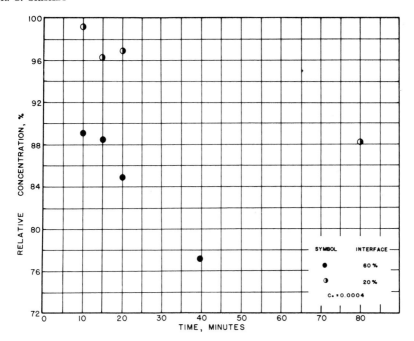

Fig. 7. Adsorption of sodium dodecyl sulfate with time for an initial detergent concentration of 0.0004 mole/liter (27).

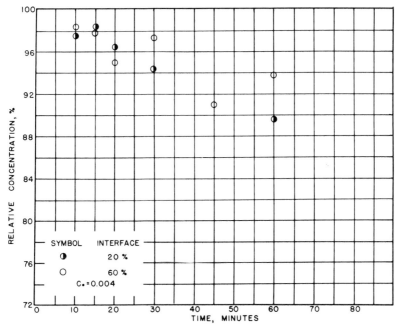

Fig. 8. Adsorption of sodium dodecyl sulfate with time for an initial detergent concentration of 0.004 mole/liter (27).

detergent levels. Likewise, in Fig. 7, with the same initial detergent concentration, decidedly different behavior results for the 20% and 60% polystyrene interfaces. On the other hand, no sizable effect of adsorbent interface appears in Fig. 8 for an initial detergent concentration level of 0.004 mole/liter.

The earlier data of Harkins (*1, 23*) are of interest in analyzing the behavior of Figs. 5 through 8, even though the experimental conditions were somewhat different. In Fig. 2, for example, the disappearance of potassium laurate at two different initial detergent levels is depicted. Although the interface changes continuously for the data shown, the trend is a faster adsorption rate and a greater relative adsorption for the lower initial detergent concentration. Furthermore, the data of Figs. 3 and 4 show that equilibrium adsorption of detergent (total adsorption at long times) and inversely the area occupied per adsorbed molecule both vary with initial detergent concentration. The equilibrium adsorption data climb to a maximum, decrease, and then level off. This then shows that adsorption behavior can vary with initial detergent concentration.

Harkins has attributed this behavior to the change in activity of the adsorbed detergent ions with changing detergent concentration. A number of authors (*29–32*) have shown theoretically that the activity of long-chain ions should go through a maximum with respect to overall concentration if the equilibrium between single ions, counterions, and aggregate is governed by a simple form of the mass action principle. Mean ion activity measurements for dodecyl sulfonic acid (*33*) have supported these theories. Hence, if only single ions are adsorbed at an interface, the maximum in the adsorption isotherm is due to a maximum in the activity of the adsorbed surfactant.

An additional point which can be used to explain the behavior of Figs. 5 through 8 is that polystyrene–styrene interfaces actually constitute a mixed adsorbent. As the percentage of polystyrene changes, the adsorbent itself is altered. Studies on mixed adsorbents other than polystyrene–styrene have been carried out by Schilow and co-workers (*34*) as well as by Chowdhury and Pal (*35*). These investigations showed that the amount adsorbed increased as the percentage of one of the components in a mixed adsorbent increased until a maximum was reached. The amount adsorbed then decreased until only one pure component remained (that is, 100% of the added component). This, therefore, means that the 20% and 60% polystyrene interfaces are actually two different mixed adsorbents.

In Figs. 5 and 6 we have data for two different interfaces (20% and 60% polystyrene). These systems would yield two different adsorption isotherms of the type shown in Fig. 4. Hence, for two different initial detergent concentrations (as with Figs. 5 and 6) possibly either the same or different amounts could be absorbed (depending on the adsorption isotherm position). For Fig. 5 the adsorption isotherm is probably such that the same adsorption process takes place at both concentrations, while for Fig. 6 the adsorption at a concentration of 0.0004 mole/liter represents a higher position on the adsorption isotherm than for a concentration of 0.004 mole/liter. Likewise, since the two interfaces represent entirely different adsorbent systems, we could at constant concentration have either the same or different amounts adsorbed. This would explain the behavior in Figs. 7 and 8.

The data of Figs. 5 and 6 were also subjected to a treatment of the adsorption rate process. The data were treated with a series of possible rate equations. One form tested was

$$\frac{df}{dt} = -k_1 C_0^{n-1} f^n \tag{1}$$

where $f = C_t/C_0$, C_t is surfactant concentration at time t; C_0 is surfactant concentration at time $t = 0$; k is the rate constant; and n is the order of the adsorption.

Another form tested was that proposed by Davies (36) for surfactant adsorption (although not adsorption on polymer interfaces):

$$\frac{df}{dt} = k_2(1 - \theta)f \tag{2}$$

where f, C_t, C_0, and t are as above; k_2 is the rate constant; and θ is the fraction of adsorbent surface covered. Although all the data of Figs. 5 and 6 fit equation 2, k_2 was found to vary with initial surfactant concentration, C_0, which should not have occurred. Hence, it was felt that Davies' equation was not suitable for describing the rate of adsorption in polymeric systems.

Tests of the same data with equation 1 yielded mixed results. The data of Fig. 5 were found to follow a first-order mechanism, while the data of the lower curve of Fig. 6 followed second-order behavior. The mechanism for the upper curve of Fig. 6 is uncertain. The differences in rate order found for the data of Figs. 5 and 6 could possibly be due to differences between the 20% and 60% interfaces.

Comment on Fig. 9 was delayed until the rate process aspect had been considered. The data on this plot represent measurements for a 60% polystyrene interface and an initial detergent concentration of 0.004 mole/liter at two different temperatures, 25°C and 70°C. The data show an increased adsorption rate at 70°C. This behavior can be readily explained by the usual temperature effect found in rate processes as per the Arrhenius equation:

$$k = A \exp(-E/RT) \tag{3}$$

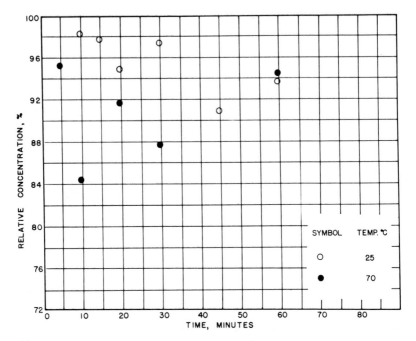

FIG. 9. Adsorption of sodium dodecyl sulfate with time for a 60% polystyrene interface and an initial detergent concentration of 0.004 mole/liter (27).

where k is a rate constant, A is a constant, E is the activation energy, R is the gas constant, and T is the absolute temperature.

Conclusions

Surfactant adsorption is a complex function of the polymer interface and the initial surfactant concentration. The former parameter is a consequence of the composition of the interface which can result in a mixed adsorbent, while the latter is related to the changing activity of the surfactant ions.

The rate of adsorption appears to be governed by the equation

$$\frac{df}{dt} = -k_1 C_0^{n-1} f^n$$

with adsorption being first or second order and in some cases of uncertain order.

The effect of temperature on adsorption of surfactants is that generally found for rate processes, namely, an Arrhenius type of relation.

References

1. W. D. Harkins, M. L. Corrin, E. L. Lind, and A. Roginsky, *J. Colloid Sci.*, **4,** 485 (1949).
2. E. I. Valko, *Ann. N.Y. Acad. Sci.*, **46,** (6), 451 (1946).
3. J. H. Baxendale, M. G. Evans, and G. S. Park, *Trans. Faraday Soc.*, **42,** 155 (1946).
4. J. H. Baxendale, M. G. Evans, and J. K. Kilham, *Trans. Faraday Soc.*, **42,** 668 (1946).
5. M. G. Evans, *J. Chem. Soc.*, 266 (1947).
6. F. A. Bovey and I. M. Kolthoff, *J. Polymer Sci.*, **5,** 487 (1950).
7. T. Bito and H. Yamakita, *Yukagaku*, **8,** 22 (1959).
8. P. M. Khomikovskii, L. G. Senatorskaya, and Z. G. Serebryakova, *Khim. Nauka i Promy.*, **4,** 598 (1959).
9. C. E. McCoy, *Offic. Dig. Federation Soc. Paint Technol.*, **35,** 327 (1963).
10. A. I. Yurzhenko and V. A. Vilshanskii, *Dokl. Akad. Nauk. SSSR*, **148,** 1145 (1963).
11. T. Motoyama and Y. Mitsuoka, *Kogyo Kagaku Zasshi*, **65,** 1303 (1962).
12. K. Chujo, Y. Harada, and H. Morita, *Kolloid-Z.*, **201,** (2), 155 (1965).
13. S. Muroi, *Kogyo Kagaku Zasshi*, **68,** (9), 1773 (1965).
14. B. Dogadkin, V. Balandina, K. Berezan, A. Dobromyslowa, and M. Lapuk, *Bull. Acad. Sci. URSS*, 397, 423 (1936).
15. B. Dogadkin, K. Berezan, and A. Dobromyslowa, *Bull. Acad. Sci. URSS*, 409 (1936).
16. H. Fikentscher, *Angew. Chem.*, **51,** 433 (1938).
17. G. Gee, C. B. Davies, and W. H. Melville, *Trans. Faraday Soc.*, **35,** 1298 (1939).
18. W. P. Hohenstein and H. F. Mark, *J. Polymer Sci.*, **1,** 127 (1946).
19. C. F. Fryling and E. W. Harrington, *Ind. Eng. Chem.*, **36,** 114 (1944).
20. E. W. Hughes, W. M. Sawyer, and J. R. Vinograd, *J. Chem. Phys.*, **13,** 131 (1945).
21. W. D. Harkins, *J. Chem. Phys.*, **13,** 381 (1945).
22. W. D. Harkins, *J. Chem. Phys.*, **14,** 47 (1946).
23. W. D. Harkins, *J. Am. Chem. Soc.*, **69,** 1428 (1947).
24. W. D. Harkins, *J. Polymer Sci.*, **5,** 217 (1950).
25. W. D. Harkins and R. S. Stearns, *J. Chem. Phys.*, **14,** 214 (1946).
26. C. H. Nestler, Ph.D. Dissertation, University of Akron, Akron, Ohio, 1964.
27. R. G. Griskey and C. E. Woodward, *J. Appl. Polymer Sci.*, **10,** 1027 (1966).
28. L. du Nouy, *Science*, **69,** 251 (1929).
29. G. S. Hartley and R. C. Murray, *Trans. Faraday Soc.*, **31,** 183 (1935).
30. K. A. Wright and H. V. Tartar, *J. Am. Chem. Soc.*, **61,** 544 (1939).
31. A. B. Scott, H. V. Tartar, and E. C. Lingafelter, *J. Am. Chem. Soc.*, **65,** 698 (1943).
32. K. Wohl, *Ann. N.Y. Acad. Sci.*, **46,** (6), 402 (1946).
33. L. L. Neff, O. L. Wheeler, H. V. Tartar, and E. C. Lingafelter, *J. Am. Chem. Soc.*, **70,** 1989 (1948).
34. N. Schilow, M. Dubinin, and S. Torapow, *Kolloid-Z.*, **49,** 120 (1929).
35. J. K. Chowdhury and H. N. Pal, *J. Indian Chem. Soc.*, **7,** 451 (1930).
36. J. T. Davies, in "Surface Phenomena in Chemistry and Biology," ed. by J. F. Danielli, K. G. A. Pankhurst, and A. C. Riddiford, Pergamon, New York, 1957, pp. 55–69.

Factors in Interface Conversion for Polymer Coatings*

F. R. EIRICH

Polytechnic Institute of Brooklyn
Brooklyn, New York

Introduction

Conference Theme. In this review and summary of this conference it would be impossible and presumptuous to evaluate adequately the many individual contributions made in the form of prepared papers or during the discussions. Instead, I shall endeavor to place these contributions into a coherent context, underline some of the main points as I see them, and add some comments and draw some conclusions.

This conference has been organized into areas dealing with the nature of phase boundary interactions with surface preparations, and with the processes of wetting, adsorption, and penetration of surfaces. The papers presented correspond to these divisions: Michaels and Bolger on the forces at interfaces, Hansen and Bobalek on surface activation for chemisorption, Stromberg and Zettlemoyer on the thermodynamics of adsorption of polymers and ions, respectively, Laukonis and Machu on the preparation and structure of phosphate coatings, and Mark and Cheever on porous surface penetration. The topics of the formation and setting of the coating as a function of surface structure and wetting, and the consequent performance of the coating, have been touched upon only tangentially, and a few additional remarks should be in order.

The conference title, "Interface Conversion for Polymer Coatings," not only defines the subject quite succinctly, but also indicates that the conversion should serve a specific purpose—that of optimizing the bond between a polymeric coating and a metal base. Metals and polymers are not particularly miscible or compatible, and experience has shown that their bond can be greatly enhanced by metal surface conversions which serve the purpose of enhancing the future bond between the two adherends. The task of promoting this bond is a most exacting one, since the surface preparation must

---

* This contribution discusses only the lectures presented at the Interface Conversion Symposium and does not include the papers noted as Contributed Discussions because these papers were submitted after the Symposium.

not be conducive to metal corrosion; it must be easy to apply, and be mechanically, chemically, and temperature- and water-resistant. The protective coating itself has to satisfy the same requirements, but in addition it must be light- and weather-resistant, decorative, and easy to repair. These requirements show that the use of polymers is indeed indicated, because no class of material is known which combines the needed properties to a better degree. On the other hand, the nature of what constitutes a suitable conversion is not equally well predicated, and it is one of the interesting points

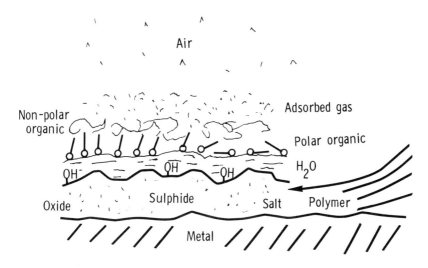

FIG. 1. "Hierarchy" of spontaneously adsorbed layers on a "metal" surface.

brought out at this conference that phosphate compounds appear to be superior to all others in forming a durable interlayer.

Interface Structures. I might begin the discussion with the axiom that there are no free, or pure, surfaces in nature. Every so-called surface, in actuality an interface between two physically recognizable phases, covers itself spontaneously with a hierarchy of adsorbed layers such that the free surface energy becomes minimized. A metal surface thus will coat itself first by a layer of sulfides, oxides, and salts, and then by hydrates followed by water, organic polar compounds, other organic compounds, mostly hydrophobics, and eventually adsorbed layers of gases (Fig. 1). If we want to "coat" a surface, this means really that we wish to establish a different hierarchy of adsorbed layers which has to compete with, and stand up against, the natural order. It follows from this that the endeavored structure should more or less simulate the natural order—that is, it should consist of layers of diminishing surface energy and establish minimal gradients of free energy. This reduces the chance that extraneous molecules interlope at strata or energy levels which are especially hospitable to their structure, as, for instance, fatty acid molecules would at the interface between metal and oil. Obviously, the interlayer should have adequate mechanical strength so as not to be easily disturbed or displaced. It is thus advantageous to have areas of chemical bonding between coating and substrate and to have some depth to the interface—that is, an interphase—in order not to be dependent on the delicate leaflet structures of monolayers or a few multilayers. I became forcibly aware of this when investigating the structure and mechanism of action of the so-called zinc oxychromate phosphate wash primer some years ago (1). Its essential feature turned out to be firmly anchored layers of ferrous–ferric oxychromate and ferric and zinc phosphates, chelated to polyvinylbutyral and overlaid with the same resin.

Adhesiveness of Polymers. For good adherence, the polymer itself must wet either the interlayer or the substrate and act as its own adhesive to the remainder of the polymer coating. In view of all the mechanical and chemical vicissitudes of an interface, it requires a material which is capable of undergoing extensive mechanical and thermal deformations in a more or less reversible fashion—in other words, a material which combines reasonable strength with the ability to absorb mechanical energy with little fatigue. This is necessary on account of the inevitable differences across any interface boundaries in the coefficients of thermal expansion and isothermal compressibility and in the moduli and Poisson ratios which create tangential stresses of great magnitude with every stress or temperature cycle. Again, creating interlayers—that is, interphases instead of interfaces—will help to reduce these stresses.

Last, but not least, it will help stress reduction during the establishment of the solid bond if one side of the boundary or interlayer consists of a material with wide softening ranges instead of sharp phase transitions. The danger from differential stresses can be further reduced by dispersing high modulus material as filler throughout the adhesive or softer adherend.

Small volume changes during glass transitions or gradual melting ranges are also helpful in view of the inevitable difficulties related to the phase changes during interfacial bond formations. All adhesives have to be applied in a rather fluid state so that they can make molecular contact by wetting and by flowing into all surface roughnesses. Subsequently, they have to increase in rigidity. With few exceptions, adhesives undergo solidification processes by drainage or drying out, by freezing, by glass transition, or by polymerization or cross-linking—phase changes which are accompanied by greater or lesser volume changes. The substrate, or the adherends, however, have as a rule the lower thermal expansion coefficients; in addition they do not undergo phase transitions. Thus, when the adhesive needs to shrink along the interface with the adherend, it is restrained in doing so by the difficulties in the way of lateral diffusion of adsorbed molecules along a solid wall. Unfortunately, the more strongly the molecules are adsorbed or bonded, the greater will be the activation energy for lateral migration and the slower the relaxation rate. As a result, the polymer experiences large tangential stresses during the process of "setting." This leads to frequent dewetting or crack formations along the interface—that is, to its substantial weakening— because interfacial flaws act as stress raisers and/or attract environmental impurities which settle in suitable niches and establish a different hierarchy. Moreover, the presence of surface active impurities may act as the entrance port from which they spread to other areas and interpose themselves in other sections of the interfaces. It is the major art of all interfacial bond formation to create substantial wetting and to maintain it during and after solidification. In many cases this is achieved by interposing an energetically fitting interlayer capable of large deformations which has the function of mitigating tangential stresses and keeping out intruding impurities. In the following discussion I wish to enlarge on and illustrate these comments and relate them to the presentations made at the Symposium.

Some Comments on Conference Papers

Phase boundaries are loci of special forces and geometries; they occur in so many varieties and are determined by so many factors that any single case studied is found to have its own individuality. Figure 2 shows a simple schematic diagram which emphasizes the deep energy trough that accompanies every phase boundary. Physical adsorption means the nonspecific collection of molecules in that trough, and chemisorption means the specific bond formation between resident and attached molecules. In any case, for adsorption to occur, the new trough on the outside of the adsorbed layer must be shallower than the old one. A series of such adsorbed strata constitutes the hierarchy mentioned above. The cases discussed during the presentations by Michaels and Bolger and by Zettlemoyer refer to specific surface–adsorbate interactions which can serve as models, since their behavior is likely to reflect the actual situation rather closely. Their discussions offer the advantage that they refer not only to original pure metal or metal oxide layers but also to molecular layers that could be located anywhere as internal boundaries within an interphase. Hansen's paper on the action of activated gases deals with an important method for the preparation of large, uniform, well-defined surfaces. To this extent these authors go beyond the considerations of forces and processes related to a single, idealized step or stratum in the interconversion of the original phase boundary. The total strategy, however,

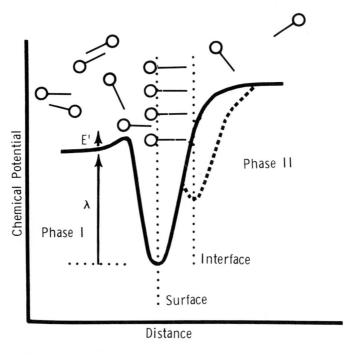

FIG. 2. Chemical potential of polar molecules at an interface. The polar heads are shown as circles, and the hydrocarbon tails by ———. The second potential trough shows the energetics at the monolayer–phase II interface.

must remain concerned with the destruction of the original hierarchy of layers and the establishment of a new and more desirable one from the point of view of the specific purpose of anchoring a polymer layer. Bobalek's paper is of particular significance, therefore, because it prepares the ground for a direct, covalent bonding of polymers in depth to almost any variety of metal oxide or metal salt surface, while Stromberg referred to recent work that showed that physically adsorbed polymer also forms an interphase several hundred angstroms in thickness.

The papers on the formation of phosphate coatings, the phosphating of oxidized surfaces, and the wetting of the resulting highly porous maze of phosphate crystals provide essential background knowledge for the optimal handling of these primer coatings. The latter proved to be of such value precisely because they provide what one might call a reinforced interphase.

Even more valuable, in my opinion, is the fact that these coatings are epitactically grown, or chemically welded, to the basic phase boundary and therefore serve like protruding iron rods in the binding of a new concrete layer on an old one. The porous nature helps in addition to hold the overlayer not only by physical—or chemisorption, but by extensive interlocking. This changes the prospect of interphase failure from that of an adhesive or locally cohesive rupture to that of an intricate composite material.

On the basis of the evidence presented at this Symposium, of my own experience with primers (1), and of other data quoted at this conference, it appears that the epitactic growth of phosphates is due to the ability of phosphoric acid to form a wide variety of salts with a wide variety of metal ions. It is also important that many of these phosphates are isomorphic or capable of forming solid solutions, or at least of having rather similar structures and surface free energies permitting ready seeding and overgrowth on metals or metal oxide layers. By virtue of the same mechanism, further phosphates can grow on the first layer and so forth and thereby build up a structure of "welded" surface layers of high energy and reactivity. Because of the many accidental crystal orientations, twinnings, dislocations, and impurities, these layers grow extremely haphazardly and full of pores and channels. They form a large and eminently suitable bed for the adsorption of ions, salts, acid and basic groups, and polar organic compounds.

Many points could be made with respect to the wetting and penetration of such surfaces. Suffice it to say that, to penetrate into the phosphate maze within an acceptable cooling or setting time, the first advancing layer of polymer must have a rather small wetting angle. A number of papers at the recent Bristol conference on spreading and wetting (2) dealt extensively with this problem of penetration, re-emphasizing the fact that even under most favorable conditions porous media such as fabrics, sponges, and powders may wet with difficulty. The geometry of widening pores especially leads to an effective increase in wetting angles, a decrease in wettability, and the formation of pockets, as discussed further below.

The papers by Mark and by Cheever have paid attention to some of these important points. Mark suggests that porous surfaces can be looked upon as filter beds and can be treated by the known laws of bulk penetration of porous solids by liquids. He points out that the finely divided surface may have also catalytic advantages in cases of surface polymerizations or for the hardening of polymers. Parenthetically, I should like to mention that all aspects of porous bed–liquid interactions have been greatly advanced under the impetus of studies on more effective flushing of fossil oils from submerged stratas (3), or on the mechanics of frost heaving (4). Papers on these topics contain valuable parallels to our problem and reflect great improvements and refinements in our understanding of the thermodynamics of penetration of porous media.

The study by Cheever is concerned directly with the problem of the wetting and penetration of phosphate-primed surfaces. On the basis of his characterized coated surfaces, using model liquids and an adaptation of the Washburn-Rideal theory, he could calculate a "degree of porosity" and was able to grade liquids with respect to their actual "penetration power." Cheever's results could possibly be refined and reinterpreted by reviewing them in the light of the Bristol Symposium. However, at this point some special aspects of Cheever's work merit comments in conjunction with recent studies carried out by Schonhorn, Frisch, and Kwei (SFK) (5).

The Kinetics of Wetting

Intrigued by the results of earlier workers, SFK set out to study the nature of the basic physical variables which must be considered in describing wetting processes, the actual physical mechanism of wetting, the way in which it can be defined by dimensional analysis, and how the kinetics of the wetting process of polymers are related to the kinetics of the contact angle. The simplest form of the wetting process would be that between two particles, or between a surface and a droplet of the same material, when Frenkel's law (6) or a modification of it, should apply. One might write in general terms: $r_t^2/r_0^2 = K(\gamma t/r_0 \eta)$, where r_t^2 and r_0^2 correspond to the areas covered by polymer at times t and zero, γ is the surface free energy, η is the viscosity, and K is a function of the dimensionless ratio in the bracket. Much will depend on whether the liquid will spread by way of a thin advancing layer, or "foot," with a central "cap," as shown by Cheever, or whether the spreading liquid will more or less maintain a definite contact angle. Using aluminum, mica, and FEP Teflon as substrates, and polyethylene and polyvinyl acetate melts as liquids, SFK studied simultaneously the progress of wetting and the changes in angle (Fig. 3a). They found that the dimensional relationship was indeed obeyed with respect to the wetting rate and that, rather surprisingly, a contact angle was maintained which permitted the shape of the fluid drop to remain that of a spherical segment. Further, a frequency a_T could be formulated: $a_T = \gamma/L_w\eta$, which represents a characteristic quantity, provided the value of L_w, which has the dimension of a length, can be defined. This turns out to be difficult, because L_w does not depend on the original dimensions or mass of the droplet or on the temperature (Fig. 3b), nor on surface porosity or roughnesses. The value of L_w, however, is determined by the nature of the substrate and the wetting material. It can be compared with Cheever's term $d \cos \theta$, but there are significant differences.

It must be emphasized that Cheever (and Washburn-Rideal) describes pore penetration under a constant contact angle, θ—that is, a steady-state process occurring so slowly as to allow the use of equilibrium values for γ and θ.

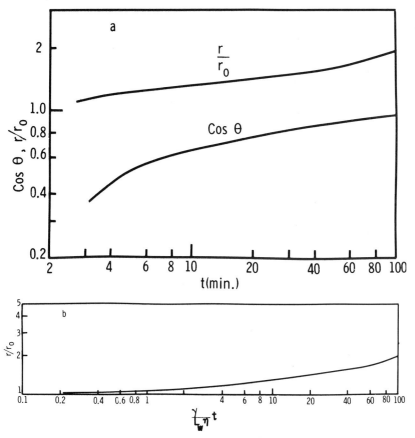

FIG. 3. (a) Relative spreading and changes of contact angle (5). (b) Superposability of spreading rates when shifted along the $a_T[\approx f(1/\eta)]$ axis.

SFK study kinetics; their θ is tied to the progress of wetting and eventually assumes values close to zero. Cheever's results show by the variability of the exponent, n, that his pore penetration is not at steady state. It would be better not to use the empirical expedient of varying n, since this destroys the dimensional correctness of his equation, but to apply a fractional exponent to cos θ or, better, to allow γ and/or θ to be time-dependent. This would be more appropriate in view of his observations of the shape of the advancing droplet. Except for a trivial normalization factor $1/R_0^2$, it would bring his equation in line with SFK's. Then Cheever's equation would have the advantage of containing the experimentally accessible pore dimension, d, as compared to the hard-to-define parameter L_w in SFK's approach. Most likely L_w is a measure of the distance of spreading for a unit change in free surface energy.

From these discussions we derive an insight into the structural features of the ratio $(\gamma \cos \theta)/\eta$, which has the dimension of a velocity, and as such is determined by the balance between the decrease of the (driving) surface free energy and its dissipation via the mechanism of the resisting viscous friction: $[\gamma/l] = [\eta/t] = [\Delta F/V]$, with l instead of r to show the dimension of length.

In view of the thoroughness of Cheever's and SFK's papers, little needs to be added or re-emphasized. Of the four relevant terms, t is the one of concern; l stands for surface geometry, $\gamma \cos \theta$ for substrate–coating interaction, and η for the coating mobility. The terms γ, η, and l are the quantities which one has to manipulate for the desired fast, and therefore likely complete, wetting penetration. The direction of manipulation is immediately apparent from the equations: One should aim at as large a surface tension of the fluid polymer as is compatible with a small wetting angle, as small a viscosity as is compatible with good mechanical properties of the solidified coating, as small a pore size as is possible without creating too much viscous resistance, and a pore geometry not detrimental to developing small wetting angles.

Some Important Surface Details

Porosity and Wetting. As stated, pore geometry is of major importance. The "capillaries" of the pores of phosphate coatings have been shown to be extremely irregular. The quantity d is therefore an average of a wide distribution. It is perfectly possible to encounter two equal averages which may differ significantly in the details of the pore structure, and it is, therefore, important to obtain specific information on the nature of the contact angle or of the apparent free surface energy as a function of the details of the porosity. Cheever has already discussed this point and has also quoted the work of

Wenzel. Johnson and Dettre at the Bristol Conference (2) derived the contact angles for given surface tensions for a number of model surfaces. Figures 4 and 5 illustrate a few of the basic models. Particularly well documented is the fact that adverse entrance angles into the capillaries can change small

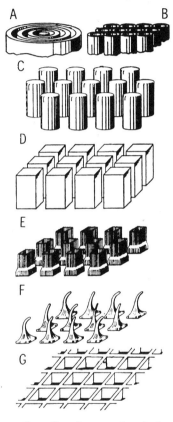

FIG. 4. Surface configurations for several typical porous surfaces (2).

contact angles to beyond 90°, a fact well known to the glass fiber laminate industry. The conference proceedings should be consulted with respect to many other aspects also.

Roughness. This discussion permits us to arrive at a rather multisided answer to the old question of whether smooth or rough surfaces are more conducive to good adhesion. Firstly, we have just seen that the geometry of the roughness can help or hinder wetting. Secondly, I have postulated that the

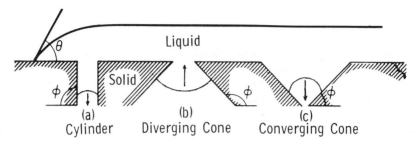

Fig. 5. Effect of pore shape on capillary pressure. Arrows indicate direction in which liquid wants to flow (2).

porosity of phosphate coatings permits interlocking with the polymer to take place, changing the locus of any failure process and rendering it a mostly cohesive one; this ought to be true for any rough porous surface. Thirdly, it has been pointed out (7) that plane surfaces have the disadvantage that all existing wetting flaws are lined up in one plane, so that any crack (Fig. 6) starting from one wetting flaw can readily propagate via other flaws in the same plane. This cannot be the case for a rough surface. Finally, the argument that roughnesses enlarge the surface areas in contact, wetting permitting, is a largely fallacious one because, although the deployment of intermolecular forces is enlarged, slip planes inclined toward the direction of separation have been introduced which reduce the areas of the strong dilative resistance to joint separation. Summarizing these factors, one can say that, provided that roughnesses are not conducive to pocket formation, and especially when there can be interlocking, rugosity of a surface can be an effective factor in the formation of a strong interphase in depth.

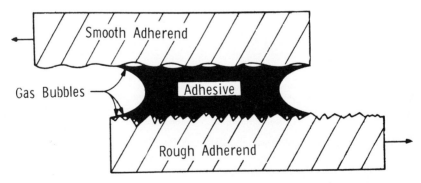

Fig. 6. Effect of surface roughness on coplanarity of gas bubbles (8).

Abhesion. Yet another important factor has not been referred to at this Symposium. By virtue of the special force field at interfaces, adsorbed molecules become as a rule strongly oriented, always with the lowest-energy groups in the molecules toward the less polarizable phase, usually the outer, or gas, phase boundary. This means, as follows from the concept of strata hierarchy, that purposely or accidentally adsorbed molecules will give rise to low-energy surfaces. Thus, if a fluid advances a molecular foot with, say, the paraffinic portion of the molecules uppermost, this will amount to a layer over which the very same liquid will find it difficult to advance because of the adverse contact angle of the bulk liquid on a paraffin layer (Fig. 7). Such processes have

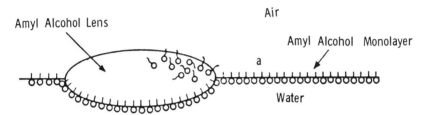

FIG. 7. Amyl alcohol does not finally spread over film-covered (9) water at a because the nonpolar CH_3— groups, oriented upward, are slightly less polar than is the amyl alcohol: the latter coheres to itself more readily than it will adhere to the oriented film at a.

been called auto-abhesive by Zisman (10), and they may seem to defy all rules of wetting. The best help in such cases is small amounts of additives, which are more strongly adsorbed but present less forbidding surfaces when applied, or when advancing, as monolayers.

Not only abhesive, but other strongly preadsorbed impurities as monolayer or multilayer, very often water, can prevent the spreading of an otherwise suitable coating. Zisman (10) has shown that it is often possible to remove this barrier layer by placing another material on top that will spread over it. The sidewise advance of this substance will exert sufficient tangential friction on the abhesive layer to drag it also with it and thereby replace the abhesive layer by one of the spreader (Fig. 8). If the latter is more lightly adsorbed, or volatile, or more soluble in the intended coating, it has then effectively acted as a cleaner.

Polymer Adsorption as Basis of Adhesion

The Structure of the Adsorbed Layers. If we assume a polymer that wets a surface and advances over it, it is important to understand on a molecular

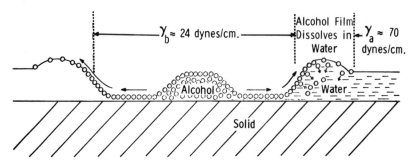

FIG. 8. Displacement of adsorbed water layers from a solid surface by spreading one of the higher alcohols over it.

scale the manner by which the polymer becomes attached. Stromberg's paper has presented a comprehensive review, so I shall emphasize merely a few particular points. Stromberg has stated, and our work (*11*) has given the same results by different methods, that polymer molecules are adsorbed from solution more or less as random coils; that is, they are anchored to the interface by a number of segments and with the intervening loops of the

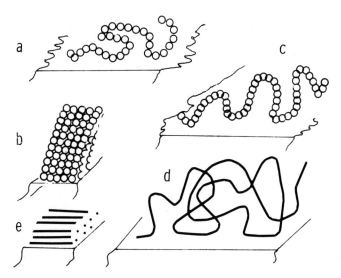

FIG. 9. Hypothetical conformations of chain molecules (*11*) adsorbed on solid–liquid interfaces: (*a*) Lying on surface. (*b*) Standing on end. (*c*) Looping. (*d*) Coiled. (*e*) Flat multilayer. *a*, *c*, *d*, single chains adsorbed with decreasing affinity; *b*, *e*, condensed surface layers.

polymer extending into the solution (Fig. 9). No one has as yet determined, or even found a way of determining, the adsorption from a polymer melt. We know, however, that the average conformation and therefore also the average energetics of polymer coils in melts are those of molecules dissolved in theta solvents (*12*). There is no reason to believe that the adsorption steps from a melt will be qualitatively different from what was found on adsorption from theta solvents. However, this first monolayer will lack the usual specific structures of condensed monolayers of small molecules, and moreover, since the adsorbed molecules and the "solvent" molecules are of the same kind, they will freely interpenetrate. Consequently, we can assume that there is not merely one first adsorbed monolayer of single theta coils, but really a system of interpenetrating coiled molecules in an adsorbed layer of approximately bulk density. Figure 10 gives a schematic impression of the difference in polymer adsorption from solutions and from a melt. Last, but not least,

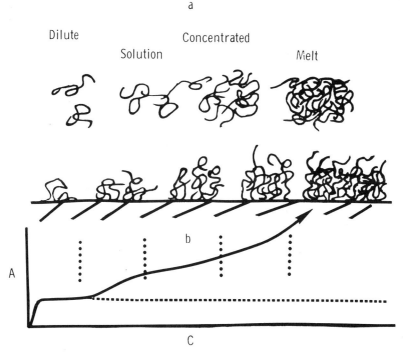

FIG. 10. (*a*) Conformation and coil interpenetration of polymer chains at the interface as a function of polymer solution concentration, or from melt (basic conformation corresponds to *d* of Fig. 9). (*b*) Corresponding adsorption isotherms.

the multipoint adsorption of polymer molecules on sites up to 100 or more angstroms apart renders their attachment to the interface particularly insensitive to surface variations. Similarly, if the interface offered is not smooth or not quite compact, partial diffusion of chain molecules into most adsorbed substrates, with parts remaining protruding, allows chain polymer molecules to become very intimately tied, on a molecular scale, into any kind of surface structures.

There is another interesting aspect. The first dense polymer monolayer has no definite boundary against the next layer by virtue of the very variable extension of individual molecules away from the surface. Extended molecules reach into the next layer, or even into the third layer, and, vice versa, molecules from the "interior" of the melt will penetrate into the "adsorbed" layer. This continuous interweaving and intermeshing over several hundred angstroms' distance, between molecules which have segments truly absorbed and others which are merely entangled with adsorbed molecules, is unique with polymers and helps to build a substantial interphase.

Reversibility and Displacement. A special aspect of polymer adsorption is its very slow reversibility (*11*) due to the required simultaneity of desorption of many anchoring points. The desorption rate will decrease further during the cooling or setting stage, and, since polymers have smaller volume contraction during these transitions than most other organic substances, they can be counted on to maintain better molecular contacts during solidification than do wetting layers of better packing and faster equilibrating materials.

Turning to yet another aspect, we have seen again at this conference how turbulent are phase boundaries and adsorbed layers. Even though damped by high viscosity and modified by viscoelasticity, the adsorbed polymer layers will be no exception. We can therefore count on a continual interchange of adsorbed and adjacent molecules and on a moving toward equilibrium, such that molecules of greater affinity to the surface will become enriched there. Thus, notwithstanding the pronounced adsorption hysteresis, polymers do exchange at interfaces. The case of polystyrene on chromium surfaces where the self-exchange was measured with the help of labeled polymers has been mentioned by Stromberg. Other workers (*13*), and we ourselves, found that exchanges take place between different molecular-weight classes and also between different polymers. The former shows as a rule a small bias in favor of larger molecular weights on smooth surfaces (*11*) and the reverse on microporous surfaces (*14*). Displacement by other polymers depends on all the factors active in polymer adsorption—that is, on all the energy and entropy changes accompanying the required desorption of solvent from surface and polymer, and the various adsorption steps of the latter. As a general rule, the polymers which are less soluble in the liquid phase, and those more capable of forming polar or hydrogen bonds with the solid phase, will displace the incumbent molecules. For instance, acrylics carrying carboxylic acid groups will displace their esters on polar surfaces which in turn will displace polystyrene. The reverse will be true on nonpolar surfaces and more polar solvents. The fundamental role played by the dielectric constant in determining adsorption affinities, at least in dilute systems, has been worked out by Frisch and Stillinger (15).

Effects of Affinity. An important aspect of polymer adsorption is that the thicknesses of the adsorbed layers vary in an unexpected manner with adsorption affinity as established by the initial slopes of the adsorption isotherms (16). As the affinity increases from zero to very high values, the adsorbed thicknesses go from zero through an initial maximum, where only a few segments per molecule are anchored and very long loops extend into the fluid. The thicknesses then decrease as the affinity goes up and/or the polymer solubility decreases, until the theta situation is reached when the thickness is at its low point notwithstanding the large amounts adsorbed (Fig. 11), equivalent to about three to eight complete monolayers of individual segments.

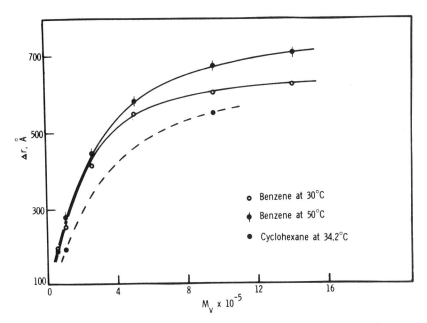

FIG. 11. Film thickness as a function of molecular weight: polystyrene in benzene and cyclohexane. The amounts adsorbed increase from the upper to the lower curve by about 50%. The dashed curve refers to a theta solvent (11). The adsorbent is Pyrex No. 7740 glass.

This is therefore also the point at which the layer adsorbed from solution is densest. For still greater affinities the situation becomes unstable and precipitation on the surface takes over.

Again, there is little reason to doubt that a similar situation occurs during adsorption in the bulk state, where we will find not so much variations of density, which is rather fixed by molecular forces, but varying degrees of molecular interpenetration between layers as a function of polymer affinity. A medium affinity should produce the best results (smallest discontinuity between adsorbed and entangled molecules) from the point of view of coatings.

Effects of Impurities. Exchanges on adsorption sites will take place not only between molecular-weight classes and different polymer species, like varying degrees of branching, blocking, or for different end groups, but also for any impurities which are present in the polymer. If a bulk polymer wets a surface, it does so always in competition with whatever impurities (including the solvent) are lodged at the original interface and now become partly dissolved in the polymer. The latter contains its own impurities. These may not possess a large affinity for the substrate surface proper, yet have a large affinity for the polymer–substrate interface, collect there, form an interlayer, and thus sever the polymer–substrate bond. This process in itself would not damage the joint even if the interlayer were abhesive, as long as molecular contact between this interlayer and the polymer, or substrate, were perfectly maintained. It might even help to relax frozen mechanical strains. An abhesive layer provides, though, ready self-lubrication on all slip and shear planes, which allows the migration and merging of flaws and, during mechanical work, the separation of the faces at points of stress concentration, such as at corners, edges, cracks, and the outer rims.

External impurities, say water or organic vapors, which are soluble in the polymer, will be replenished from the surroundings including the atmosphere while migrating to, and collecting at, the interphase. Thus, poor barrier properties of a coating will always lead to rapid destruction of the bond even if it was strong initially. The presence at, or access to, the interface of certain detergents, or of complexing agents such as nitriles and ketones, and of multihydric acids, alcohols, or amines, may prevent polymer adsorption completely.

Chemisorption. It is in this respect that scattered areas of chemisorption, say the presence of a percentage of polymers with acidic or complexing functions, help strongly. First layers of polyelectrolytes, polyalcohols, or polybases form particularly good primers. This has been the principle of the polyvinylbutyral-based wash primer, of most coupling agents, and for preparing tooth surfaces for improved adhesion of dental resins (*17*). Here phosphonated or sulfated polymers are supposed to bond and chelate with the calcium ions of the hydroxyapatite (Fig. 12). Thus, priming by chemisorbed polyelectrolytes, of one sign or by pairs of opposite charge (Fig. 13), opens the possibility of depositing water-borne organic coatings or lacquers much in the manner of latex paints.

The action of chlorinated rubbers as primers on metal or metal oxide surfaces is somewhat different. There, the electron-rich chlorine atoms act sufficiently well as Lewis bases to become strongly adsorbed on surfaces containing metal ions. Such a rubbery overlayer offers space for interdiffusion from the next layer of polymers.

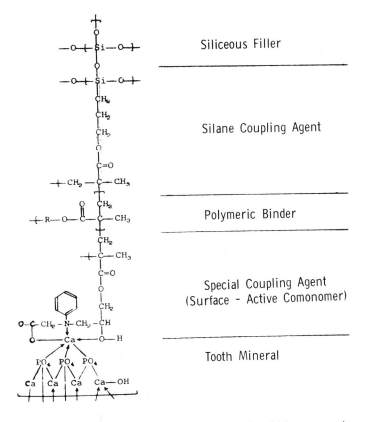

FIG. 12. Hypothetical, schematic diagram of manner in which a composite coating might be held to substrate. The figure shows the example of a filled polymer attached to a hydroxyapatite surface (*17*).

It follows that there are at least two effective principles of surface conversion: (1) a preadsorbed or chemisorbed polymeric undercoating which either reacts with the polymeric overylayer, or allows interdiffusion to a similar extent as if the polymer proper had become the primary adsorbate; (2) an extensive chemical attack on, or conversion of, the original substrate surface to turn it

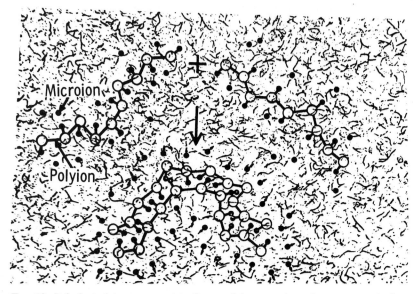

Fig. 13. Polyelectrolytes may react to form polyion complexes in a manner similar to the meshing of a zipper (*18*). Shown in plane view is a thin polyion complex layer over a surface. The ions shown in black are negative, the others positive. In three dimensions chains not only intermesh but twist and will complex with substrate and polar (polymer) overlayer.

into a porous, firmly anchored, fixed bed which holds the polymer by interlocking. There may also, of course, be the combination between the two, namely, a priming of the pore walls such as to reduce the contact angle and enhance interpenetration. Similarly, and even more critically, there should not be any abhesive contamination within the pores, nor should any arrive there during the wetting process.

The Strength of Interphases

It remains to discuss one more of the important group of factors in the formation and the life of a coating, namely, the stresses and strains which develop in the interphase during aging and usage conditions. It is important to understand these stresses and to anticipate their effects in building preventive or strengthening elements from the beginning into the interphase. Concerning rather flat interphases, I mentioned the many sources of mechanical and thermal stresses and might add now the consequences of chemical attack which usually leads to a cross-linking and/or a degradation of the polymer molecules and therefore to a mechanical weakening. We are further faced with the contradiction that the bonds between the various strata should be strong, yet they should permit tangential stress relaxation. This can be combined, as mentioned before, by having an elastomeric interphase which is covalently anchored on the substrate side, is almost equally permanently anchored by entanglements on the polymer side, and is capable of tolerating substantial shear and biaxial tension and undergoing subsequent relaxation which dissipates the energy stored in the interphase. In other words, I wish to emphasize the essential advantages of a viscoelastic dissipation in some joint interfaces. A second possibility would be to have an interphase layer of strong, but rapidly exchanging bonds as they exist in the glide planes of a metal in plastic deformation. In polymers, the mechanical rupture and re-forming of largely polar cross-links has been observed at higher temperatures (19), and this principle has been utilized extensively when creating the class of so-called ionomers. There, polar, ionic, and salt bonds are bunched in such a way as to permit interchange of linkage partners during mechanical gliding. One might consider accordingly a superficial ionic double layer of polyacids and bases, rather similar to the ionic silicate layers, or some layer crystals, and in contrast to the nonlocalized bonding to an abhesive slip layer.

In the case of the anchoring of the polymer layer by interlocking, the situation is still more complex, and our growing knowledge of the micromechanics, stress analysis, and mechanical performance of multiphase systems has to be brought to bear on the problem. Once again the boundary layer between the penetrating polymer and the pore walls ought to be viscoelastic. The penetrating polymer itself should not be notch-sensitive. The pores should funnel out away from the substrate boundary, and there should be as few sharp corners, edges, etc., as possible that could give rise to stress concentrations. For the same reason the pore penetration should be as complete as possible and free from air pockets, and the polymer should set as slowly as possible to reduce thermal stresses.

It may be permissible to formulate these postulates rather than to give technical descriptions of materials and situations, since they have been rather extensively explored—for example, by the glass fiber laminate industry

(*20*), in the cases of metal-fiber reinforcement (*21*), by *in situ* fiber formation (*22*), and in the cementing or coating of foams and fabrics (*23*). The recent book by Kraus (*24*), "The Reinforcement of Elastomers," and that by Holliday, "Composite Materials" (*25*), and the new periodicals, *Journal of Composite Materials* and *International Journal of Fracture Mechanics* (*27*) and others, should be scanned to bring the most up-to-date theory and data to bear on the overall problem of the stress concentrations at primed and coated surfaces and on the nature and performance of "bonding."

Polymer Property Requisities

Finally, I wish to refer to the least understood factor. A successfully constructed interphase moves the failure locus into the polymer, the strength of which now becomes the limiting factor. We should then consider the type of polymer which not only has good adhesive properties, good barrier properties, and slowly varying properties with temperature, but also exhibits the required moduli; deformability and strength. Again, there is extensive literature on any of these properties or functions of polymers, but little on such delicately balanced, often conflicting, multiple-purpose responses as are needed for a good "coating." Much of the challenge of the future in the coatings field will indeed come in the form of designing such special-purpose resins. These can hardly be simple, because simple materials cannot fulfill a multitude of requirements. Thus, one has to weigh the properties of tactic, atactic, block, and graft copolymers and heteropolymers, effects of degrees of cross-linking and crystallinity, of polyblending, of filled resins, etc.

After the foregoing requirements, the most essential requisites for bonding include, firstly, the presence of some types of molecules or portions of molecules with the ability to adsorb and flow readily, to set slowly, to remain amorphous, and thereby to mediate the adsorption–adhesion stresses. Next, there must be present domains which provide the modulus and strength and thus consist of either crystalline areas or filler particles. Next, there must be molecular sections permitting rather close packing without leading to true crystallization, but rather to paracrystalline or nematic structures, so as to provide barrier properties without brittleness. To provide chemical stability, there must be a minimum of unsaturation, of labile hydrogens, of reactive acid, basic, amide, and ester groups, and of reactive carbonyls. All of this indicates the application of either polymers of the desired chemistry—say, some blends of chlorinated rubber, butyl rubber, and a partly crystalline EPR rubber, or blends of a butylmethacrylate and ethylmethacrylate, plus a a copolymer of acrylic acid, styrene, and methylmethacrylate with addition of atactic methylmethacrylate—or blends of block or graft polymers—say, of an acrylonitrile grafted on styrene with blocks of isobutylene on styrene. Alternatively, one will have to resort to new combinations such as chlorhydrin–polyphenoxy blocks, vinylidene cyanide–novolak, silicone–polycarbonate, epoxy–phenazine, and similar combinations.

The recent spate of advanced physical testing of a wide variety of polymers in efforts to correlate structures with properties has shown convincingly, at

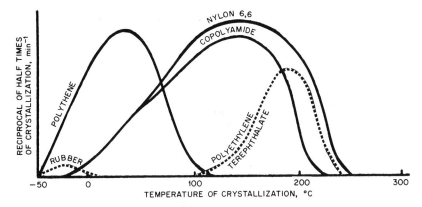

FIG. 14. Reciprocal crystallization half-times for various polymers (28).

least to me, that one of the most important bases to give a broad variety of desirable properties is the ready ability of a polymer to undergo cold flow. This ability renders a material capable to work-harden at large strains rather than to fail, to absorb energy rather than to store it elastically. This material then, able to relax, will not suffer from the stress concentrations at the boundary between the porous primed surface and at the tip of flaws. In its ability to organize into ordered domains, such a material also suffers relatively less from impurities. These materials can be recognized best by spinnability, large areas under their stress–strain curves, and wide ranges of crystallization temperatures and rates, the latter exemplified by Fig. 14.

Conclusion

In conclusion, I have tried to show in this survey, first, how the individual papers presented at this conference have contributed to our knowledge of interface conversion for polymer coatings and where, within the wide area of problems encompassed by this theme, they make their individual contributions.

Next, I have tried to provide some of the rationale which makes it more readily apparent how the individual papers contribute to a major progress in our thinking. In addition, I have touched upon a few areas which could not possibly be treated fully within the time limits of this conference. It may now be permitted by means of a number of qualitative statements and tolerable oversimplifications to render a sketch of what we know, and what we would like to know and do, to obtain stronger and better interfacial bonds and coatings.

All artificial coatings have been stated to compete with indigenous, or subsequently spontaneously forming, surface strata and structures. Polymers are well suited as coating materials because of their viscoelastic nature and ability to dissipate accumulated stresses. Cross-linked elastomeric polymers are particularly good interlayers because with their help one can build a substantial interphase which may be strongly anchored on either boundary and yet prevent stress concentrations. The tendency of polymers to attach themselves by adsorption of extended coils which become entangled with nonadsorbed bulk polymer is recognized as being particularly helpful for the formation of strong interphases.

The practice of phosphate priming earns its success from the fact that it yields highly porous surface structures which, besides favorable adsorption, permit also an interlocking of the adjacent polymer phases and a reinforcement of the interphase thus created. However, the porous phosphate coating should not contain too many bottle-shaped or diverging pores, and it should not be too rugged geometrically. The rate of penetration into porous beds has been found to be proportional to the wetting energy, and inversely proportional to the viscosity and to a characteristic length defined by the pore structure Accordingly, one should aim to select polymers with surface tensions in the fluid state as large as is compatible with a small contact angle, with as small a viscosity as is compatible with good mechanical properties of the solidified coating, with as small a pore size of the converted surface as is possible without creating too much viscous resistance, and with a pore geometry not detrimental to developing small wetting angles.

Particular attention must be paid to autoabhesion of the polymer and to impurities inimical to polymer adsorption by competition or abhesion. Preferably, the polymer should have good barrier properties to prevent additional access of abhesives and also should have good solubility for surface contaminations that are thereby removed by what might be termed a self-cleaning process.

The polymer should be allowed to "set" as slowly and with as little volume change as possible, and, preferably, the coating should be annealed. The adsorption process can advantageously be strengthened by polymeric additives which carry strongly chemisorbing groups and help the polymer to compete successfully with, or to displace, abhesive competitors.

The mechanical properties desired for a successful coating come close to those found in so-called spinnable resins, which exhibit also appreciable cold flow. The many contradictory performances often demanded from a coating should best be fulfilled by partly rubbery, partly glassy, and partly crystallizable block copolymers, or by polyblends. The formation of the interfacial bond as well as the subsequent behavior of the coating is best understood by considering the coating as a partly composite, partly laminated material, and by drawing from the extensive research in all areas of composite systems.

References*

1. F. R. Eirich, NONR-1129(00), Project NR 036-012 (June 1957).
2. "Wetting," *Soc. Chem. Ind. (London), Monograph* **25** (1967); (see especially the paper by R. H. Dettre and R. E. Johnson, Jr.).
3. J. C. Melrose and W. H. Wade, in preparation; J. C. Melrose, *Soc. Petrol. Engrs. J.*, **234**, 259 (1965); N. Hackerman and W. H. Wade, *J. Phys. Chem.*, **69**, 314 (1965).
4. D. H. Everett, *Trans. Faraday Soc.*, **57**, 1541 (1961).
5. H. Schonhorn, H. L. Frisch, and T. K. Kwei, *J. Appl. Phys.*, **37**, 4967 (1966).
6. J. Frenkel, *J. Phys. (USSR)*, **9**, 385 (1945).
7. "Surfaces and Interfaces," ed. by J. J. Burke, N. L. Reed, and V. Weiss, Sagamore Army Materials Research Conferences, 1966; "Adhesion and Cohesion," ed. by P. Weiss, Elsevier, Amsterdam, 1962; J. J. Bikerman, "The Science of Adhesive Joints," Academic Press, New York, 1961; see articles on structural adhesive bonding, "Applied Polymer Symposia, No. 3," ed. by M. J. Bodnar, Interscience, New York, 1966; M. L. Williams, "Stress Singularities, Adhesion and Fracture," *Proc. 5th U.S. Natl. Congr. Appl. Mech., A.S.M.E.* (1966); J. R. Huntsberger, in "Treatise on Adhesion and Adhesives," Vol. I, ed. by R. Patrick, M. Dekker, New York, 1967; R. Irvin, *ibid.*
8. W. A. Zisman, *Ind. Eng. Chem.*, **57** (1), 27 (1965).
9. J. T. Davies and E. K. Rideal, "Interfacial Phenomena," Academic Press, New York, 2nd ed., 1963, p. 24.
10. W. A. Zisman, *Advan. Chem. Ser.*, **43**, 1 (1964); H. R. Baker, P. B. Leach, C. R. Singleterry, and W. A. Zisman, *Ind. Eng. Chem.*, **59** (6), 29 (1967).
11. F. Rowland, R. Bulas, E. Rothstein, and F. R. Eirich, *Ind. Eng. Chem.*, **57** (9), 46 (1965); F. Rowland and F. R. Eirich, *J. Polymer Sci.*, **A1**, 2033, 2401 (1966); R. Stromberg, see references in his paper, this volume.

* I wish to acknowledge the benefits derived from attendance at many conferences on adhesion, and the contributions from many colleagues to the evolution of the ideas here presented. Even a very incomplete list of references would fill many pages; therefore the rather arbitrary number quoted here should be consulted further for the pertinent literature.

12. P. J. Flory, "Principles of Polymer Chemistry," Cornell University Press, Ithaca, New York, 1953, p. 602.
13. C. Thies, P. Peyser, and R. Ullman, *Proc. 4th Intern. Congr. Surface Activity, Brussels* (1964); G. Steinberg, *J. Phys. Chem.*, **71**, 292 (1967).
14. S. Claesson, *Discussions Faraday Soc.*, **7**, 321 (1949); H. Jellinek and H. Northey, *J. Appl. Polymer Sci.*, **3**, 26 (1960).
15. H. Frisch and F. Stillinger, Jr., *J. Phys. Chem.*, **66**, 823 (1962).
16. A. Adamson, "Physical Chemistry of Surfaces," 2nd ed., Interscience, New York, 1967.
17. "Adhesive Restorative Dental Materials," Proc. 1st Workshop, National Institute Dental Research, 1961; "Adhesive Restorative Dental Materials," Proc. 2nd Workshop, National Institute Dental Research, 1965.
18. A. S. Michaels, *Ind. Eng. Chem.*, **57** (10), 32 (1965).
19. A. Tobolsky, "The Properties and Structure of High Polymers," Wiley, New York, 1960; E. Otocka and F. R. Eirich, *J. Polymer Sci.*, **A2**, 6, 895, 913, 921, 933 (1968).
20. "Glass Fiber Reinforced Plastics," ed. by A. de Dani, Interscience, New York, 1960; K. L. Loewenstein, in "Composite Materials," ed. by L. Holliday, Elsevier, New York, 1966.
21. N. J. Grant, in "The Strengthening of Metals," ed. by D. Peckner, Reinhold, New York and London, 1964, p. 163; R. W. Guard, in "Strengthening Mechanisms in Solids" American Society of Metals, 1962, p. 253.
22. G. C. Smith, *Powder Met.*, **11**, 102 (1963).
23. G. C. Grimes, in "Applied Polymer Symposia, No. 3," ed. by M. J. Bodnar, Interscience, New York, 1966, p. 157.
24. "The Reinforcement of Elastomers," ed. by G. Kraus, Interscience, New York, 1965.
25. "Composite Materials," ed. by L. Holliday, Elsevier, New York, 1966.
26. *Journal of Composite Materials*, ed. by S. W. Tsai, Technomic Publishing Co., Stamford, Connecticut.
27. *International Journal of Fracture Mechanics*, ed. by M. L. Williams, P. Nordhoff, Ltd., Groningen.
28. H. Leaderman, *Lecture Notes on High Polymer Physics*, Tokyo, 1957.

Discussion

COWLING: The very properties that are desirable for achieving a strong bond between the matrix of the adhesive and the substrate, especially pertaining to displacing and resisting subsequent attack with water, frequently render that matrix susceptible to attack by water over the long term. To make strong adhesives generally involves introducing more polarity, and this makes them more susceptible to water. Therefore, it is necessary to look at the total problem of the bonded joint, rather than the interface.

EIRICH: Not only that, but, as one makes the bond stronger, one enlarges the inherent and latent stresses, strains, and cracks. Water will find its way through the slightest fissures and destroy the bond. Epoxies contain polyesters and additional polar groups and are susceptible to hydrolysis, just as all polyesters are unless highly crystalline. Epoxies cannot be crystallized, only cross-linked, and, therefore they are basically susceptible to hydrolysis and degradation unless the cross-link density is high, while at the same time residual stress relaxation can prevent the development of cracks. In other words, epoxies owe their resistance to the fact that they are highly cross-linked and yet somewhat elastic.

KUMINS: We studied crystallite fillers dispersed in a polymeric binder in order to increase the surface area, and we actually showed a decrease in the density of the polymer. If one determines the density of the composite, and if one assumes that the density of the crystallites remains constant, it turns out that the polymer is actually lower in density when it is in contact with the substrate than it is by itself. Now, I do not deny that there is a contraction in the bulk of the adhesive or coating film, but I would like to point out that at the interface it is possible to have an expansion of volume, and this can, in some cases counteract the stresses that are induced. Now, the question is, what type of surfaces do this? In our experience, we have investigated hydrophilic and hydrophobic surfaces, and we found this expansion phenomenon to occur on both types of surfaces. We have not studied metal surfaces.

EIRICH: What you say is in many cases true, but you will find that those are usually cases of rather high volume loadings of the filler. If you go to only 1 or 2% volume, this effect is not readily found. At high volume loadings, above 10%, the material cannot flow and/or contract as it wants during the mixing process, and therefore one creates a certain amount of voids. Moreover, the polymer is in a state of strain and is slightly expanded because the filler disturbs the normal packing of the chains. Thus, there are all kinds of effects which increase the volume at high filler loadings, because one replaces bulk volume of the resin by strained interfaces between the resin and the particles.

KUMINS: We have noticed this effect with loadings as low as 1% by volume. I believe that, as the adhesive bond thickness approaches the thickness of the "interphasial" layer, both the modulus is increased by the organization of matter which occurs there, and the number of stresses and strains around that interface is reduced.

ASBECK: We have done a little work in this area and find that another mechanism can readily explain these effects. This is due to the fact that when an adhesive bond is tested, for instance in a pure tensile test, the sample is tested at some arbitrary jaw separation rate. Now, it is obvious that, the thinner the adhesive joint, the shorter will be the actual test piece (the adhesive) and the more rapid will be the true extension rate. In other words, the thinner the joint, the faster the stress is applied, and it is proportional to the arbitrarily chosen machine speed. There has been a lot of work done and reported in the literature on the correlation between time and temperature effects on the tensile strength of polymers. Usually, the higher the stress rate, the higher is the tensile strength. This can encompass several orders of magnitude of tensile strength over a broad range of applied stress velocities [J. J. Lohr, *Trans. Soc. Rheol.*, **9** (*1*), 65 (1965)]. Exactly the same thing seems to be true with the addition of pigment or filler to the adhesive. As pigment is added, the effective sample length which is stressed (the polymer matrix) is shortened. As a consequence, if the highly pigmented system is stressed at the same jaw separation rate as an unfilled sample of the same polymer adhesive matrix, it is stressed at a very much higher rate. As a matter of fact, at the critical pigment volume content, the stress rate of the polymer binding the pigment particles together can become infinitely high [W. K. Asbeck, Preprint Booklet, Division of Organic Coatings and Plastics Chemistry, American Chemical Society, **26,** No. 2, 13 (1966)].

KUMINS: This should be taken into account by running the tests at different strain rates.

ASBECK: This is not necessarily so, since the effect of pigment addition on true strain rate can be much greater, up to five or six orders of magnitude, than the strain rate variations achievable on a conventional tensile machine. For instance, if the tensile strength of a series of coating compositions, loaded with increasing pigment content, is determined at a fixed jaw extension rate, the tensile strength, particularly of the softer polymer compositions, often is found to vary by as much as 1000:1 in going from no pigment up to the critical pigment content. Beyond this point, air becomes a part of the coatings composition and a decrease in tensile strength is encountered to a logical value of zero at 100% pigment. As a matter of fact, the PVC (pigment volume concentration) ladder test is commonly used to determine the CPVC (critical pigment volume concentration) of coatings compositions such as latexes which cannot be readily measured by the filter cell technique.

KUMINS: This still does not explain the role of the structure around an interface which can be as much as 8000 Å in thickness.

ASBECK: You do not need the concept of structure to explain this particular phenomenon. Although the structure is undoubtedly there, as is well-documented by your experiments, this cannot alone account for the effect, since it would require, under some circumstances, that the structured polymer be one hundred or more times as strong as its unstructured counterpart.

GUTH: Dr. Eirich did not mention the possible significance of theories of polymer adsorption, which he himself pioneered. Adsorption of polymers is, supposedly, the primary step in the formation of polymeric interfacial bonds, be it in organic adhesion or in coatings, and even in biological membranes. The first theory by Frisch et al. [H. L. Frisch, R. Simha, and F. R. Eirich, *J. Chem. Phys.*, **21**, 365 (1953); *J. Phys. Chem.*, **57**, 584 (1953); H. L. Frisch and R. Simha, *ibid.*, **58**, 507 (1959)] implies partial adsorption of the polymer. Short sequences of segments are held on the surface, and random loops extend into the solvent. The later theories by Silberberg [A. Silberberg, *J. Phys. Chem.*, **66**, 1872 (1962)], DiMarzio et al. [C. A. J. Hoeve, E. A. DiMarzio, and P. Peyser, *J. Chem. Phys.*, **42**, 2558 (1965); E. A. DiMarzio, *ibid.*, **42**, 2101 (1965); E. A. DiMarzio and F. L. McCrackin, *ibid.*, **43**, 539 (1965)], Roe [R. J. Roe, *Proc. Natl. Acad. Sci. U.S.*, **53**, 50 (1965); *Bull., Am. Phys. Soc.*, **10**, 381 (1965)], and Rubin [R. J. Rubin, *J. Chem. Phys.*, **43**, 2392 (1965)] assume almost total adsorption of the polymer and are random-walk lattice models. Long sequences of segments are held on the surface with comparatively little looping into the solvent. While these and related theories do not yet permit one to predict which coatings will be best in particular cases, it is hoped that they will contribute to the understanding of the mechanisms of interface conversion for polymer coatings.

AUTHOR INDEX

Numbers in parentheses indicate the numbers of the references when these are cited in the text without the names of the authors.

Numbers set in *italics* designate the page numbers on which the complete literature citation is given.

Acrivos, A., 84(6), *88*
Adamson, A., 365(16), *374*
Albrecht, E., 156(31), *176*
Alfrey, T., 69(14), *79*
Anderson, J. H., 13(28), 15(28), *51*, 222, 223(8), *232*
Andrew, K. F., 240(7), 241(8), 243(7), *256*
Arevalo, A., 144, *145*
Armstrong, R. A., 219, *232*
Artmann, K., 207, *207*
Asbeck, W. K., 376, *376*
Atlas, S. M., 64(11), *79*

Baird, W., 37, *52*
Baker, H. R., 11(23), *51*, 361(10), *373*
Bakker, C. A. P., 87, *88*
Balandina, V., 338(14), *349*
Bangham, D. H., 152(17), *176*
Bardolle, J., 238(1), 240, 241, *256*
Barrer, R. M., 61(1), *79*
Bartell, F. E., 150(1), *176*, 223(13), *232*
Bascom, W. D., 152(18), 153, 172, *176*
Bass, R. L., 43, *52*
Bastick, J., 221(4), *232*
Baxendale, J. H., 338, *349*
Beek, W. J., 87(15), *88*
Bénard, J., 238(1, 2), 240, 241, 242, 243, *256*
Benedict, M., 62(6), *79*
Berezan, K., 338(14, 15), *349*
Berg, J. C., 84, *88*
Berghausen, P. E., 224(16), *232*
Bernett, M. K., 11(24), *51*
Bielak, E. B., 152(15), *176*
Bikerman, J. J., 3(1, 2), 4, 13(29), *50*, *51*, 153(23), 157, *176*, 360(7), *373*
Bito, T., 338, *349*
Blokhuis, G., 156(33), *177*
Bloor, D. W., 182(2), 183, *200*
Bogen, R. H., 63(9), *79*

Bobalek, E. G., 290(1), 296(6), *313*
Bolger, J. C., 3, 20, *52*, 81, *81*
Bondi, A., 151(11), *176*
Bonetskaya, A. K., 226(21), *233*
Bosanquet, C. H., 156(29), *176*
Boucher, E. A., 150(9), *176*
Boudart, M., 84(6), *88*
Bovey, F. A., 338, *349*
Bowden, F. P., 13, 14, *52*
Boyd, G. E., 14, 20, *52*
Brian, P. L. T., 87, *88*
Bridgeford, D. J., 291(4), 297(4), 311(4), 312, *313*
Brinkman, H. C., 76, *80*
Brittin, W. E., 221(5), *232*
Broetz, W., 63(9), *79*
Broge, E. C., 222(6), *232*
Brooks, C. S., 13, *51*
Brooks, L. H., 84, *88*
Brunauer, S., 66, 67(13), *79*, 225, *232*
Bulas, R., 362(11), 364(11), *373*
Buob, K. H., 185(8), *200*
Bursh, T. P., 223, 225, *232*
Byrne, B. J., 76(24), *80*

Calderwood, G. F. N., 77(26), *80*, 156(36), 171(36), *177*
Carman, P. C., 13(30), *51*, 61(1), *79*
Casey, J. P., 78(28, 30), *80*
Caule, E. J., 185(7, 8), *200*
Cheever, G. D., 14, *52*, 77, *80*, 136, *145*, 195(10), 196(12), 197, *200*
Chessick, J. J., 15(53), *52*, 229(30, 32), 230(33, 34, 35, 37), *233*
Cholnoky, L., 62(5), *79*
Chowdhury, J. K., 346, *349*
Chujo, K., 338, *349*
Cibulka, B., 126(2, 3, 4), *127*
Claesson, S., 364(14), *374*

Clark, M. W., 87, *88*
Cohen, M., 185(6, 7), *200*
Collie, B., 150(4), *176*
Colson, J. P., 324, 325, *335*
Columbo, A. F., 34(59), *52*
Cooke, S. R. B., 34(59), *52*
Copan, T. P., 238(5), 243, 247, 254, *256*
Corrin, M. L., 338(1), 343(1), *349*
Cote, W. A., Jr., 290(3), *313*
Cottington, R. L., 152(18), 153(18, 21), 172(18), *176*
Cupr, V., 125(1), 126(2, 3, 4, 5, 6), *127*

Darcy, H., 72, *80*
Davies, C. B., 338(17), *349*
Davies, J. T., 347, *349*, 361(9), *373*
De Benedictis, T., *283*
Debye, P. J. W., 13, 20, 22, *52*
de Mendoza, A. P., 295, 311(5), *313*
Den'shchikova, G. I., 152(19), *176*
Dettre, R. H., 355(2), 359, *373*
DiMarzio, E. A., 323(4, 7), *335*
Diserens, L., 63(8), *79*
Dobromyslowa, A., 338(14, 15), *349*
Doede, C. M., 3(3), *50*
Dogadkin, B., 338, *349*
du Nouy, L., 339, *349*
Drain, L. E., 229(31), *233*
Dubinin, M., 346(34), *349*
Durer, A., 108(5), *119*
Duret, M., 242(11), 243(11), *256*

Egorov, M. M., 223(12, 14), *232*
Einstein, A., 76(25), *80*
Eirich, F. R., 322, 328, 333, *336*, 355(1), 362(11), 364(11), 369(19), *373*, *374*
Elder, M. E., 290(1), *313*
Eley, D. D., 156(32), *177*
Emmett, P., 66(13), 67(13), *79*
Epik, A. P., 100(4), 101(4), *119*
Epstein, N., 76(24), *80*
Evans, M. G., 338(3, 4, 5), *349*
Everett, D. H., 356(4), *373*
Eyring, H., 71, *80*

Fikentscher, H., 338, *349*
Fitch, B., 81, *81*
Fleishmann, R. M., 41(61), *52*

Flory, P. J., 363(12), *374*
Fok, S. M., 290(1), *313*
Fontana, B. J., 331, *336*
Fowkes, F. M., 7, 9, 10, 11, 14, 21, *51*, *52*, 209, *232*, 330, *336*
Frenkel, J., 356, *373*
Frisch, H. L., 153(22), 164(22), *176*, 322, *335*, 337, *337*, 356, 358, 365, *373*, *374*
Fryling, C. F., 338, *349*

Ganichenko, L. G., 226(22), *233*
Gardon, J. L., 9(16), *51*
Garrett, H. E., 150(10), *176*
Gebhardt, M., 91(2), 93(2), 95(2), 96(2), 97(3), *119*, 132(7), *145*, 194(10), 195(10), 196(10), *200*
Gee, G., 338, *349*
Ghali, E., 134, *145*
Gilbert, L. O., 183, *200*
Gillespie, T., 156(37, 38), *177*
Girifalco, L. A., 7, 9, 10, *51*
Good, R. J., 7, 9, 10, 11, 20, 22, *51*, 224(16), *232*
Goodwin, E. T., 207, *207*
Goryunov, Yu. V., 152(19), *176*
Grant, N. J., 370(21), *374*
Grant, W. H., 333(21), 334(22), *336*
Gray, V. R., 150(6), 151(6), 153(6), *176*
Gregg, S. J., 150(7), 152(7), *176*
Greinacher, H., 156(39), 157, 171(39), *177*
Grimes, G. C., 370(23), *374*
Griskey, R. G., 338, 339(27), 343(27), 344(27), 345(27), 348(27), *349*
Groger, K., 87(16), *88*
Grönlund, F., 242(11), 243(11), *256*
Guard, R. W., 370(21), *374*
Guderjahn, C. A., 224(16), *232*
Gulbransen, E. A., 238(5, 6), 239, 240(6, 7), 241(8, 10), 243, 247(14, 15), 248, 254, *256*

Hache, A., 134(9), *145*
Hackerman, N., 14, 20, *52*, 223(10), 226(19), *232*, 356(3), *373*
Hagen, 70(16), *79*
Hansen, R. H., *283*
Happel, J., 76(24), *80*
Harada, Y., 338(12), *349*
Hardy, W. B., 152(16), *176*

Harkins, W. D., 152, *176*, *213*, 338(1, 21, 22, 23, 24, 25), 339(23), 341(23), 343(1), 346(1, 23), *349*
Harr, M. E., 61(1), *79*
Harrington, E. W., 338, *349*
Hartley, G. S., 346(29), *349*
Healey, F. H., 229(30), 230(33, 34), *233*
Heller, W., 337, *337*
Hickman, J., 238(6), 239, 240(6), *256*
Hildebrand, J. H., 7, *51*
Hirschfelder, J. O., 7
Hobden, J. F., 44(64), *52*
Hobson, J. P., 219, *232*
Hoeve, C. A. J., 323, 330, *335*
Hohenstein, W. P., 338, *349*
Holliday, L., 370, *374*
Holmes, B. G., 221(5), *232*
Hosler, C. L., 230(36), *233*
Hughes, E. W., 338, *349*
Huntsberger, J. R., 5, *51*, 150(2), 153(2), *176*, 360(7), *373*

Iler, R, K., 13(31), *51*
Ilkka, G., 152(12, 13), *176*
Irvin, R., 360(7), *373*
Irwin, G. R., 14, *52*
Ito, T., 114, *119*
Iwasaki, I., 34, *52*
Iyengar, R. D., 227, *233*

Jaeger, E. L., 306, 307(8), *313*
Jellinek, H. H. G., 44(64), *52*, 364(14), *374*
Jenckel, E., 321, *335*
Jimeno, E., 144, *145*
Johannson, O. K., 41(61), *52*
Johnson, R. E., Jr., 150(5), *176*, 355(2), 359, *373*
Jumpertz, E., 194(10), 195(10), 196(10), *200*

Kanamura, K., 153(24), *176*
Kantro, D. L., 225(17, 18), *232*
Keller, D. V., 13, *52*
Khomikovskii, P. M., 338, *349*
Kidd, D. J., 15(51), *52*
Kingdon, K. H., 228(27), *233*
Kintner, R. C., 87, *88*
Kiselev, V. F., 223(14), 226(7, 20, 22, 23), 227, *232*, *233*

Knudsen, M., 70(17), *80*
Kolthoff, I. M., 338, *349*
Komagata, S., 156(31), *176*
Konopicky, K., 130, *145*
Kozeny, J., 75, *80*
Krasilnikov, K. G., 223(12, 14), 226(21, 22), *232*, *233*
Kraus, G., 9(17), *51*, 370, *374*
Krieger, I. M., 290(1), *313*
Kwei, T. K., 153(22), 164(22), *176*, 356, 358, *373*

Lafratta, F. G., 311(9), *313*
Laidler, K. J., 71(18), *80*
Langmuir, I., 228(27), *233*
Lapuk, M., 338(12), *349*
Lasater, J. A., 221(5), *232*
Lavelle, J., 211, *213*
Leach, P. B., 361(10), *373*
Leaderman, H., 371(28), *374*
Lester, G. R., 150(8), *176*
Levine, J. D., 207, *207*
Levine, M., 152, *176*
Lewis, J. B., 84, *88*
Liebau, F., 113, *119*
Lind, E. L., 338(1), 343(1), *349*
Linde, H., 87, *88*
Lingafelter, E. C., 346(31, 33), *349*
Livingston, H. K., 14, 20, *52*
Lloyd, W. G., 69(14), *79*
Loesser, E. H., *213*
Loewenstein, K. L., 370(20), *374*
Lohr, J. J., 376, *376*
Lower, W. K., 222(6), *232*

McCandless, E. L., 241, *256*
McCoy, C. E., 338, *349*
McCrackin, F. L., 323, 324, 325, *335*
Machu, W., 128, 129, 131(5, 6), 132, 136, 137, 138, *145*, 182(1), 183, *200*
Mack, G. W., 153(25), 171(25), *176*
McKelvey, J. M., 290, *313*
McLean, D. A., 156(30), *176*
McMillan, W., 240(7), 243(7), *256*
Madow, B. P., 290(1), *313*
Maier, C. G., 62(6), *79*
Makrides, A. C., 223(10), *232*

Manegold, E., 156(31), *176*
Marangoni, C., 83, 84, 87, *88*
Mardles, E. W. J., 77(26), *80*, 152(15), 156(36), 171(36), *176*, *177*
Mark, H. F., 20(57), *52*, 64(11), *79*, 338, *349*
Mark, P. H., 207, *207*
Maron, S. H., 290(1), *313*
Marshall, S. N., 299, 301(7), 304, 307(7), 311(7), *313*
Marshall, W. R., Jr., 62(7), *79*
Martin, W. M., *283*
Matiatos, 87(19), *88*
Maurer, J. I., 189(9), *200*
Mayer, M., 115, *119*
Mehl, R. F., 241, *256*
Melrose, J. C., 356(3), *373*
Melville, W. H., 338(17), *349*
Michaels, A. S., 3, 20, *52*, 81, *81*, 368(18), *374*
Mitsuoka, Y., 338, *349*
Moilliet, J. L., 150(4), *176*
Montillon, G. H., 62(4), *79*
Montonna, R. E., 62(4), *79*
Mori, H., 114, *119*
Morita, H., 338(12), *349*
Morrison, J. A., 229(31), *233*
Motoyama, T., 338, *349*
Müller, W. J., 130, 139, *145*
Murina, V. V., 223(14), 226(22), *232*
Muroi, S., 338, *349*
Murphy, C. M., 153(21), *176*
Murray, R. C., 346(29), *349*
Muskat, M., 61(1), 62(3), *79*

Neff, L. L., 346(33), *349*
Nestler, C. H., 338, *349*
Neuhaus, A., 91(1, 2), 93(2), 95(2), 96(2), 111(1), *119*, 132(7), *145*, 194(10), 195(10), 196(10), *200*
Noller, C. R., 22, 29, *52*
Northey, H., 364(11), *374*

Olander, D. R., 87, *88*
Orell, A., 87, *88*
Orowan, E., 14, *52*
Otocka, E., 369(19), *374*
Osterhof, H. J., 150(1), *176*
Oudar, J., 242(11), 243(11), *256*

Pal, H. N., 346, *349*
Park, G. S., 338(3), *349*
Parks, G. A., 16, 17, 18, 19, 48, *52*
Pascale, J. V., *283*
Passaglia, E., 325(11, 14), 333(21), *335*
Patrick, R. L., 3(3), *50*
Patton, T. C., 83, *88*
Pauling, L., 8, 22, 23, 32, *51*
Paynter, D. A., 224(16), *232*
Pearson, J. R. A., 83(4), *88*
Peek, R. L., Jr., 156(30), *176*
Pelikan, J. B., 125(1), 126(5, 6), *127*
Pepper, D. C., 156(32), *177*
Pertsov, N. V., 152(19), *176*
Peyser, P., 323(4), 325(13), 328(15), 331(13, 15, 19, 20), *335*, 364(11), *374*
Pfefferkorn, G., 238(3), 243, *256*
Phelps, R., 238(6), 239, 240, *256*
Porter, M. R., 43, *52*
Poshkus, D. P., 227, *233*

Reddy, L. B., 87, *88*
Rentzepis, P. M., *283*
Rhead, G. E., 243, *256*
Rhodin, T. N., 228, 229, *233*
Rideal, E. K., 77, *80*, 156, 171, *176*, 361(9), *373*
Roe, R. J., 323, *335*
Roginsky, A., 338(1), 343(1), *349*
Rothstein, E., 362(11), 364(11), *373*
Rowland, F. W., 328, 333, *336*, 362(11), 364(11), *373*
Ruka, R. J., 241, *256*
Rumbach, B., 321, *335*
Ruth, B. F., 64(4), *79*

Saison, J., 108(6), *119*, 136, 139, 140(12), 144, *145*, 183(3), 194, 195(3), 196, 198, *200*
Salomon, G., 13(32), *51*
Samsonov, G. V., 100(4), 101(4), *119*
Saweris, Z., 152(17), *176*
Sawyer, W. M., 338(20), *349*
Scheidegger, A. E., 61(1), 64(12), 70(15), 71(19), 75(21), *79*, *80*
Schilow, N., 346 *349*
Schmid, E., 108(5), *119*

AUTHOR INDEX

Schonhorn, H., 4, *50*, *51*, 153(22), 164, *176*, 217, *232*, *283*, 356, 358, *373*
Schwab, G. M., 63(9), *79*
Schwarz, E., 87(16), *88*
Scott, A. B., 346(31), *349*
Scott, R. L., 7, *51*, 155, *176*
Scriven, L. E., 84, 85, 87, *88*
Senatorskaya, L. G., 338(8), *349*
Serebryakova, Z. G., 338(8), *349*
Sewell, P. B., 185(6), *200*
Shafran, E. G., 13, *52*
Sharpe, L. H., 4, *50*
Shchukin, E. D., 152(19), *176*
Sherwood, T. K., 84, *88*
Shockley, W., 207, *207*
Silberberg, A., 322, 323, *335*
Simha, R., 322, *335*
Singleterry, C. R., 11(23), *51*, 152(18), 153(18, 21), 172(18), *176*, 361(10), *373*
Smith, F., 15(51), *52*
Smith, G. C., 370(22), *374*
Smith, L. E., 325(12), 329(12), *335*
Snow, C. W., 156(40), 160, *177*
Spence, R., 63(9), *79*
Spengler, H., 63(9), *79*
Srinivasen, G., 230(37), *233*
Stark, F. O., 41(61), *52*
Stearns, R. S., 338, *349*
Sterman, S., 43, *52*
Sternling, C. V., 84, 85, 87, *88*
Stillinger, F., Jr., 365, *374*
Stockbridge, C. D., 185(6), *200*
Strain, H. H., 62(5), *79*
Stromberg, R. R., 12, *51*, 325(11, 12, 13, 14), 329(12), 331(13, 20), 333(21), 334(22), *335*, *336*, 362(11), *373*
Studebaker, M. L., 156(40), 160, *177*
Suggit, R. M., 223(13), *232*
Suhrmann, R., 228(28), *233*
Summ, B. D., 152(19), *176*
Sysoev, E. A., 223(12), *232*

Tabor, D., 13, 14, *52*
Takagi, R., 238(4), 243, *256*
Tanaka, W., 337, *337*
Tartar, H. V., 346(30, 31, 33), *349*
Taylor, H. S., 63(9), *79*
Tcheurekdjian, N., 230(35, 36), *233*

Teller, E., 66(13), 67(13), *79*
Ter-Minassian-Saraga, L., 13, *52*
Terzaghi, K., 62(2), *79*
Thies, C., 328, 331(15), *335*, 364(13), *374*
Thomas, J. R., 331, *336*
Thompson, J., 83, *88*
Throssell, W. R., 13, *52*
Tobolsky, A., 369(19), *374*
Tollenaar, D., 156(34, 35), *177*
Toogood, J. C., 43, *52*
Torapow, S., 346(34), *349*
Turkalo, A. M., 155, *176*
Tutas, D. J., 325(11, 13, 14), 331(13), *335*

Ulevitch, I. N., 290(1), *313*
Ullman, R., 328(15), 331(15), *335*, 364(11), *374*

Valko, E. I., 338(2), *349*
Van Beek, L. K. H., 13, *52*
van Buytenen, P. M., 87(15), *88*
Varley, C., 83, *88*
Vaughan, W. A., Jr., 3(3), *50*
Vickerstaff, T., 63(8), *79*
Vilshanskii, V. A., 338, *349*
Vinograd, J. R., 338(20), *349*
Vivian, J. E., 87(19), *88*
Voeltzel, J., 134(9), *145*
Vogel, G. E., 41(61), *52*
Voyutski, S. S., 12, *51*

Wade, W. H., 14, 20, *52*, 226(19), *232*, 356, *373*
Ward, F. H., 84, *88*
Washburn, E. W., 156, 160, *176*
Waters, G. W., 162(14), *177*
Watt, J. A. C., 44(65), *52*
Wei, J. C., 84, *88*
Weise, C. H., 225(17, 18), *232*
Weiser, H. B., 15(52), *52*
Weiss, P., 152(12, 13), *176*
Wenzel, R. N., 152, *176*, 359, *373*
Westwater, J. W., 87, *88*
Whalen, J. W., 223(9), 226, 227, *232*
Wheeler, O. L., 346(33), *349*
Wickersheim, K. A., 13(28), 15(28), *51*, 222, 223(8), *232*
Williams, D. J., 296(6), *313*

Williams, M. L., 360(7), *373*
Wohl, K., 346(32), *349*
Woodward, C. E., 338, 339(27), 343(27), 344(27), 345(27), 348(27), *349*
Wright, K. A., 346(30), *349*

Yamakita, H., 338, *349*
Yeh, S. J., 337, *337*
Young, G. J., 13(33), *51*, 216, 223(11, 15), 225, 230(33), *232*

Yu, Y. F., 230(34), *233*
Yurshenko, A. I., 338, *349*

Zechmeister, L., 62(5), *79*
Zettlemoyer, A. C., 15(53), *51*, *52*, 227, 229(30, 32), 230(33, 35, 36, 37), *233*
Zimmerman, J. R., 221(5), *232*
Zisman, W. A., 4, 5, 9, 11, 13, 20, *50*, *51*, *52*, 150, 152, *176*, 360(8), 361(8, 10), *373*
Zuidema, H. H., 162(41), *177*

Subject Index

Abhesion,
  definition of, 361
Accelerators,
  role of during phosphating, 140–141
Active centers,
  role of during phosphating, 131–136
Adatom, 115–116
Adhesiveness,
  of polymers, 352–353
Adsorbed layers,
  structure of, 361–363
Adsorption,
  of polymers, 81
  polymers and physical, 3–6
  reversibility of, 334–335
Adsorption isotherms,
  and heats of immersion, 218–219
Adsorption rate processes, 347–348
Adsorption rates,
  of polymers, 333
Adsorption theories,
  monolayer, 289–290
  active site, 290
Affinity,
  of polymers, 365–366
Alternating current,
  during phosphating, 128
Aluminum oxide,
  IEPS, 17
  surface of, 14–16, 19
Anodic areas,
  of metals, 128–129, 148
Arithmetic mean averaging, 211–213
Attenuated total reflection measurements,
  of adsorbed polymer films, 331–333

Bond strength, 3–6
  polyethylene, 4
Bonding forces,
  on metal surfaces, 121–123
Bond types and energies, 8
Bridgeford concept, 297–298
Brushite, 108, 110
Burnishing layers,
  on iron and iron alloys, 95–100

Capillary penetration, 57
Capillary pressure, generated in,
  a slit, 157
  zinc phosphate coatings, 174
Carbide, nitride, and boride layers,
  on metal substrates, 100–102
CASING, 267, 277–285
Cathodic areas,
  of metals, 128–129, 148
Cellulose,
  composites, 291–297
  nature of, 290–291
Chemical potential,
  of polar molecules at interfaces, 354
Chemisorption, 366–367
Chromic acid treatment,
  of zinc phosphate coatings, 161, 174
Chrysotile asbestos,
  structure, 20
Cohesive energy density, 8, 21
Collective picture,
  of surfaces, 205–207
Conduction band, 206
Conversion coatings,
  adhesion on metal surfaces, 126
  formation on metal surfaces, 125–126
Coordinate interfacial bonding, 45–46, 60
Coupling agents,
  silane and titanate, 40–45
Covalent interfacial bonding, 37–40
Critical surface tension,
  of wetting, 4–5, 9–10
Crystal growths,
  model for, 251–253

Diffusion, 76–79
  of additives, 44, 50, 57
  during phosphating, 115–118
  of water, 6–7
Dipole and ionic interactions, 22–37

Electronic state,
  adsorption, 230–231
Ellipsometry measurements,
  of adsorbed polymer films, 324–331, 336

Emulsification,
  of organic materials, 24–25, 57–58
Emulsion polymerization,
  effect of surfactants in, 338–342
Enthalpies,
  of adsorption, 14
Epitaxial corrosion layers,
  on iron, cobalt, and nickel, 92–95, 118–119
Epitaxy,
  concept of, 91–92
Equation,
  Arrhenius, 347–348
  BET, 66–68, 80
  Darcy, 72–75
  Fowkes, 10–12
  Good-Girifalco, 9–12
  Hagen-Poiseuille, 62, 156, 158
  Lagrange, 69
  Maxwell, 69
  Parks, 16–19
  Poisson, 207
  Rideal-Washburn, 77
  Schrodinger, 205
  Stokes-Navier, 69–70
  Young, 151
  Young-Laplace, 153
Etching,
  during phosphating, 117–118, 128–129

Ferrous ions,
  displacement from fibers, 306–311
Fermi level, 207

Gas reactions,
  with high-melting metals, 102–104
Gases,
  flow of, 70–71
  kinetic theory of, 71
Geometric mean averaging, 209–210
Grain boundaries, 128, 136, 147–149
Graphite,
  adsorption of sodium dodecyl sulfate on, 343–344

Harkins' spreading coefficient, 152

Heat of dehydration,
  of silanol groups, 225–226
Heat of immersion,
  based on water areas, 227
  calculations, 213–217
  and heterogeneities, 219–221
  measurements, 214–215
  of oxides in organic liquids, 226–227
  of polystyrene, 233–235
Helium,
  effect on,
    fluorocarbons, 270, 272
    hydrocarbons, 270
    polyethylene, 267–270, 273, 275
Heterogeneities,
  from adhesion and heat of immersion measurements, 227–230
  determination of, 219–221
  on solid surfaces, 217–221
High-energy surfaces,
  isoelectric point of, 16–19
  model of hydrated, 12–14
  physicochemical nature of, 12–19
Hopeite, 108–113, 139–140, 146, 162, 194–196, 199
Hureaulite, 108–110, 132
Hydrogen bonding, 10, 20–25, 59
Hydroxyl groups,
  surface, 12–19, 38

Impurities,
  effects of, 366
Infrared measurements,
  of adsorbed molecular segments, 331
Interaction efficiency, 9–12, 20–21
Interface structures, 351–352
Interfacial forces,
  additive of forces, 8
  free energy change, 6–7, 14, 17, 20–22, 57–58
  inverse square relationship in, 8–12
  methods for estimation of, 6–12
Interfacial turbulence, 83–88
Interphases, strength of, 368–370
Ion exchange,
  on pulp fibers, 299–302, 313–316
  reactions,
    kinetics of, 302–306

SUBJECT INDEX

Iron,
    dissolution of during phosphating, 183, 200–202
Iron oxide surfaces,
    phosphating of, 189–192
    preparation of, 185
    survival of,
        during phosphating, 186
Isoelectric point,
    solid oxides, 16–20, 52–53, 56, 60

Joint strength, 276–282

Kinetics, of wetting, 356–358

Liquids,
    displacement of, 11–12
    flow of, 69–70, 72–75, 77, 80–81
    Newtonian, 69–70, 80–81, 156, 177
    non-Newtonian, 177–178
London forces, 5–6, 8–10, 20–22, 210

Marangoni effect, 83–84, 87
Mass transfer,
    in interfacial turbulence, 84–87
Metal chelates, 45–50, 53
Metal pretreatments, 58, 183–184
Molybdenum oxides,
    IEPS, 16
Monetite, 108

Nylon,
    effect of CASING on, 277–278
Nylon, 6, 6,
    bonding to oxide surfaces, 22

Ore flotation, 34–35, 53
Oxalation layers,
    on iron, zinc, manganese, magnesium, and aluminum, 104–107
    phase stability of, 104–105
    textures, epitaxies, and adhesive bonds, 105–107
Oxide formation,
    model of, 241–243
Oxide layers,
    formation and significance of, 123–125

Oxide nucleation,
    thin film range,
        normal pressure, 238–239
        low oxygen pressures, 239–241
Oxide nuclei,
    orientation of, 241
Oxide platelets,
    blunted-shaped, 247–251
    pointed blade-shaped, 244–247
    rounded, 247–248
Oxide surfaces, 221–231
    interactions with, 25–32
Oxide whiskers, 243–244, 246

Paint films,
    drying of, 83–84
Permeability constants,
    uses of, 75–76
    values of, 74
Permeation,
    emperical approach to, 72–75
    physical models for, 75–77, 81
Phosphate coatings,
    determination of,
        electrochemical, 136–139
        gravimetric, 136–137
        x-ray, 136, 139–140, 146
    formation of, 120–127, 131–136, 141–144
    kinetics of formation, 136–139
Phosphate layers,
    on iron, zinc, copper, brass, and aluminum, 107–115
    mechanism of growth on metals, 115–118
Phosphating processes,
    electrochemical nature of, 128–131
Phosphophyllite, 108–113, 132, 136, 139–140, 146, 162, 194–196, 199
Polar organic compounds,
    interaction between oxide surfaces and, 20–46, 48–50
Polyethylene,
    oxidation of, 257–259, 271
    reaction with atomic oxygen, 262, 265–267
Poly(ethylene o-phthalate) adsorbed,
    on chrome, 330–331

Polymer adsorption,
  on substrates, 321–337
  theory of, 321–323
Polymer liquids,
  gel permeation chromatographs of, 165
  molecular weights of, 164
  physical properties of, 163
  wetting of zinc phosphate coatings, 151–180
Polymer requisites, 370–371
Polymer structure,
  ease of oxidation and, 261
Polymerization,
  on cellulose, 311–312
Polymers,
  effect of excited gases on, 267–283
  oxidation of,
    atomic oxygen, 262–267
    thermal, 257–262
Polypropylene,
  oxidation of, 260–261
Polystyrene adsorbed,
  on chrome, 325–328, 336
  on mercury, 328–329
Polystyrene-styrene surfaces,
  adsorption of sodium dodecyl sulfate on, 343–346
Porosity,
  definition, 64
  measurement of, 65–69
  uses of, 75–77
  values of, 66
Porosity and wetting, 358–359
Porous systems,
  characterization of, 64–69
  chemical reactions in, 63–64, 77–79
  examples of,
    catalysts, 63
    concrete, 63
    distillation columns, 64
    filters, 62
    molecular sieves, 62
    paper, 63
    paper chromatography, 62
    printing, 63
    seepage of gas and oil, 62
    soil, 62
  Pulping of wood, 77–78

resin treatment,
  of fabrics, 79
  of paper, 78
Potential shift,
  during phosphating, 129–131
Pretreatment,
  of metal surfaces, 131–136
Printability, 277, 279

Rate-dependent processes, 6–7, 57
Redox system,
  in polymerization reactions, 291–297
Reversibility and displacement,
  of polymer layers, 364–365
Roughness and wetting, 359–360
Rubber,
  reaction with atomic oxygen, 264

Schlieren photography, 86–87
Scholzite, 108–110, 132
Silicates,
  isoelectric point of, 16–20
  nature of surface, 12–19
Solute transfer, 84–85
Specific surfaces,
  measurement of, 65–69
  porous systems, 64–69
  values of, 67
Spontaneous emulsification, 83
Spreading liquids,
  kinetics of, 167–169, 172
  shape of droplets of, 164–166
Strengite, 108
Stress corrosion, 254–255
Surface charges,
  in adhesion, 60
Surface hydroxyls, 221–231, 235–237
Surface states, 206–207
Surfaces,
  general nature of, 56
  interaction of fluids with, 71–72
  pretreatment of metal, 19–20
  wettability of, 72
Surface tension,
  variation of, 83–84
Surface treatment,
  of metals, 120, 131–136

SUBJECT INDEX

Surfactant adsorption,
  experimental methods for studying, 339
  on polymers, 338–349

Titanium dioxide,
  IEPS, 17
  surfaces of, 14–15, 17, 19
Titanium phosphate, 132, 134, 136, 190–192
Topotaxy,
  concept of, 92

Vivanite, 108–111, 114, 132, 136

Water,
  interactions of surfaces with, 12–19, 21, 47–48, 57
Wavefunctions, 205–207
Wetting,
  adhesion and, 3–6, 21
  critical surface tension for, 4–5, 9–10
  oxides and, 13–14
  spreading and, 5, 21, 57–58
Wetting fundamentals,
  rough surfaces, 152–155
  smooth surfaces, 150–152

Wetting model,
  for zinc phosphate coatings, 155–161
Wetting processes,
  in adhesion, 208
  on oxide surfaces, 231
Work of adhesion, 215–216

Zinc phosphate baths,
  ingredients of, 182–184
Zinc phosphate coatings, 59, 63–64, 77, 81–82
  constituents of, 195–196
  corrosion resistance of, 190
  growth of, 196–199
  nature of, 186–188
  penetration of, 170
  porosity of, 155–157, 163, 170–174
  structure of, 154–155
  wetting of, 355–356
Zinc phosphate crystals,
  form and density of, 192–196
Zinc phosphating,
  immersion versus spraying, 188
  kinetics of formation of, 188–196
Zinc phosphating mechanisms, 182–183